SAC CRAIBSTONE
29 NOV 2004
LIBRARY

Humic Substances
Versatile Components of Plants, Soil and Water

Humic Substances
Versatile Components of Plants, Soil and Water

Edited by

Elham A. Ghabbour
Soil, Water and Environmental Research Institute, Alexandria, Egypt and Northeastern University, Boston, USA

Geoffrey Davies
Northeastern University, Boston, USA

Based on the proceedings of the fourth Humic Substances Seminar held on 22–24 March 2000 at Northeastern University, Boston, Massachusetts

The front cover illustration is taken from the contribution by G. Haberhauer, W. Bednar, M. H. Gerzabek and E. Rosenberg, p. 144

Special Publication No. 259

ISBN 0-85404-855-3

A catalogue record for this book is available from the British Library

© The Royal Society of Chemistry 2000

All rights reserved.

Apart from any fair dealing for the purpose of research or private study, or criticism or review as permitted under the terms of the UK Copyright, Designs and Patents Act, 1988, this publication may not be reproduced, stored or transmitted, in any form or by any means, without the prior permission in writing of The Royal Society of Chemistry, or in the case of reprographic reproduction only in accordance with the terms of the licences issued by the Copyright Licensing Agency in the UK, or in accordance with the terms of the licences issued by the appropriate Reproduction Rights Organization outside the UK. Enquiries concerning reproduction outside the terms stated here should be sent to The Royal Society of Chemistry at the address printed on this page.

Published by The Royal Society of Chemistry,
Thomas Graham House, Science Park, Milton Road,
Cambridge CB4 0WF, UK

For further information see our web site at www.rsc.org

Printed by MPG Books Ltd, Bodmin, Cornwall, UK

"you cannot prove a vague theory wrong."
>
> Richard Feynmann, in
> 'The Character of Physical Law', M.I.T. Press,
> Cambridge, MA, 1965, p. 158,
> quoted by Robert L. Wershaw, p. 5

Preface

This book is a companion of the volumes 'Humic Substances: Structures, Properties and Uses' and 'Humic Substances: Advanced Methods, Properties and Applications' published by the Royal Society of Chemistry in 1998 and 1999, respectively. These books report the best and latest research on humic substances (HSs), the remarkable brown biomaterials in animals, coals, plants, sediments, soils and waters. HSs functions include water retention, pH buffering, photochemistry, redox catalysis, solute sorption and metal binding. HSs are chameleons whose behavior depends on the circumstances. The structures responsible for this remarkable range of natural functions still need to be understood, but real progress is being made thanks to more and more discriminating analytical and physical measurements on reproducible HSs samples.

This book and its companion volumes are based on remarkable meetings called the Humic Substances Seminars, which have been held at Northeastern University in the Spring of each year since 1997. The Seminars have become the major forum for discussion of HSs as biomaterials of paramount importance to the productivity, health and safety of the world's land and water.

The proper uses of analytical chemistry and emphasis on good physical chemistry of HSs have intensified over the past few years. Another great development is the increasing participation of young men and women from many disciplines in fundamental HSs research with many potential applications.

Some burning issues in HSs science are addressed in this book. First is a strong desire to understand HSs structures from knowing HSs precursor structures (for example, in senescent leaves) and from working out humification mechanisms. Spectroscopic and mass based approaches (especially quantitative NMR) are the instruments of progress on this front. The need for every published NMR spectrum of an HSs sample to be quantitative can be answered now the tools to make this possible have been assembled. Underlying the push to quantitation is the prospect that HSs from different locations have more than a few common properties, perhaps because they have similar molecular structures. An example is the narrow range of solubility parameters derived for HSs from many different sources, which should be useful in understanding HSs sorption of endocrine disruptors, pyrethroids and other hydrophobic environmental poisons. Driving the enthusiasm and new effort is the desire to make HSs research a 'hard' science, as so eloquently expressed by Dr. Robert Wershaw.

A second issue is the 'molecularity' of HSs. Dr. Wershaw reminds us that HSs are a continuum of molecules that differ in molecular structures and sizes. So-called humic acids (HAs) dominate HSs in soils and the current debate on whether HAs are macromolecules or supramolecular assemblies makes fascinating reading and gives much food for thought. Field flow fractionation (FFF) has been revived as an HSs research technique that is chromatography without a solid phase, an advantage in studying 'sticky' HSs. FFF theory is well established and the FFF methods are simple and reliable. Perhaps there is no great difference between the molecular weights of 'fulvic' and 'humic' acids?

A third issue is the 'inner' and 'outer' parts of HSs molecules, micelles and particles. More and more evidence is appearing on 'fast' and 'slow' components of solute desorption and metal release for both dissolved and solid HSs. In fact, more and more common properties of dissolved and solid HSs support the notion of an HSs molecular suite with related properties in the 'wet' solid state and in solution.

Another spotlight is on the photochemistry of HSs, in which fluorescence is used to understand HSs primary interactions with solutes like amines and is being identified with very aromatic 'impurities' in soil derived HAs.

Also covered is the use of HSs for soil enhancement, water purification and soil remediation. The growing coal based HSs products industry offers real alternatives to mass transportation of healthy soils and too-slow composting for soil improvement or replacement, expensive monofunctional sorbents and metal binders for water purification and wasteful incineration of soil for remediation. These coal-based HSs products are being quantitatively compared with soil HSs, standardized and put to good use in the field.

We have entered an era in which HSs increasingly are acknowledged as important materials and where random measurements on HSs are downplayed. Hard science has the power of prediction and good, comprehensive data lead to predictive models like those in the fine paper by Buckau *et al.* on radionuclide migration. There is no doubt that the Humic Substances Seminars are contributing to the 'hardening' of humic substances science, thanks to authors who bring their best work to Northeastern for discussion.

This book is derived from Humic Substances Seminar IV, which was held at Northeastern University, Boston, Massachusetts, USA on March 22–24, 2000. We were honored by the presence of Drs. James Alberts, Michael Hayes and Nicola Senesi (Past Presidents) and Yona Chen (President-Elect) of the International Humic Substances Society, together with other IHSS officers and nearly 100 eminent authors from sixteen countries. It was an excellent meeting and a clear indication that HSs research has new impetus and is going from strength to strength.

ACKNOWLEDGEMENTS

The authors and reviewers did a great job and the co-operation of everyone involved in the production of this collection of hallmark papers is appreciated. Financial support from Arctech, Inc., the Barnett Institute of Chemical and Biological Analysis, the Seminar IV Exhibitors and our other sponsors is gratefully acknowledged. Northeastern University provides excellent facilities for the Humic Substances Seminars and Michael Feeney ably manages the Seminar presentations. The staff of the Barnett Institute is a model of efficiency and the undergraduates and graduates of the Humic Acid Group are warm-hearted hosts. It is a real pleasure to acknowledge the hard work of Janet Freshwater and her staff at the Royal Society of Chemistry for timely publication of the best in humic substances research. It's here, right in this book.

Boston, Massachusetts
July, 2000

Elham A. Ghabbour
Geoffrey Davies
Editors

Contents

The Study of Humic Substances – In Search of a Paradigm   1
    Robert L. Wershaw

Humic Substances and Humification   9
    W. Ziechmann, M. Hübner, K. E. N. Jonassen, W. Batsberg, T. Nielsen,
    S. Hahner, P. E. Hansen and A.-L. Gudmundson

Humification of Duck Farm Wastes   21
    M. Schnitzer, H. Dinel, H.-R. Schulten, T. Paré and S. Lafond

Catalytic Effects of Hydroxy-aluminum and Silicic Acid on Catechol
Humification   37
    C. Liu and P. M. Huang

Effect of Cover Crop Systems on the Characteristics of Soil Humic
Substances   53
    Guangwei Ding, Dula Amarasiriwardena, Stephen Herbert,
    Jeffrey Novak and Baoshan Xing

Structural-group Quantitation by CP/MAS ^{13}C NMR Measurements of
Dissolved Organic Matter from Natural Surface Waters   63
    R. L. Wershaw, G. R. Aiken, J. A. Leenheer and J. R. Tregellas

Structural Investigation of Humic Substances Using 2D Solid-State Nuclear
Magnetic Resonance   83
    Jingdong Mao, Klaus Schmidt-Rohr and Baoshan Xing

Procedures for the Isolation and Fractionation of Humic Substances   91
    Michael H.B. Hayes and Colin L. Graham

Differences in High Perfomance Size Exclusion Chromatography between
Humic Substances and Macromolecular Polymers   111
    A. Piccolo, P. Conte and A. Cozzolino

Characterization of the 'Fluorescent Fraction' of Soil Humic Acids   125
    M. Aoyama, A. Watanabe and S. Nagao

Investigations of Humic Materials Aggregation with Scattering Methods   135
    James A. Rice, Thomas F. Guetzloff and Etelka Tombácz

Application of MALDI-TOF-MS to the Characterisation of Fulvic Acids 143
  G. Haberhauer, W. Bednar, M. H. Gerzabek and E. Rosenberg

Sorption of Aqueous Humic Acid to a Test Aquifer Material and Implications for Subsurface Remediation 153
  D. R. Van Stempvoort, J. W. Molson, S. Lesage and S. Brown

Validation of a One-Parameter Concept to Elucidate the Sorption of Hydrophobic Organic Compounds into Humic Organic Matter and Bioconcentration Processes 165
  Juergen Poerschmann

The Interaction between Esfenvalerate and Humic Substances of Different Origin 177
  L. Carlsen, M. Thomsen, S. Dobel, P. Lassen, B. B. Mogensen and P. E. Hansen

Adsorption–Desorption Interactions of Environmental Endocrine Disruptors with Humic Acids from Soils and Urban Sludges 191
  E. Loffredo, M. Pezzuto and N. Senesi

Binding of Organic Nitrogen Compounds to Soil Fulvic Acid as Measured by Molecular Fluorescence Spectroscopy 205
  C. L. Coolidge and D. K. Ryan

Flow Field-Flow Fractionation-Inductively Coupled Plasma-Mass Spectrometry (FLOW-FFF-ICP-MS): A Versatile Approach for Characterization of Trace Metals Complexed to Soil-Derived Humic Acids 215
  Dula Amarasiriwardena, Atitaya Siripinyanond and Ramon M. Barnes

EXAFS and XANES Studies of Effects of pH on Complexation of Copper by Humic Substances 227
  Anatoly I. Frenkel and Gregory V. Korshin

Main Conclusions of the EC-Humics Project: 'Effects of Humic Substances on the Migration of Radionuclides: Complexation and Transport of Actinides' 235
  G. Buckau, P. Hooker, V. Moulin, K. Schmeide, A. Maes, P. Warwick, Ch. Moulin, J. Pieri, N. Bryan, L. Carlsen, D. Klotz and N. Trautmann

Natural Organic Matter from a Norwegian Lake: Possible Structural Changes Resulting from Lake Acidification 261
  J. J. Alberts, M. Takács and M. Pattanayek

Organoclays Remove Humic Substances from Water 277
  George R. Alther

Mass Spectrometry and Capillary Electrophoresis Analysis of Coal-derived Humic Acids Produced from Oxihumolite. A Comparison Study 289
  D. Gajdošová, L. Pokorná, A. Kotz and J. Havel

Analysis and Characterization of a 'Standard' Coal Derived Humic Acid 299
  L. Pokorná, D. Gajdošová, S. Mikeska and J. Havel

Performance Improvement and Applications of Humasorb-Cs™: A Humic Acid-Based Adsorbent for Contaminated Water Clean Up 309
  A. K. Fataftah, H. G. Sanjay and D. S. Walia

Humic Acid Products for Improved Phosphorus Fertilizer Management 321
  K. S. Day, R. Thornton and Harry Kreeft

Subject Index 327

THE STUDY OF HUMIC SUBSTANCES—IN SEARCH OF A PARADIGM

Robert L. Wershaw

U.S. Geological Survey, Denver Federal Center, Denver, CO 80225

1 INTRODUCTION

The successful pursuit of any field of scientific research requires that all the practitioners in the field share a common set of basic premises and definitions. For example, the Kuhn[1] model of 'normal science' requires that the research of all the scientists in a particular field be based on a shared set of conceptual models called paradigms. Or, as Medawar[2] has described the Kuhn model: "That which the scientist measures his hypotheses against is the current 'establishment' of scientific opinion—the current framework of theoretical commitments and received beliefs—the prevailing 'paradigm' in terms of which the day-to-day problems arising in a science tend to be interpreted." I shall show that progress in the field of humic substance research in soils and natural waters is impeded by the lack of a shared paradigm on which the research is based. The shortcomings in the present research regime will be outlined below and suggestions will be made for a more logical approach to the study of the natural organic compounds in soils and natural waters.

2 SHORTCOMINGS OF THE PRESENT REGIME

The first shortcoming arises from the fact that there are no universally accepted definitions of the terms humic substance and humus that are widely used to refer to the nonliving, natural organic matter in soils and waters. Naturally occurring organic compounds in soils have been the object of scientific research for more than 200 years because agricultural scientists early recognized the importance of natural organics in enhancing soil fertility and water-holding capacity.[3] The most general term for the natural organic compounds in soils, sediments, and natural waters is natural organic matter (NOM); however, a number of other terms have been applied to different NOM fractions. The most commonly used of these are: humus, humic substances, humic acid, fulvic acid and humin. The dissolved fraction of the NOM in natural waters is commonly called dissolved organic carbon (DOC). Historically, the term humus has been applied to the dark-colored, "rotted" organic matter in soils, and the terms humic acid, fulvic acid and humin have been used to designate different fractions of humus.[3]

Early workers recognized that soil humus arises mainly from the degradation of dead plant tissue with a lesser contribution from decaying animal remains. Many of these early workers assumed that humus was composed of the end products of synthetic reactions that alter the structures of plant degradation products. Other workers, however, maintained that humus is a complex mixture of plant degradation products. This controversy has persisted. Stevenson[4] defined soil humic substances as "A series of relatively high-molecular weight, yellow to black colored substances formed by secondary synthesis reactions," and humus as the "Total of the organic compounds in soil exclusive of undecayed plant and animal tissues, their 'partial decomposition' products, and the soil biomass." In contrast, a number of workers recently have proposed that humic substances consist mainly of the partial degradation products of plant polymers.[5-7] However, these workers generally do not explicitly state, as Stevenson[4] does, what NOM compounds are excluded from the category of humic substances; therefore, their definition of humic substances is necessarily incomplete. Most workers would agree that soil NOM consists of humified components, which have been called humus or humic substances and the nonhumified components of soil organic matter. Those workers who do not accept Stevenson's[4] definitions must, therefore, distinguish between those NOM components that have been altered enough to be called humic substances and those that have not. It probably is useful to exclude completely undecomposed plant and animal tissue and living microorganisms from the definition of humus; however, there is no practical way to exclude "partial decomposition" products from humus isolates, and indeed, it is likely that soil humus consists mainly of such products.[7,8]

The second shortcoming of the present regime of humic substance research arises when one tries to isolate a material that fits whatever definition one chooses for a humic substance. Any attempt to exclude a group of NOM components from the definition of humic substances or humus requires that methods exist that allow one to remove the nonhumic components from the crude humic isolates normally obtained. If one cannot reliably separate humic substances from nonhumic substances, then whatever analytical technique one uses will not uniquely characterize the humic substance of interest. If one accepts Stevenson's[4] definitions of humic substances and humus, then the first step in any study of humic substances is to separate those substances formed by secondary synthesis reactions from all other NOM components. If one cannot accomplish this task then, of necessity, one cannot claim to have analyzed or characterized a humic substance.

Most workers apparently have assumed that the humic-nonhumic separation is accomplished by the commonly used extraction procedures; however, there is no evidence that this is the case. In fact, no reliable procedures have been proposed that exclude nonhumic components from the soil extracts. Hayes[9] has reviewed procedures that have been used for extraction of humic substances from soils. He concluded that humic substances are most efficiently extracted with strongly basic aqueous solutions. Unfortunately, these solutions can degrade macromolecules and produce artifacts. Much less degradation can take place when neutral salt solutions are used for extraction, but they generally do not solubilize as much material as the basic solutions. Polar organic solvents have also been tried, but much less material is generally extracted than with aqueous bases. All of these solvent systems will also extract monomeric and oligomeric plant and animal components such as amino acids, proteins, saccharides, and plant polyphenols, which are excluded from the Stevenson definitions of humus and humic substances.

Soil humic extracts are generally fractionated into two fractions by reducing the pH to 1. By definition, the fraction that precipitates at pH 1 is humic acid; the fraction that remains in solution is called fulvic acid. The fraction of soil humus that remains in the soil after extraction with a basic solution is called humin. These definitions are purely operational and are not based on any well-defined chemical structural criteria.

3 HUMIC SUBSTANCE PARADIGM

The research of those scientists who accept the Stevenson[4] definition of humic substances is predicated on the basic assumption or paradigm (henceforth called the humic substance paradigm) that humic substances constitute a distinct group of natural organic macromolecular species in soils and natural waters, which are composed of structural units that are different from the structural units of the chemical components of intact plant and animal tissue. Acceptance of the humic substance paradigm has led many scientists to assume that all of the isolates of a specific humic substance, for example all humic acids, are composed of a single characteristic set of monomeric units. Numerous chemical structural diagrams have been published purporting to represent the chemical structures of humic acids.[4] In a recent paper Sein et al.[10] used a molecular modeling program to develop a series of molecular structural representations of "humic acid monomers." However, Sein et al.[10] present no evidence that humic acids are made up of repeating monomeric units. Schulten[11] has generated a much different chemical structural model of humic acid using pyrolysis-mass spectrometry data obtained from humic acids isolated from three different soils.

After 100 years or so of research using classical chemical structural techniques, we appear to be no closer to being able to write a single structural diagram for humic acid than scientists were at the beginning of the period. The reason for this is very simple: humic acid isolates are mixtures of components that do not possess a single chemical structure. Wershaw et al.[12-14] fractionated humic acids isolated from a number of different environments on dextran gels and measured the ^{13}C NMR spectra of the different fractions in solution. The spectra of the fractions from a single humic sample were all different, and the spectra of corresponding fractions from different humic acids were similar, but not identical. These results indicate that humic acid isolates are composed of mixtures of different chemical species. The fractionation on dextran partially separates the chemical components of the mixtures, but the separations are far from complete. Wershaw et al.[14] proposed that their NMR results indicate that humic acids are mixtures of the partially degraded chemical constituents of plant tissue that retain much of their original chemical structure. Hayes et al.[15] also have pointed out that heterogeneous, multicomponent samples such as humic isolates are intrinsically unsuitable for chemical structural analysis. For a pure sample of a single chemical species, a chemical structural representation is normally developed from detailed data on the bonding of all of the atoms in a molecule such as those obtained from single-crystal x-ray diffraction or multidimensional NMR analysis. In the case of humic acid isolates, the best that has been done so far is to identify the chemical structural units present in a given isolate and perhaps to estimate the relative concentrations of the units in the sample. In other instances, the structural units identified from a number of different samples have been used. Therefore, even under the best of circumstances, a published structure of a humic acid isolate represents only one of the many possible configurations that will account for the structural-unit distribution in a given sample. In addition, because a given model is based on the structural unit distribution in the mixture of compounds that make up a particular humic acid isolate, there is no way to transfer the structure to any other humic acid isolate.

Other humic substance isolates have also been shown to be heterogeneous mixtures. Leenheer et al.[16,17] used pH gradient chromatography to fractionate Suwannee River fulvic acid into approximately 30 fractions of different metal-ion binding characteristics. They have postulated two possible structures for the fraction that exhibits the strongest binding for Cu(II).[16] These structures were derived from the structures of well-characterized plant polymers.

4 HUMIFICATION PARADIGM

I would like to suggest that a more fruitful approach to the study of NOM in soils and natural waters than that based on the humic substance paradigm is to study the chemical reactions that the chemical components of plant tissue undergo during and after senescence. That is to say, to concern oneself with the humification process rather than with ill-defined intermediates in the continuum from well-characterized plant components to carbon dioxide.

Laidler[18] has distinguished between "hard" and "soft" science. "Hard" science is that "science that can be formulated mathematically, and can be tested by experiment." In contrast, 'soft' science is more descriptive in nature. In the past, biology and geology could be described as "soft" sciences, physics as a 'hard' science, and chemistry as intermediate between the two categories. However, much of the recent progress in all of these fields would be classified as "hard" science. For example, the introduction of the "hard" scientific techniques of molecular biology has truly resulted in a Kuhnian scientific revolution in biology and medicine.[1]

A "hard" science approach to the study of humification would start out with well-defined molecular species from the precursors of NOM, such as plant tissue, soil biomass, or plant exudates and then follow the degradation of these compounds in soils and natural waters. As a first step in such an approach, Wershaw et al.[19] used solid-state ^{13}C NMR to follow the changes in the structures of the chemical components of leaves of different species that take place during senescence. They found evidence of oxidation of lipids and cutins, hydrolysis of lignin methyl ether groups, lignin depolymerization, and hydrolysis of peptides. Leaching of the senescent leaves with water yields degradation products of lignin, hydrolyzable tannins, nonhydrolyzable tannins, lipids, carbohydrates, and peptides.[20] The NMR and infrared spectra of fractions of leaf leachates provide evidence of oxidative degradation of the polyphenolic components of leaf tissue. The oxidative degradation appears to follow the sequence of O-demethylation and hydroxylation followed by ring fission, chain shortening and oxidative removal of substituents. Oxidative ring fission leads to the formation of carboxylate groups on the cleaved ends of the rings. The carbohydrate components are broken down into aliphatic hydroxy acids and aliphatic alcohols.[21]

There are two possible approaches to the elucidation of the changes that the organic components of leaf leachates undergo in soils and natural waters. One can attempt to infer the reactions from a retrospective study of the structures of the components of NOM isolates or a prospective study in which one follows the changes that a known compound or group of compounds undergo when introduced into a soil or natural water system. Crude soil organic matter extracts are generally far too complex to allow one to infer with any assurance the reactions that have altered the precursor molecules. Soil NOM is a mixture consisting of organic coatings on mineral grains, organic precipitates, partially decomposed plant fragments, microbial biomass and charcoal.[7,22-24] Each of these components, with the possible exception of the charcoal, will contribute to the mixture of NOM compounds that will be extracted in a standard basic solvent extraction.[9] Each of these NOM sources should yield a different suite of organic compounds when extracted with strong base. The organic compounds in the NOM of organic coatings and precipitates originates from compounds that were leached by rain water from plant litter and from exudates of living plants and microorganisms. The partially decomposed plant fragments, on the other hand, contain those compounds that were resistant to leaching, and a still different suite of compounds will be derived from living and dead microorganisms.

In many soils a substantial part of the NOM extract may be derived from partially decomposed leaves and other plant fragments. Wershaw and Kennedy[25] demonstrated that isolates that fit the definitions of fulvic and humic acids can be extracted with sodium hydroxide solutions from senescent leaves that had previously been extracted with distilled water. Solid-state ^{13}C NMR spectra of the isolates indicated that the fulvic acid from senescent *Acer campestre* L. leaves was composed mainly of gallotannins in which some of the carboxylate groups were attached to carbohydrate groups by ester linkages. Most of the carboxylate groups, however, were present as free acids. The humic acid isolate consisted mainly of lignin units and polyflavonoid structures linked to carbohydrates. Long-chain polymethylene structures derived most likely from lipids and cutins were also important components of the humic acid. These results suggest that humic and fulvic acids extracted by sodium hydroxide or other strong bases from senescent leaves are fragments of plant polymers that are released by basic hydrolysis. Published solid-state ^{13}C NMR spectra of soil fulvic and humic acids have bands that are generally in the same positions as those of the senescent leaf extracts; however, the bands are broader and less well-resolved in the soil isolates. These differences are probably due to heterogeneity of the source material of the soils isolates and to diagenetic changes that take place after incorporation into the soil matrix.

Prospective studies have been used very successfully for the elucidation of biochemical reactions. For example, Krebs and coworkers established the reactions involved in the citric acid cycle by following the chemical changes that took place when known compounds were added to minced muscle tissue suspensions.[26] In a similar fashion, one could introduce isotopically-labeled model precursor compounds into natural soil and water systems and then follow the changes that these compounds undergo. Such an approach has been used by Bollag and coworkers to follow the reactions of isotopically-labeled xenobiotic compounds in model soil systems.[27]

5 CONCLUSIONS

Progress in scientific research requires the constant testing of hypotheses and the occasional reformulation of a paradigm to accommodate new information. The study of NOM is at the stage where the introduction of a new paradigm is necessary. The prevailing humic substance paradigm is irreparably flawed and should be replaced by a humification paradigm in which the chemical and biochemical reactions that convert precursor molecular species into NOM are elucidated. Elucidation of these reactions will allow one to predict the types of chemical structural units that will be present in NOM derived from a particular set of precursors. Techniques can then be devised for the identification of these structural units. In this way, one can test the specific reaction mechanisms that one postulates for the formation of NOM components in a way that is not possible for those working with the uncertainties inherent in the current humic substance paradigm. For example, scientists who assume that a humic substance such as humic acid is an identifiable entity that possesses some vaguely defined general chemical structure cannot rigorously test their chemical structural hypotheses because all humic acid isolates are heterogeneous mixtures that are intrinsically unsuitable for chemical structural analysis. As Feynman[28] has pointed out "you cannot prove a vague theory wrong."

References

1. T. S. Kuhn, 'The Structure of Scientific Revolutions', University of Chicago Press, Chicago, 1996.
2. P. B. Medawar, 'Advice to a Young Scientist', Harper & Row, New York, 1979.
3. M. M. Kononova, 'Soil Organic Matter-Its Nature, Its Role in Soil Formation and in Soil Fertility', Pergamon Press, New York, 1961.
4. F. J. Stevenson, 'Humus Chemistry. Genesis, Composition, Reactions', 2nd Edn. Wiley, New York, 1994.
5. J. A. Baldock, J. M. Oades, A. G. Waters, X. Peng, A. M. Vassallo and M. A. Wilson, *Biogeochem.*, 1992, **16**, 1.
6. P. G. Hatcher and E. C. Spiker, in 'Humic Substances and their Role in the Environment', F. H. Frimmel and R. F. Christman, (eds.), Wiley, Chichester, 1988, p. 59.
7. R. L. Wershaw, 'Membrane-Micelle Model for Humus in Soils and Sediments and Its Relation to Humification', U.S. Geological Survey Water-Supply Paper 2410, 1994.
8. H. Knicker, H. -D. Lüdemann and K. Haider, *Eur. J. Soil Sci.*, 1997, **48**, 431.
9. M. H. B. Hayes, in 'Humic Substances in Soil, Sediment, and Water', G. R. Aiken, D. M. McKnight, R. L. Wershaw and P. MacCarthy, (eds.), Wiley, New York, 1985, p. 329.
10. L. T. Sein, Jr., J. M. Varnum and S. A. Jansen, *Environ. Sci. Technol.*, 1999, **33**, 546.
11. H. -R. Schulten, *Environ. Toxicol. Chem.*, 1999, **18**, 1643.
12. R. L. Wershaw, K. A. Thorn, D. J. Pinckney, P. MacCarthy, J. A. Rice and H. F. Hemond, in 'Peat and Water-Aspects of Water Retention and Dewatering in Peat', C. H. Fuchsman, (ed.), Elsevier, London, 1986, p. 133.
13. R. L. Wershaw, K. A. Thorn and D. J. Pinckney, *Environ. Technol. Lett.,* 1988, **9**, 53.
14. R. L.Wershaw, D. J. Pinckney, E. C. Llaguno and V. Vicente-Becknett, *Anal. Chim. Acta,* 1990, **232**, 31.
15. M. H. B. Hayes, P. MacCarthy, R. L. Malcolm and R. S. Swift, in 'Humic Substances II: In Search of Structure', M. H. B. Hayes, P. MacCarthy, R. L. Malcolm and R. S. Swift, (eds.), Wiley, Chichester, 1989, p. 4.
16. J. A. Leenheer, G. K. Brown, P. MacCarthy and S. E. Cabaniss, *Environ. Sci. Technol.*, 1998, **32**, 2410.
17. J. A. Leenheer, P. A. Brown and T. I. Noyes, in 'Aquatic Humic Substances-Influence on Fate and Treatment of Pollutants', I. H. Suffet and P. MacCarthy, (eds.), Amer. Chem. Soc. Advances in Chemistry Series 219, Washington, D. C., 1989, p. 25.
18. K. J. Laidler, 'To Light Such a Candle', Oxford University Press, Oxford, 1998.
19. R. L. Wershaw, K. R. Kennedy and J. E. Henrich, in 'Humic Substances: Structures, Properties and Uses', G. Davies and E. A. Ghabbour, (eds.), Royal Society of Chemistry, Cambridge, 1998, p. 29.
20. R. L. Wershaw, J. A. Leenheer and K. R. Kennedy, in 'Humic Substances: Structures, Properties and Uses', G. Davies and E. A. Ghabbour, (eds.), Royal Society of Chemistry, Cambridge, 1998, p. 47.
21. R. L. Wershaw, J. A. Leenheer and K. R. Kennedy and T. I. Noyes, *Soil Sci.,* 1996, **161**, 667.

22. R. L. Wershaw, E. C. Llaguno and J. A. Leenheer, *Colloids Surfaces A: Physicochem. Eng. Aspects,* 1996, **108**, 213.
23. J. O. Skjemstad, P. Clarke, A. Golchin and J. M. Oades, in 'Driven by Nature: Plant Litter Quality and Decomposition', G. Cadisch and K. E. Giller, (eds.), Cab International, Wallingford, U.K., 1997, p. 253.
24. J. O. Skjemstad, L. J. Janik and J. A. Taylor, *Aust. J. Exp. Agric.*, 1998, **38**, 667.
25. R. L. Wershaw and K. R. Kennedy, in 'Humic Substances: Structures, Properties and Uses', G. Davies and E. A. Ghabbour, (eds.), Royal Society of Chemistry, Cambridge, 1998, p. 60.
26. F. L. Holmes, *Fed. Proc.*, 1980, **39**, 216.
27. J. Dec and J.-M. Bollag, *Soil Sci.*, 1997, **162**, 858.
28. R. Feynman, 'The Character of Physical Law', M.I.T. Press, Cambridge, Mass., 1965, p. 158.

HUMIC SUBSTANCES AND HUMIFICATION

W. Ziechmann,[1] M. Hübner,[2,3] K. E. N. Jonassen,[3] W. Batsberg,[3] T. Nielsen,[3] S. Hahner,[4] P. E. Hansen[5] and A. -L. Gudmundson[5]

[1] Kiefernweg 2, D-37085 Göttingen, Germany
[2] Montecatini CRA, I-48023 Marina di Ravenna, Italy
[3] Risø National Laboratory of Biogeochemistry, DK-4000 Roskilde, Denmark
[4] Bruker-Daltonik, D-28359 Bremen, Germany
[5] Department of Life Science and Chemistry, Roskilde University, DK-4000 Roskilde, Denmark

1 INTRODUCTION

Unlike any other group of natural products, humic substances (HSs) have remained a mystery to structural analysis. The obstacles involved are the result of a complex, virtually chaotic genesis.[1] Therefore, it is reasonable to focus on elucidating the process of humification and, in doing so, gain insight into HSs structural properties.

The circumstances and characteristics of humification reveal many features of the process itself. Usually, biosynthesis of organic material in living systems is subject to enzymatic, compartment and transport control. These conditions hardly are involved in the formation of natural HSs. Moreover, a seemingly unlimited range of reaction partners is available, providing an even broader variety of reactive sites. An important feature of humification is the presence of parallel reaction paths with many arbitrary cross linkings ('open genesis'). This principle is completely opposite to the conditions for conversions in organisms ('closed genesis'). In the past, attempts have been made to simulate natural humic substance formation in restricted systems, especially with a restricted selection of conditions. These experiments have largely contributed to what we know about HSs genesis today.

Among the first studies was Miller's experiment (1955)[2] on chemical evolution, which – besides the building blocks for life – produced HSs. As a matter of fact, these primary HSs ('archeo-HSs') consisted of all those compounds which are not 'fit' for becoming life. Contemplating chemical evolution from a different point of view, HSs may be regarded as buffers for entropy that assume a state of lower order to compensate for a state of higher order in living systems (Figure 1).[3] Analogous capabilities for internal transmission are observed with respect to energy and electrical charge.[3]

A simple experiment simulating the formation of HSs is the autoxidation of polyphenols in alkaline solution in the presence of oxygen.[4] Phenyl radicals possess different reactive sites due to formation of mesomeric forms. This gives rise to diverse possibilities of polymerization. Aerobic/oxidizing conditions necessary for autoxidation are frequently met, especially in the first phase of humification. Autoxidation is a comparatively fast process and therefore is capable of profoundly influencing subsequent steps. Evidence has been found that core structures are formed that provide a macromolecular basis for the incorporation of other molecules.[5] Humic acid precursors (HAP) themselves possess enough reactive sites to allow further transformations, e.g.,

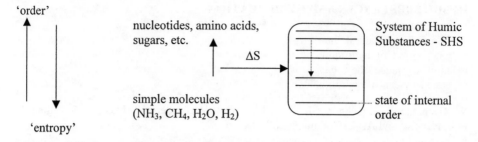

Figure 1 *'Entropy' in systems of humic substances*

condensation reactions. Each substance taking part in humification has a favored stage to enter – for example, amino acids, lignins, pectins or carbohydrates enter at a rather early stage, whereas aromatic hydrocarbons tend to be incorporated at a later stage (Table 1). Assembled by intermolecular forces (donor-acceptor, ionic, hydrophilic, hydrophobic), a so-called 'system of humic substances' (SHSs) eventually is formed. This system is comprised of different fractions specified by the degree in humification. The degree of humification is determined not only by solubility but also by reactivity (especially the

Table 1 *Conversions in the process of humification*

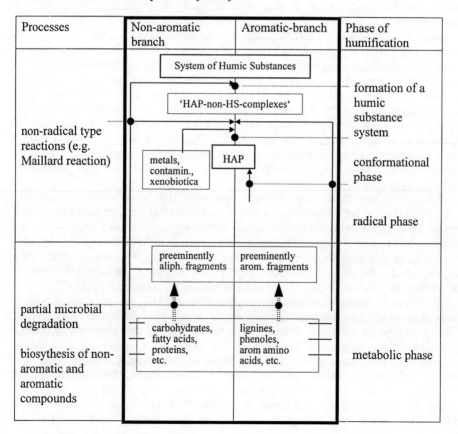

ability to consume oxygen), particle size, electric conductivity and spectroscopic characteristics.

Similar to natural humification, the base-promoted (e.g., NaOH) autoxidative polymerization of hydroquinone yields different fractions.[6] Addition of acid stops the reaction. Because they have distinct physico-chemical properties, fractions can be separated by centrifugation, adsorption of the products on solid materials and extraction with organic solvents. Different fractions dominate the reaction mixture at different points of time.

The time-dependent changes within the system of humic substances (SHSs) can also be depicted by a mathematical model involving Markoff processes. The term 'Markoff process' is a statistical expression. Markoff processes are based on the concept that each state within a system only depends on the previous states. Such a concept proves true for a chemical system with a) a defined substance in each reaction step or b) a plethora of substances in each reaction step, as found in a SHSs.

With the assumption of a specific reaction sequence, a specific time-dependent development of each fraction is to be expected mathematically. As an additional expression, the Shannon entropy (H) has been defined to characterize the state of humification of a SHSs. The Shannon entropy indicates to what extent the whole material is distributed among different fractions. Presupposing different reaction sequences leads to different behavior of the Shannon entropy. Mathematical predictions based on a corresponding formula agree very well with the experimental results for the hydroquinone system. Different behavior would have been observed for alternative reactions and reaction sequences. Markoff processes and the Shannon entropy are effective tools for describing humification as a radical dominated process.

Since data for base catalyzed oxidation of hydroquinones allowed definite conclusions about model humification, the same approach was applied in the present study to natural HSs from both terrestrial and aqueous origins. Our main objective was to investigate changes in molecular size and functional moieties on autoxidation. Moreover, the experiments were aimed at characterizing natural HSs with respect to future investigations of interactions between HSs and polycyclic aromatic hydrocarbons (PAH) for purposes of soil and ground water bioremediation. In this context, HSs are to be applied in solution, not in the solid state. Therefore, extracts of HSs have been subjects of our research. Solvents with different polarity were used to model hydrophobic pollutants in a hydrophilic medium.

2 MATERIALS AND METHODS

2.1 Chemicals and Reagents

All chemicals were of analytical grade and purchased from Merck, Germany unless specified otherwise. Solutions were obtained by diluting or dissolving chemicals in high purity water (Milli-Q system, Millipore). Deuterated solvents for NMR spectroscopy, $CDCl_3$ at 99.8 atom% D purity with 0.03 %wt TMS and D_2O at 99.9 atom% D purity with 0.05 %wt 3-(trimethylsilyl)propionic-2,2,3,3-d_4-acid were obtained from Aldrich. All reactions and preparations were carried out under standard conditions unless specified otherwise.

2.2 Isolation of Humic Substances

2.2.1. Aqueous Humic Substances. Aqueous HSs were sampled from a brook in Northern Germany (Braker Sieltief, Ovelgönne) in October 1999 (pH 7.1 ± 0.1; T (water) = 10°C; T (air) = 15°C) in a volume of 700 L surface water. Pre-filtration was carried out through a pre-rinsed single use polypropylene wound fibre-filter with 20-40 μ pore size (Beckum, Germany). TOC content after pre-filtration was 19.0 ± 0.5 mg L^{-1}. The sample was stored in pre-rinsed 50 L vanadium steel beer kegs (Beck, Germany). 5 mL 0.046 M silver nitrate solution was added to each keg to give a final concentration of 500 ppb silver to suppress microbial growth. The samples were brought to 4°C within 2 h.

Isolating a fraction with a nominal molecular weight between 1 and 100 kDa and simultaneously concentrating the sample was achieved by two-step ultrafiltration with a Sartocon II crossflow unit equipped with Sartocon polysulfone casettes (0.7 m^2 surface) with 1 and 100 kDa nominal molecular weight cutoffs, respectively (Sartorius AG, Germany). The system included a rotary pump (Grundfos type CRN 50), two valves for regulating inlet and outlet pressure, a protective metal sieve filter and a cooling device for regulating the temperature in the reservoir (200 L). In the first step, all material with a nominal molecular weight exceeding 100 kDa was removed. The initial inlet pressure was set at 200 kPa with an outlet pressure of 0 kPa. A tangential flow of 11.8 L min^{-1} and a penetrate flow rate of 2.5 L min^{-1} at 17°C resulted (the penetration flow rate with dimineralized water was 8 L min^{-1} with the same pressure conditions). The final inlet pressure (after 700 L) was 350 kPa at an outlet pressure of 0 kPa with a penetrate flow rate at 25°C of 0.6 L min^{-1}.

Then, all material with nominal molecular weight lower than 1 kDa was removed and the retentate was concentrated to give a volume of 12.5 L. The initial inlet pressure was chosen to be 350 kPa with an outlet pressure of 300 kPa. A tangential flow of 4 L min^{-1} and a penetrate flow rate of 89 mL min^{-1} at 17°C resulted (the penetration flow rate with dimineralized water was 300 mL min^{-1} at an inlet pressure of 400 kPa and an outlet pressure of 300 kPa). Once during the process it was necessary to rinse the membrane with 30 L 1 N NaOH for 2 h (subsequent neutraliztion was with 30 L 0.5 % citric acid and 30 L demineralized water). The final inlet pressure (after 700 L) was 380 kPa at an outlet pressure of 300 kPa with a penetrate flow rate at 22°C of 14 mL min^{-1}.

The concentrate was found to have a DOC content of 0.47 ± 0.01 mg L^{-1}. Partial freeze drying of the product gave 3.08 ± 0.05 g L^{-1} dry substance. According to elemental analysis (instrument: LECO CS200) the dried sample contained 17.6 ± 0.1 % carbon and 3.3 ± 0.2 % sulfur (IR determination of CO, CO_2 and SO_2 after combustion).

Three portions of the freeze dried product (200, 232 and 232 mg) were extracted by means of ultrasound (Branson 2210E-DTH, 90 Watt, 47 kHz) at 50°C with water or methanol or ethanol (4 x 100 mL, each time over 6 h and with subsequent centrifugation with a Sigma Laboratory Centrifuge Model 4-15) to give (after evaporation of the solvent under reduced pressure) 192, 79 and 54 mg dry product, respectively.

2.2.2. Soil Humic Substances. A commercial peat preparation (isolated from Danish 'Pottemuld', a bog peat provided by Djursland/Sphagnum A/S; controlled by Dansk Erhvervsgartnerforenings Laboratorium and containing 3.5 kg m^{-3} $CaCO_3$) that is recommended for plant nutrition was freeze dried within 48 h (free water content 65%). Four portions of the freeze dried product (1.265, 0.11, 1.200 and 1.200 g) were extracted with ultrasonification at 50°C with water or 0.1 M KH_2PO_4 or methanol or ethanol (4 x 100 mL each, each time over 6 h and with subsequent centrifugation) to give (after filtration through Pyrex 4 and Pyrex 5 glass filters, 0.45 μ filtration with solvent resistant Sartorius syringe filters and evaporation of the solvent under reduced pressure) 85.6, 7,

44.4 and 48.0 mg of dry product, respectively.

2.3 Oxidation of Freeze Dried Humic Substances

20 mg freeze dried product was redissolved in 40 mL of the corresponding solvent (water or methanol or ethanol). Base mediated oxidation was achieved by adding 20 µL 1 M NaOH and exposing the samples for 2 h to air with ultrasound at 40°C. Subsequently, 20 µL1 M HCl were used to neutralize the solution. Control samples were prepared by adding 20 µL 1 M NaCl solution.

2.4 MALDI-TOF MS Experiments

MALDI-TOF mass spectrometry was performed with a Bruker Daltonik Biflex III instrument with the SCOUT 384TM target. The instrument is equipped with a N_2 laser (337 nm) for ionization/desorption and pulsed ion extraction capabilities (PIETM). The initial acceleration potential was 25 kV. A dual microchannel plate detector was used for detection. The samples (concentration: 10 mg mL^{-1}) were mixed with the matrix (3-hydroxypicolinic acid (50 g L^{-1}) in 15 % acetonitrile) forming a microcrystalline layer upon drying on the probe target. Prior to sample preparation, the matrix solution was incubated with an appropriate amount of NH_4^+-loaded cation exchange beads (Bio-Rad, München, Germany) to reduce the amount of alkali counter ions. Oligonucleotides were used for system calibration.

2.5 HPSEC Experiments

HPSEC investigations were carried out on a Shimadzu LC-10 HPLC system with a photo-diode array detector, thermostatted column oven and auto-injection facility. The samples (injection volume 150 µL; concentration of HSs 0.5 mg mL^{-1} in the mobile phase; internal standard 2 µL mL^{-1} acetone) were passed through a SEC column (TSK G 3000 PW; 7.5 mm ID x 30 cm, Toso Haas) with 1:1v/v methanol:0.05 M sodium chloride as the mobile phase at a flow rate of 0.5 mL min^{-1} at 40°C. Detection was at 254 nm. Calibration was performed with polystyrenesulfonate standards (1200, 780, 100, 74, 35, 18, 8 kDa; Polymer Laboratories, peak detection at 254 nm).

2.6 IR Spectrometry

FTIR spectra of freeze dried HS extracts were recorded on a Perkin-Elmer model 1760x FTIR spectrometer as KBr pellets (3 mg sample mixed with 300 mg KBr) with a standard procedure recommended by the manufacturer.

2.7 Proton NMR Spectrometry

Proton NMR spectra were recorded on a Varian INOVA 600 instrument at 300 K with an acquisition time of 1.024 s, a relaxation delay time of 0.01 s and optional suppression of the water signal. NMR samples were prepared from freeze dried HSs by adding 300 µL of D_2O and subsequent freeze drying, followed by dissolution in 700 µL solvent with varying concentrations (10 mg in D_2O for water extract of aqueous HSs, 7 mg in D_2O for water extract of peat HSs, 5 mg samples in D_2O for methanol and ethanol extracts of aqueous HSs, and 7 mg in CD_3OD for methanol and ethanol extracts of peat HSs).

2.8 DOC Determinations

Total organic carbon (TOC) content was measured with a Shimadzu Model TOC 5000 analyzer fitted with a ASI-5000A autosampler and configured in combined combustion and non-dispersive IR gas analysis modes. 50 µL samples were injected. Acidification with 25 wt% phosphoric acid (stimulating the complete release of all inorganic carbon) allowed distinction between organic and inorganic carbon.

3 RESULTS AND DISCUSSION

3.1 Determination of Molecular Sizes – MALDI-TOF MS and HPSEC

The molecular sizes or molecular weights (Mw) of HSs are assumed to be crucial indicators of the degree of humification. However, reliable methods of determining absolute molecular sizes or weights of HSs still have not been found. Methods based on entirely different principles have been applied in the present study.

Conventional mass spectrometry has a narrow range for mass detection and moreover suffers from fragmentation of the molecules. MALDI-TOF mass spectrometry does not have these disadvantages but, as in other forms of mass spectrometry, signal intensity does not allow quantification of the Mw distribution. In MALDI-TOF experiments, small molecules may be desorbed preferentially, leaving heavier molecules partly undetected. Furthermore, interaction with matrix molecules may occur, causing artefacts. For this reason, HSs themselves have been used as a matrix.[7] However, formation of molecular clusters still can be a problem in MALDI-TOF spectroscopy.

Since 3-hydroxypicolinic acid performs well as a matrix for DNA or peptide MALDI-TOF analysis, it appeared worth investigating the water extract of the freeze dried aqueous HSs in the same manner. The mass spectrum revealed a series of quite regularly recurring maxima. The height of the maxima decreased towards higher masses (Figure 2).

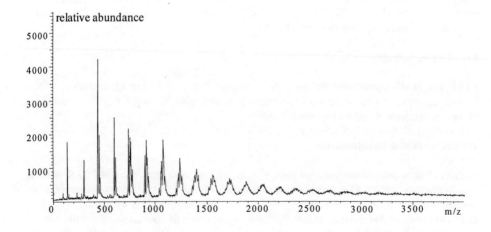

Figure 2 *MALDI-TOF mass spectrum of water extract from freeze dried aqueous HSs*

Similar observations were made for MALDI-TOF mass spectra of lignins,[8] substances thought to be involved at the onset of humification. The corresponding fine structure in those spectra was assigned to oligomeric lignin molecules.

Evidence from combined small-angle neutron scattering (SANS), small-angle x-ray scattering (SAXS) and x-ray microscopy with synchrotron radiation showed cluster formation of fulvic acids at concentrations higher than 5 mg mL^{-1} to give spectra with similar accumulations in molecular size distribution around certain maxima.[9] In that case the results were explained in terms of chain formation of globular structures at higher concentrations (the so-called ball-chain model).

A close correlation between molecular weight and size usually is assumed. Therefore, other methods (besides scattering experiments) have been developed to measure molecular sizes instead of weights. Among these are ultracentrifugation,[10-14] field flow fractionation (FFF),[15] and especially size exclusion chromatography (SEC),[16-23] all of which have been used to characterize HSs.

In view of the widespread application of size exclusion chromatography, it was used here to detect (at 254 nm) the molecular size distribution in different extracts of freeze dried HSs materials and to monitor changes in molecular sizes caused by base mediated oxidations with oxygen. Conditions that suppress both repulsive and attractive interactions with the column material were found to be optimum with a mobile phase of 1:1v/v methanol:0.05 M sodium chloride. Calibration with polystyrenesulfonates was chosen to allow comparison with literature data.

Generally speaking, the chromatograms (Figure 3) show a high content of nominal molecular masses in the region of 1 kDa for extracts with an aqueous medium, whereas extracts obtained especially with ethanol contain molecules smaller than 0.5 kDa and larger than 100 kDa.

Exposure to oxygen in a basic medium induced changes in all the samples. Considerable changes were observed for phosphate buffer extracts (in contrast to water extracts) and also for extracts with methanol and ethanol. Based on the fact that both increases and decreases of molecular masses were observed on autoxidation, it became clear that both formation and degradation processes were involved. Degradation processes in alkaline media have been registered previously and were attributed to cleavage of ester bonds and so on.[24] On the other hand, formation processes are assigned to radical reactions, as described for polyphenols.[25]

Even if MALDI-TOF spectrometry and SEC do not reflect the full truth about the sizes of HSs, important tendencies are clear from our experimental measurements.

3.2 IR and NMR Spectrometry

IR spectra (Figure 4) provided further evidence that significant alterations in molecular size, functional groups and primary structures of HSs occur in alkaline media containing oxygen. A precise assignment of functional group modifications appears difficult due to overlapping spectral features. It is obvious, however, that different moieties in different extracts became altered on autoxidation.

IR spectra (Figure 4) and NMR spectra (Figure 5) show an increasing content of C-H functionalities (IR: ν = 2800-3000 cm^{-1}; NMR: δ = 0.5-1.5 ppm) with increasing hydrophobicity of the extraction medium. Also, more hydrophobic extracts appear to be more susceptible to base catalysed alterations.

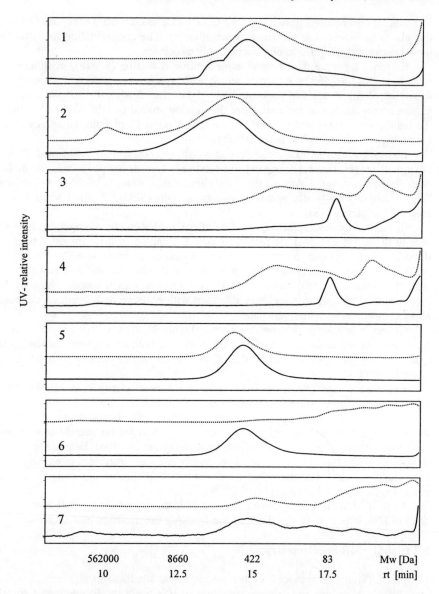

Figure 3 *SEC spectra of extracts of HSs – 1: peat HS with water, 2: peat HS with 0.1 M KH_2PO_4 buffer, 3: peat HS with methanol, 4: peat HS with ethanol, 5: aqueous HS with water, 6: aqueous HS with methanol, 7: aqueous HS with ethanol; solid line: original sample; dashed line: exposed to oxygen in basic medium under ultrasound; rt: retention time; Mw: molecular weight. Molecular weight calibration carried out with polystyrene sulfonate standards (1200, 780, 100, 74, 35, 18, 8 kDa); acetone as internal standard (rt: 20.5 min); absorbance measured at 254 nm.*

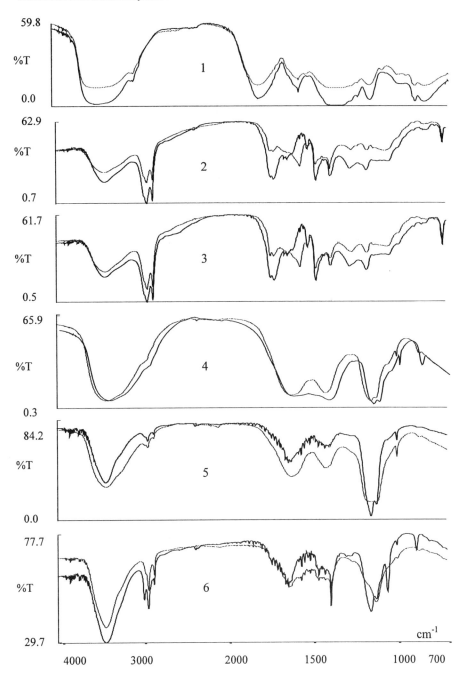

Figure 4 *FTIR spectra of extracts of HSs – 1: peat HS with water, 2: peat HS with methanol, 3: peat HS with ethanol, 4: aqueous HS with water, 5: aqueous HS with methanol, 6: aqueous HS with ethanol; solid line: original sample, dashed line: exposed to oxygen in basic medium under ultrasound*

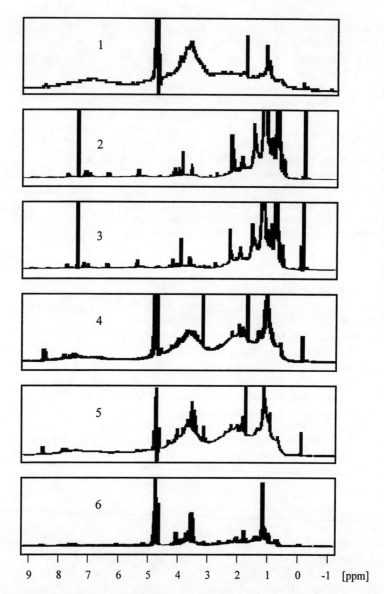

Figure 5 *NMR spectra of extracts of HSs – 1: peat HS with water (D$_2$O), 2: peat HS with methanol (CDCl$_3$), 3: peat HS with ethanol (CDCl$_3$), 4: aqueous HS with water (D$_2$O), 5: aqueous HS with methanol (D$_2$O), 6: aqueous HS with ethanol (D$_2$O); D$_2$O spectra measured with suppression of water signal*

4 CONCLUSIONS

Indications of considerable size differences between polar and non polar HSs were found in this study. Both degradation and formation processes are induced by oxygen in a basic reaction medium. Moreover, some functional moieties are altered. These results are evidence of the fact that HSs are susceptible to radical reactions and that radical reactions play a major role in HSs formation.

The results presented also support another hypothesis: there cannot be a common molecular formula nor a common structural constitution for HSs. Radicals have innumerable ways of reacting (even without radical reactions, formation of heterogeneous products can be observed, e.g. melanoidins). It hardly will be possible to find any two HSs molecules with the same structure. Even if the determination of a molecular structure for one single molecule could be achieved, the information would be irrelevant because of the lack of information about all the other molecules. The seemingly infinite variability of HSs structures is the most outstanding characteristic of HSs.

HSs have been applied in medicine for a long time with amazing success,[26] yet no biologically active compound responsible for the effects has been isolated. Effects in medicine, bioremediation and other areas may be correlated with the degree of humification of a HSs pool because of structural properties and moieties present at particular humification stages. For this reason, characterization of HSs should be carried out in the state (solid, solution) in which they will be applied.

ACKNOWLEDGEMENTS

Financial support from Stiftung Industrieforschung (Köln, Germany, in cooperation with BREGAU Institutes (Bremen, Germany), Gesellschaft für Angewandte Geologie und Biologie, Rainer Hartmann (Göttingen, Germany) and the Centre of Biological Processes in Contaminated Soil and Sediment under the Danish Environmental Research Program (Århus, Denmark) is gratefully acknowledged. Equipment for conducting the ultrafiltration generously was made available by Sartorius AG (Göttingen, Germany). Technical assistance and advice were contributed by Anja Nielsen, Ingelis Larsen, Frank Laturnus (Risø National Laboratory, Denmark), the group of Wolfgang Thiemann (University of Bremen), Wirtschaftsakademie Schleswig-Holstein (Lübeck, Germany), Wasserwirtschaftsamt (Bremen, Germany) and the Niedersächsisches Landesamt für Ökologie (Hannover, Germany).

References

1. W. Ziechmann, 'Humic Substances', Wissenschaftsverlag, Mannheim, 1994, Chapter 4, p. 70.
2. S. L. Miller, *J. Am. Chem. Soc.*, 1955, **77**, 2351.
3. W. Ziechmann, 'Humic Substances', Wissenschaftsverlag, Mannheim, 1994, Chapter 12, p. 208.
4. W. Eller, *Brennstoffchemie*, 1921, **2**, 192.
5. W. Ziechmann, 'Humic Substances', Wissenschaftsverlag, Mannheim, 1994, Chapter 4, p. 72.
6. W. Ziechmann, in 'Handbuch der Bodenkunde', H. -P. Blume et al., (eds.), Spektrum Akademischer Verlag, Heidelberg, 1997, Chapter 2.2.3.4, p. 8.
7. G. Haberhauer, W. Bednar, M. H. Gerzabek and E. Rosenberg, in this volume.
8. J. O. Metzger, C. Bicke, O. Faix, W. Tuszynski, R. Angermann, M. Karas and K. Strupat, *Angew. Chem.*, 1992, **104**, 777.
9. A. Knöchel, K. Pranzas, H. Stuhrmann and R. Willumeit, *Physica B*, 1997, **234**-

236, 292.
10. M. N. Jones, J. W. Birkett, A. E. Wilkinson, N. Hesketh, F. R. Livens, N. D. Bryan, J. R. Lead, J. Hamilton-Taylor and E. Tipping, *Anal. Chim. Acta*, 1995, **314**, 149.
11. J. W. Birkett, M. N. Jones, N. D. Bryan and F. R. Livens, *Anal. Chim. Acta*, 1998, **362**, 299.
12. N. D. Bryan, N. Hesketh, F. R. Livens, E. Tipping and M. N. Jones, *J. Chem. Soc., Faraday Trans.*, 1998, **94**, 95.
13. N. Hesketh, M. N. Jones and E. Tipping, *Anal. Chim. Acta*, 1996, **327**, 191.
14. J. R. Lead, J. Hamilton-Taylor, N. Hesketh, M. N. Jones, A. E. Wilkinson and E. Tipping, *Anal. Chim. Acta*, 1994, **294**, 319.
15. M. E. Schimpf and M. P. Petteys, *Colloids and Surfaces A.: Physicochem. Eng. Aspects*, 1997, **120**, 87.
16. H. K. J. Powell and E. Fenton, *Anal. Chim. Acta*, 1998, **362**, 299.
17. K. Johannsen, M. Assenmacher, M. Kleiser, G. Abbt-Braun, H. Sontheimer and F. H. Frimmel, *Vom Wasser*, 1993, **81**, 185.
18. G. Abbt-Braun, K. Johannsen, M. Kleiser and F. H. Frimmel, *Environ. Internat.*, 1994, **20**, 397.
19. F. H. Frimmel and G. Abbt-Braun, *Environ. Internat.*, 1999, **25**, 191.
20. D. Hongve, J. Baann, G. Becher and S. Lømo, *Environ. Internat.*, 1996, **22**, 489.
21. H. G. Barth, B. E. Boyes and C. Jackson, *Anal. Chem.*, 1998, **70**, 251R.
22. R. Artinger, G. Buckau, J. I. Kim and S. Geyer, *Fresenius J. Anal. Chem.*, 1999, **364**, 737.
23. R. Artinger, G. Buckau, S. Geyer, P. Fritz, M. Wolf and J. I. Kim, *Appl. Geochem.*, 2000, **15**, 97.
24. A. Eschenbach, M. Kästner, R. Bierl, G. Schaeffer and B. Mahro, *Chemosphere*, 1994, **28**, 683.
25. W. Ziechmann, in 'Handbuch der Bodenkunde', H. -P. Blume et al., (eds.), Spektrum Akademischer Verlag, Heidelberg, 1997, Chapter 2.2.3.4, p. 4.
26. W. Ziechmann, 'Huminstoffe und ihre Wirkungen', Spektrum Akademischer Verlag, Heidelberg, 1996, p.112.

HUMIFICATION OF DUCK FARM WASTES

M. Schnitzer,[1] H. Dinel,[1] H.-R. Schulten,[2] T. Paré[1] and S. Lafond[1]

[1] ECORC, Agriculture and Agri-Food Canada, Ottawa, Ontario K1A 0C6, Canada
[2] The University of Rostock, Rostock, Germany

1 INTRODUCTION

Composting, a widely used method for the recycling of manures and organic wastes, curtails environmental pollution, reduces landfilling and limits greenhouse gas emissions. During composting, organic substances, which often contain N, P and S in addition to C, are transformed mainly through the activities of successive microbial populations into more stable organic materials that chemically and biologically resemble humic substances. Thus, oxidative biodegradation is the major reaction mechanism underlying both composting and humification.

In recent years we have studied the chemistry governing the composting of cattle manure, chicken manure and, more recently, the co-composting of municipal solid waste and sewage sludge.[1] In this study we followed chemical changes occurring during the composting of duck excreta that had been collected in pine wood shavings which were used as bedding materials. We hoped that a more comprehensive understanding of the chemistry would lead us to practice composting more efficiently and also generate value-added products that would lower the composting costs.

2 MATERIALS AND METHODS

2.1 Feedstock

Duck excreta enriched pine wood shavings (1,500 kg - 2/3 wood shavings and 1/3 duck excreta) were collected on a duck farm south of Montreal. Separate samples of duck excreta and untreated pine wood shavings also were collected.

2.2 Composting

The duck excreta enriched wood shavings were transferred to a home-built composter, which was located inside a barn. Temperature, moisture, total weight, total C, total N, and air flow were monitored daily. The mixture was mechanically mixed for 10 min. every 2h. The moisture content was maintained at 50% throughout the composting period. The compost reached biomaturity after 29 days. The temperature rose from 29°C on day 0 to 62°C on day 2, reached a maximum of 68°C on day 12 and then fell to 29°C on day 29.

2.3 Chemical Analyses

Triplicate samples were collected, oven-dried (at 105°C) and analyzed for total C and total N by dry-combustion in a Leco combustion furnace. Lipids were extracted sequentially in an Accelerated Solvent Extractor (ASE 200, Dionex) from 2-3 g of finely ground material first with diethyl ether (DEE) and then with chloroform ($CHCl_3$). Each extract was dried and weighed. Weight ratios of DEE- and $CHCl_3$ - extractable lipids and of $CHCl_3$ - and total extractable lipids (TEL) were used as indicators of compost maturity.[2]

2.4 ^{13}C NMR Spectroscopy

Solid-state ^{13}C NMR spectra were recorded on 300 mg of each thoroughly dried compost on a Bruker CXP-180 NMR spectrometer equipped with a Doty Scientific probe at a frequency of 45.28 MHZ. Single-shot cross-polarization contacts of 2 ms were used with matching radio frequency field amplitudes of 75 MHZ. Up to 120,000 500-word induction decays were co-added with a delay time of 1 s. These were zero-filled to 4 K before Fourier transformation. Magic angle spinning rates were about 4 KHz.

2.5 Pyrolysis-Field Ionization Mass Spectrometry (Py-FIMs)

About 3 mg of each air-dry compost was transferred to a quartz micro-oven and heated linearly in the direct inlet system of the mass spectrometer from 50 to 750°C at a rate of 1°C s^{-1}. A double-focussing Finnigan MAT 731 mass spectrometer was used and about 40 magnetic scans per sample were recorded over the mass range 18 to 900 m/z. Three analytical replicates per sample were run and the FIMS signals of all spectra were integrated to produce summed spectra.

2.6 Weight Losses During Composting

The two methods used for this purpose were weighing and measuring total ion intensities (TII as counts x 10^6 mg^{-1}) during mass spectrometry. Previous studies[3] show that the TII of Py-FI mass spectra was directly proportional to the organic matter (OM) concentration.

3 RESULTS AND DISCUSSION

3.1 Chemical Characteristics of Composts

The data in Table 1 show that the percentage volatile matter (VM) decreased with composting. The total C content of the OM remaining after 12 days of composting was less but it increased on day 29 to approach the initial value (see data expressed on a moisture- and ash-free basis). By contrast, total N continued to decrease with increasing time of composting. The latter observations also are confirmed by increases in C/N ratios from day 0 to day 29. $DEE/CHCl_3$ ratios of extractable lipids decreased from 10.6 on day 0 to 2.3 on day 29 (Table 2), while

Table 1 *Analytical characteristics of composts (moisture-free)*

Days	% VM	% C	% N	C/N	% ash
0	71.0	39.2(46.1)[a]	1.7 (2.0)[a]	22.6	15.0
12	64.0	33.5(41.9)[a]	1.2 (1.6)[a]	27.0	20.0
29	58.5	34.0(43.8)[a]	1.1 (1.4)[a]	30.4	22.4

[a] ash-free

$CHCl_3$/TEL ratios increased from 0.09 to 0.30 during the same period of time. DEE/$CHCl_3$ ratios of <2.50 and $CHCl_3$/TEL ratios of >0.30 are characteristic of composts having attained maturity.[3] Thus, it appears from the data in Table 2 that the compost had reached maturity after 29 days. During the latter period of time, the compost had lost 45.5% of its initial weight as determined by weighing and slightly less as calculated from losses in total ion intensities. When we consider total weight losses during composting of 18.6% on day 12 and 45.5% on day 29, we can compute that on day 12 the residual OM had lost 24.0% of its C and 33.5% of its N, whereas on day 29 the remaining OM had lost 48.2% of its C and 62.0% of its N.

Table 2 *Analytical characteristics of composts*

Days	DEE $CHCl_3$ soluble (g kg^{-1})	$CHCl_3$	DEE/ $CHCl_3$	$CHCl_3$ /TEL	Weight losses by weighing%	TII Counts, $x10^6 mg^{-1}$	ΔTII%
0	11.5	1.1	10.6	0.1	0	2.227	0
12	9.8	2.4	4.1	0.2	18.6	1.872	-16.0
29	5.8	2.5	2.3	0.3	45.5	1.272	-42.9

3.2 Py-FI Mass Spectra of the Different Materials

Identification of the major signals in the Py-FIMS spectra are based on extensive earlier researches.[3-6] Figure 1 shows the Py-FI mass spectrum of the duck excreta. The spectrum shows mass signals at m/z 84, 96, 98, 110, 126, 132, 144 and 162, indicating the presence of polysaccharides. Other signals show the presence of a homologous series of n-fatty acids ranging from m/z 172 (n-C_{10}) to 480 (n-C_{32}). Especially prominent in this series are m/z 368 (n-C_{24}), 382 (n-C_{25}), 396 (n-C_{26}), and 424 (n-C_{28}). Other strong signals due to sterols are m/z 386 (cholesterol), 394 (ethylcholestatriene), 410 (dehydrostigmasterol), 412 (stigmasterol), 414 (β-sitosterol) and 416 (dihydro-β-sitosterol). Signals due to N-compounds are m/z 59 (acetamide), 81 (methylpyrrole), 95 (dimethylpyrrole), 103 (benzonitrile), 117 (indole), 131 (methylindole) and 167 (N-acetylglucosamine). Also, small signals indicative of the presence of proteinaceous components[7] are m/z 70, 84, 97 and 115 (acidic amino acids), m/z 57, 60, 73, 75 and 87 (neutral amino acids), m/z 74, 91, 120 and 135 (neutral aromatic amino acids), and m/z 60, 129 and 135 (basic amino acids). Judging from the contributions of the different classes of compounds to the total ion intensity, the major components of the duck excreta are: lipids (including sterols), followed by N-compounds (both heterocyclics and proteinaceous materials) and polysaccharides.

Figure 2 shows the Py-FI mass spectrum of the wood shavings (WS). Mass signals at m/z 84, 96, 98, 110, 112, 114, 126, 132, 144 and 162 are due to polysaccharide thermal

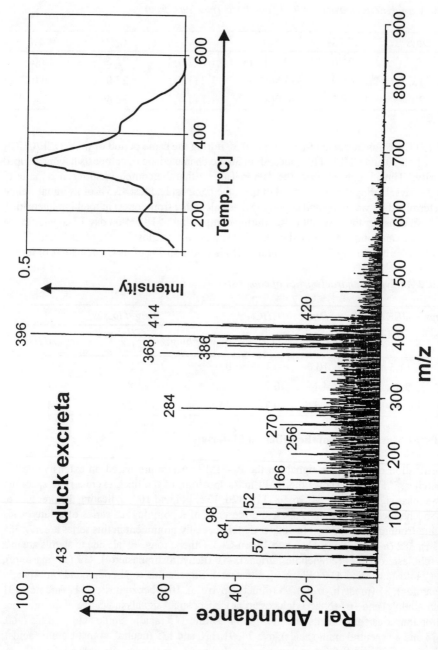

Figure 1 *Pyrolysis-field ionization mass spectrum of duck excreta*

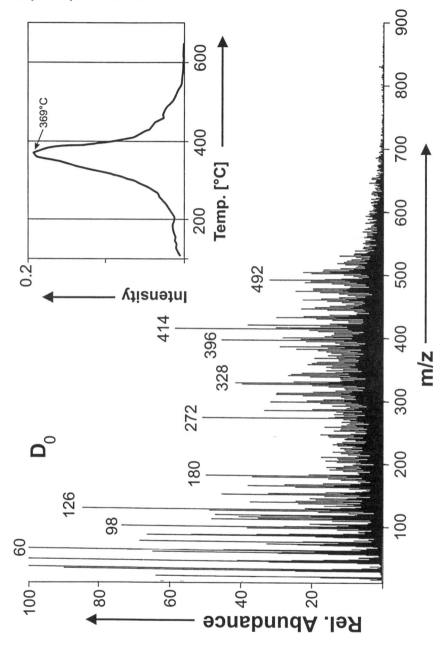

Figure 2 *Pyrolysis-field ionization mass spectrum of Pinus sp. wood shavings*

oxidation products. Strong characteristic signals originating from aldohexose subunits of cellulose can be seen at m/z 126, 144 and 162. The presence of pentoses is indicated by signals at m/z 114 and 132. Monomeric pyrolysis products of lignin with guaiacyl structures are shown by m/z 124 (I, guaiacol (from this point the Roman numerals refer to chemical structures shown in Figures 3 and 4)), 138 (II, methylguaiacol), 140 (III, hydroxyguaiacol), 150 (IV ethyleneguaicol), 152 (V, vanillin), 164 (VI, guaiacolpropene), 168 (VII, vanillic acid), 178 (VIII, coniferyl aldehyde), 180 (IX, coniferyl alcohol) and 194 (X, ferulic acid). The strongest signal in the mass spectrum is produced by coniferyl alcohol. Signals characteristic of syringyl structures are weak but can be detected at m/z 154 (XI, syringol), 168 (XII, methylsyringol), 182 (XIII, syringylaldehyde), 196 (XIV, syringylacetaldehyde), 208 (XV, sinapylaldehyde), 210 (XVI, sinapylalcohol) and 212 (XVII, syringylacetic acid). Signals due to lignin dimers can be seen at m/z 246, 260, 272, 284, 286, 296, 300, 312, 314, 316, 326, 328, 330, 340, 342, 344, 356, 358, 360, 362, 372, 374, 376, 386, 388, 402, 416 and 418. The most prominent of these signals are m/z 272 (XVIII), 284, (XIX), 298 (XX), 300 (XXI), 312 (XXII), 316 (XXIII), 326 (XXIV), 328 (XXV) and 340 (XXVI), all of which have been suggested to result from the condensation or polymerization of coniferyl alcohol (IX), a fundamental building block of coniferous lignin.[8]

Figure 2 shows additional peaks, most likely due to n-fatty acids at m/z 242, (C_{15}), 256 (C_{16}), 270 (C_{17}), 284 (C_{18}), 298 (C_{19}), 312 (C_{20}), 326 (C_{21}), 340 (C_{22}), 354 (C_{23}), 368 (C_{24}), 382 (C_{25}), 396 (C_{26}), 410 (C_{27}), 452 (C_{30}), 480 (C_{32}), 494 (C_{33}) and 508 (C_{34}). The presence of a homologous series starting at m/z 432, 446, 460, 474, 488, 502, 516 and 530 is indicative of the presence of partially oxidized phenolic esters, that is esters of phenols with n-fatty acids[5] of the type C_6H_5-O-CO-(CH_2)n-COOH with n ranging from 19 to 26.

Figure 5 shows the Py-FI mass spectrum of a mixture of wood shavings and duck feces. Similar to Figure 1, this spectrum is dominated by signals arising from polysaccharides, especially cellulose, and monomeric and dimeric lignin structures. Contributions from the duck manure are the relatively strong signals in the m/z 386 to 430 region due to sterols. The signal at m/z 386 indicates the presence of cholesterol, m/z 394 is ethylcholestatriene, m/z 410 is ethylcholesterol, and m/z 412 is stigmasterol. The strong signal at m/z 414 arises from β-sitosterol, while m/z 416 is dehydro-β-sitosterol, and m/z 430 is α-tocopherol. Other components originating from the manure are n-C_{16} to n-C_{40} fatty acids and the phenolic esters mentioned above.

The Py-FI mass spectrum of the mixture that had been composted for 12 days (Figure 6) is very similar to that of the starting materials (Figure 5). Polysaccharides, lignin monomers and dimers, sterols, n-fatty acids and phenolic esters are the major components. The only compounds that appear to have increased are the sterols m/z 412 (stigmasterol), 414 (β-sitosterol) and 416 (dehydro-β-sitosterol).

The Py-FI mass spectrum recorded on the mixture after 29 days of composting (Figure 7) is similar to the Py-FIMS spectrum of the mixture that had been composted for only 12 days. During the additional 17 days of composting, the molecular composition of the compost did not change except for increases in sterols and possibly n-fatty acids and phenolic esters. Thus, after 29 days of composting the resulting organic matter was a heterogeneous mixture of polysaccharides (cellulose), lignin residues, phenolic esters and lipids, with sterols being the most prominent members of the latter group. It is noteworthy that the Py-FI mass spectra recorded at the different stages of composting were quite similar.

a)

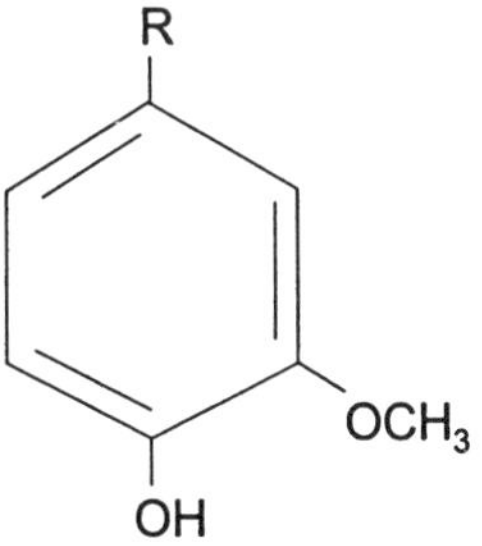

Compound No.	M/z	R
I	124	-H
II	138	$-CH_3$
III	140	-OH
IV	150	$-CH=CH_2$
V	152	-CHO
VI	164	$-CH_2-CH=CH_2$
VII	168	-COOH
VIII	178	-CH=CH-CHO
IX	180	$-CH=CH-CH_2OH$
X	194	-CH=CH-COOH

b)

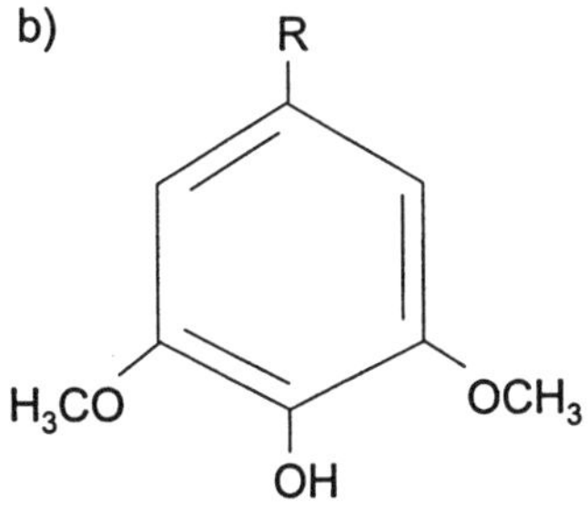

Compound No.	M/z	R
XI	154	-H
XII	168	$-CH_3$
XIII	182	-CHO
XIV	196	$-CH_2=CHO$
XV	208	-CH=CH-CHO
XVI	210	$-CH=CH-CH_2OH$
XVII	212	$-CH_2-COOH$

Figure 3 *Proposed identifications of monomeric pyrolysis products of wood shavings with guaiacyl (a) and syringyl (b) structures*

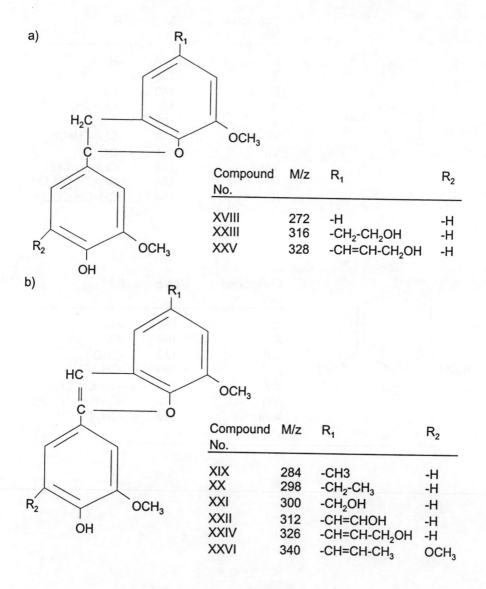

Figure 4 *Proposed identifications of dimeric pyrolysis products of wood shavings with phenylcoumaran structures*

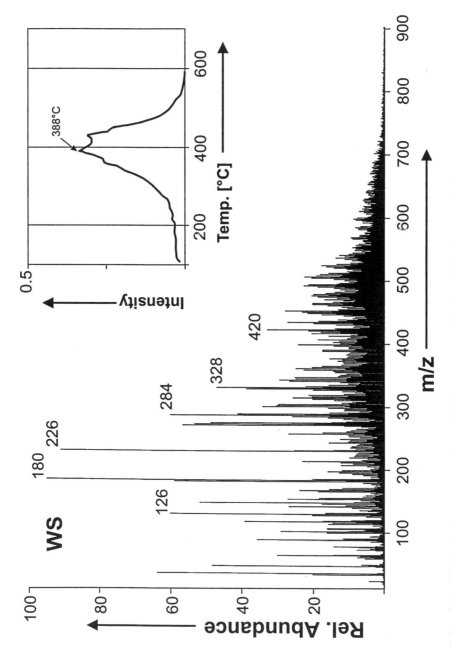

Figure 5 *Pyrolysis-field ionization mass spectrum of duck excreta enriched wood shavings on day 0*

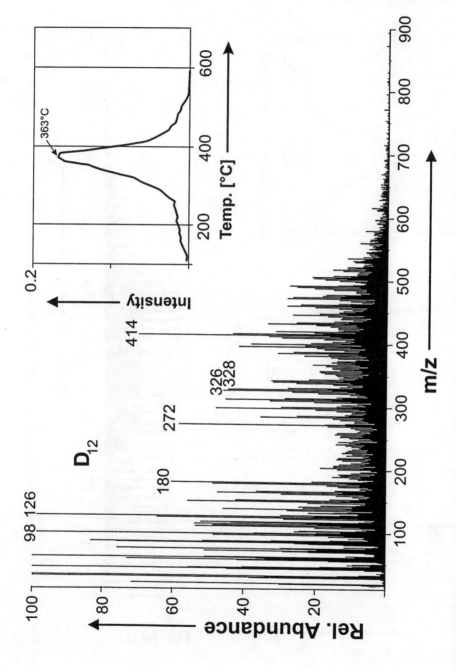

Figure 6 *Pyrolysis-field ionization mass spectrum of composted duck excreta enriched wood shavings on day 12*

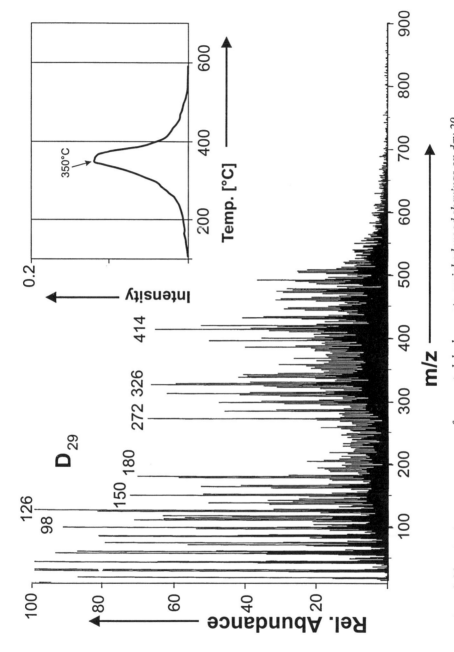

Figure 7 *Pyrolysis-field ionization mass spectrum of composted duck excreta enriched wood shavings on day 29*

3.3 ^{13}C NMR Spectra of Composts

The solid-state ^{13}C NMR spectrum of the compost on day 0 (D_0, Figure 8) shows signals at 22, 31, 37, 45, 58, 64, 72, 83, 88, 104, 118, 128, 146, 155, 160, and 170 ppm. The signal at 22 ppm arises from aliphatic CH_3, that at 31 ppm from CH_2 groups in long-chain aliphatics, and that at 37 ppm also from aliphatic structures. The resonances between 50 and 60 ppm are likely due in part to C in amino acids, peptides and proteins, and to C in OCH_3 groups of lignins whose presence is indicated by Py-FIMS. Other major contributors to resonances in the 0 to 60 ppm region are sterols whose saturated ring carbons and carbons in side chains produce signals in this area. The most prominent resonances in the spectrum are in the 60 to 105 ppm region and are due to carbohydrate-C. The strong signals at 72 and 103 ppm are characteristic of the C-2, C-3, C-5 and C-1 carbons of cellulose, while the shoulders near 64 and 88 ppm indicate the presence of the corresponding C-6 and C-4 carbons.[10] Aromatic carbons show signals at 118, 128 and 146 ppm, but these are weak except for the broad resonance near 145 ppm, which is close to the chemical shift for the C-3 and C-4 carbons of vanillyl units in lignin.[10] The signal centered near 170 ppm is due to C in COOH groups although C in amides and esters could also contribute to this resonance.

The ^{13}C NMR spectrum of the compost on day 12 (D_{12}, Figure 8) is similar to that recorded on D_0. Paraffinic aliphatics (12 to 40 ppm), sterols (0 to 60 ppm), proteinaceous materials (30-70 ppm), celluloses (60 to 105 ppm), aromatics (lignins) (105 to 150 ppm), phenolics (150 to 165 ppm) and lignins are the major components of this compost. Peaks due to aromatic C (116, 132, 142 and 146 ppm) are more prominent in spectrum D_{12} than in spectrum D_0. Compared to spectra D_0 and D_{12}, spectrum D_{29} (Figure 8) exhibits a diminution of cellulose but significant increases in aromatic and phenolic C. There are numerous distinct signals between 116 and 150 ppm (aromatic C) and between 150 and 170 ppm (phenolic C).

For a more detailed analysis of the data, the total area (0-190 ppm) under each spectrum was integrated and divided into the following regions:[11] 0-40 ppm (paraffinic C), 41-60 ppm C in OCH_3 groups, amino acids and sterols), 61-105 C ppm in celluloses, 106-150 ppm (aromatic C and C in sterols), 151-170 ppm (phenolic C), and 171-190 ppm (C in COOH, amide, and ester groups). The aromaticity was computed by expressing aromatic C (106-170 ppm) as a percentage of aliphatic C + aromatic (0-170 ppm).

Table 3 shows the C distribution of the composts on days 0, 12, and 29. While the ^{13}C NMR data are at best only semi-quantitative, they show nonetheless a number of interesting trends. As composting proceeds from day 0 to day 29 (biomaturity), there are reductions in paraffinic and proteinaceous C and certainly in cellulose C, whose concentration drops by 35.0%, but increases in aromatic and phenolic C. The aromaticity increases from 14.8 to 35.2%. Part of this increase is probably due to relatively greater weight losses of proteinaceous materials and celluloses compared to lignins, but increases in aromaticity appear to occur during composting. Further research is needed to elucidate the nature of these reactions. Py-FIMS indicates that the concentration of sterols at biomaturity remains relatively high (about 10% of the total material) so that a substantial proportion of the C in the 0-60 ppm region of ^{13}C NMR spectrum D_{29} appears to be due to C in sterols. During composting, COOH groups increased slightly.

3.4 Total Ion Intensities (TII) and Volatile Matter (VM)

Figure 9 shows that the volatility of the OM in the three mixtures is directly proportional to the total ion intensity. With increasing composting both TII and VM decrease but the total C remains relatively constant (see Table 1). These findings may be interpreted as indicating greater molecular cross-linking and more inter- and intra- molecular associations of the OM on day 29 than on day 0. The end result of composting appears to be the formation of larger and more stable molecules that are chemically and biologically less reactive than the starting materials. The overall impression that our data conveys is that the celluloses, lignins and proteinaceous materials are "glued" together by the lipids and sterols.

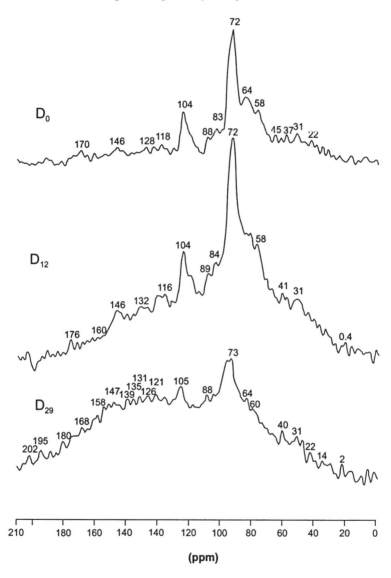

Figure 8 ^{13}C NMR spectra of composted duck excreta enriched wood shavings on days 0, 12, and 29

Table 3 *Distribution of C in D_0, D_{12} and D_{29} composts as determined by solid-state ^{13}C NMR*

Type of C	Chemical shift range (ppm)	% distribution (% of total C)		
		D_0	D_{12}	D_{29}
Aliphatic C (including C in alkanes, fatty acids, aliphatic esters, and sterols)	0-40	13.5	12.1	9.4
Protein C, C in OCH_3 and sterols	41-60	18.3	15.4	9.4
Cellulose	61-105	51.0	42.9	33.3
Aromatic C and C in sterols	106-150	12.5	23.1	30.2
Phenolic C	151-170	1.9	4.4	13.5
C in COOH groups	171-190	2.9	2.2	4.2
Aromaticity	$\frac{(106-170) \times 100}{0-170}$	14.8	28.1	35.2

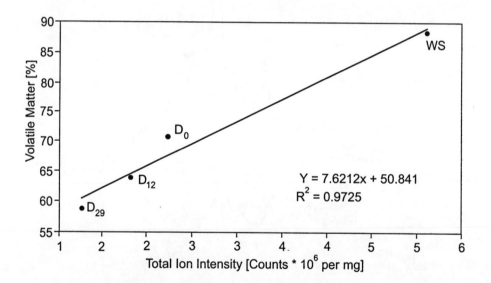

Figure 9 *Relationship between volatile matter and total ion counts for composted duck excreta enriched wood shavings on days 0, 12 and 29*

The data reported here are similar to findings of an earlier study[12] in which the effects of 80 years of cultivation on SOM were investigated. Cultivation, which brings air into the soil and enhances chemical and biological oxidation (humification) of SOM, was found to increase cross-linking of the major SOM components, leading to the formation of larger SOM molecules with higher molecular weights, stability and complexity. Also, decreases in enzyme activities involving the C, N and P cycles were observed. Thus, humification and composting appear to be based on similar chemical and biological reaction mechanisms.

4 CONCLUSIONS

Celluloses, followed by lignins, lipids (including sterols) and proteinaceous materials were the major components of the compost on day 0. After 12 days of composting, the remaining OM had lost 18.6% of its initial weight, 24.0% of its initial C and 33.5% of its N. Chemical analysis showed that the compost had attained biomaturity by day 29. The C content of the residual material was similar to that of the starting material, although by that time the compost had lost 45.5% of its initial moisture- and ash- free weight, 48.2% of its initial C and 62.0% of its initial N. Py-FI mass spectra showed that at biomaturity, the compost was a heterogeneous mixture of celluloses, lignins, phenolic esters, lipids (with sterols prominent) and proteinaceous materials. Solid state ^{13}C NMR data indicated that there were reductions in paraffinic C, proteinaceous C and especially in cellulose C with increasing composting time. While concentrations of C in sterols remained unchanged, there were increases in aromatic and phenolic C and in aromaticity. C in COOH groups increased only slightly.

With increasing composting time, both TII and VM decreased while the C content of the residual compost changed relatively little. These findings appear to indicate greater molecular cross-linking and more inter- and intra-molecular associations in the composted OM, which also becomes more aromatic. The end result of composting is the formation of larger OM molecules with higher molecular weights, greater stability and complexity, but chemically and biologically less reactive than the starting materials.

References

1. T. Paré, H. Dinel and M. Schnitzer, *Biol. Fertil. Soils*, 1999, **29**, 31.
2. H. Dinel, M. Schnitzer and S. Dumontet, *Compost Sci. Util.*, 1996, **4**, 6.
3. M. Schnitzer and H. -R. Schulten, *Adv. Agron.*, 1995, **55**, 167.
4. H. -R. Schulten and N. Simmleit, *Naturwissenschaften*, 1986, **73**, 618.
5. R. Hempfling, W. Zech and H. -R. Schulten, *Soil Sci.*, 1988, **146**, 262.
6. R. Hempfling and H. -R. Schulten, *Org. Geochem.*, 1990, **15**, 131.
7. C. Sorge, M. Schnitzer and H. -R. Schulten, *Biol. Fertil. Soils*. 1993, **16**, 100.
8. W.J. Schubert, 'Lignin Biochemistry', Academic Press, New York, 1965.
9. F. Le Roy, F. Johnson and W. Jankowski, 'Carbon-13 NMR Spectra', Krieger, Huntington, N.Y., 1978.
10. I. Kögel, R. Hempfling, W. Zech, P.G. Hatcher and H. -R. Schulten, *Soil Sci.*, 1988, **146**, 124.
11. M. Schnitzer and C. M. Preston, *Soil Sci. Soc. Am. J.*, 1987, **51**, 639.
12. H. -R. Schulten, C. M. Monreal and M. Schnitzer, *Naturwissenschaften*, 1995, **82**, 42.

CATALYTIC EFFECTS OF HYDROXY-ALUMINUM AND SILICIC ACID ON CATECHOL HUMIFICATION

C. Liu and P. M. Huang

Department of Soil Science, University of Saskatchewan, Saskatoon, SK S7N 5A8, Canada

1 INTRODUCTION

Polyphenols exist widely in soils as decomposition products of plant and animal tissues and microbial metabolites.[1,2] Transformations of polyphenols through abiotic and biotic catalytic processes are important pathways of the formation of humic substances.[3-7] Abiotic catalysts, which include a series of soil inorganic constituents, promote the transformation of phenolic compounds to humic macromolecules through their oxidative polymerization, ring cleavage, decarboxylation, and/or dealkylation.

Manganese(IV) and iron(III) oxides are powerful in causing the darkening of phenolic compounds and the subsequent formation of humic substances.[8-12] They can act as Lewis acids that accept electrons from diphenols, leading to their oxidative polymerization.[7] Aluminum oxides and silica gel were found to promote the humification of polyphenols to much less extents in a reaction period less than 2 wk compared with Mn(IV) and Fe(III) oxides.[9,13-15] This is attributable to the fact that aluminum and silicon have fixed oxidation states. Nevertheless, aluminum and silicon are among the most abundant elements in the earth crust. Further, Al and Si are always present as predominant impurities in the humic acids (HAs) and fulvic acids (FAs) extracted from soils.[4,16,17] This highlights the importance of full understanding of the role of Al and Si in the formation of humic substances.

Although the catalytic role of Al hydroxides and Si oxides in the formation of humic substances from polyphenols has been investigated,[9,13-15] there is only one report on the catalytic effect of soluble Al ions on the darkening of catechol.[15] McBride et al.[15] found that soluble Al increased the rate of oxidation of catechol by O_2, favoring the formation of highly colored solution products in the pH range 5-7 in a reaction period of less than 2 wk, although no precipitation was observed. Aerobic oxidation of hydroquinone in dilute solution can be catalyzed by the presence of silica gel in acidic media.[18] However, it has been postulated that the accelerating effect of silica gel on the formation of humic acid was mainly from metal impurities such as Fe rather than from silica.[13,18] Nevertheless, the role of silicic acid present in solutions in the humification of polyphenols is unknown. Furthermore, the residence time effect on the catalytic role of Al ions and silicic acid in the humification of polyphenols has not been established.

In the present study, the effects of aqueous hydroxy-Al ions and silicic acid at 10^{-3} M on catechol humification for reaction periods ranging from 10 to 60 days under acidic conditions were investigated. The reaction products were compared with the standard soil HA obtained from the International Humic Substances Society (IHSS).

2 MATERIALS AND METHODS

2.1 Preparation of Reaction Systems

The solution of hydroxy-Al ions at an Al concentration of 2×10^{-3} M and pH 4.4 was prepared by titrating a 0.1 M $AlCl_3$ solution with 0.1 M NaOH. A stock solution of monosilicic acid at a Si concentration of 3.56×10^{-3} M was prepared by passing Na-metasilicate through a H^+-saturated cation exchange resin column. The 2×10^{-3} M monosilicic acid was obtained by diluting the stock solution and was adjusted to pH ~ 4.4. One liter of 0.2 M catechol solution at pH 4.4 was mixed with 1 L of 2×10^{-3} M hydroxy-Al or monosilicic acid solution. A solution of 0.1 M catechol alone at pH 4.4 was also prepared as control. Two hundred milligrams of thimerosal (sodium ethylmercurithiosalicylate) was added into the 1 L catechol solution before mixing with hydroxy-Al or silicic acid solution to inhibit the growth of microorganisms. The final thimerosal concentration was 0.02% (w/v). Further, all the distilled deionized water and the containers used in the experiment were boiled for at least 10 min before solution preparation. The solutions in the polypropylene bottles were sealed with Parafilm and shaken at 298.0 ± 0.1 K for 10, 20, 40, and 60 d. The whole experiment was carried out in duplicate.

2.2 Characterization of the Reaction Products

Small quantities of the reaction solutions or suspensions were removed at the end of 10, 20, 40, and 60-d reaction periods and centrifuged for 25 min at 20,000g. The supernatants were analyzed by UV-visible spectroscopy on a Beckman Model DU 650 spectrophotometer. The solid product at the end of a 60-d reaction period was dialyzed and freeze-dried.

The supernatants of the reaction systems at the end of a 60-d reaction period were analyzed by Fourier transform infrared spectroscopy (FTIR) on a Bio-Rad Model 3240 infrared absorption spectrophotometer using a liquid cell. The solid product at the end of a 60-d reaction period and the standard soil HA (1S102H) obtained from the IHSS were also examined by FTIR on the same machine using the KBr pellet technique. One milligram of the sample was mixed with 200 mg of KBr for the FTIR analysis. Twenty milligram samples of the solid product and the standard soil HA were used for the X-ray powder diffraction (XRD) analysis on a Rigaku diffractometer with Fe-Kα radiation filtered by a graphite monochromator at 40 kV and 130 mA. The X-ray diffractograms were recorded from $2\theta = 4°$ to $80°$ with $0.005°$ steps at a speed of $0.5°$ per min. The solid state ^{13}C nuclear magnetic resonance (NMR) spectrum of solid products was obtained at 90 MHz using magic angle spinning (MAS) and cross polarization on a Model Avance 360NMR spectrometer. The ^{13}C CPMAS-NMR spectrum of the standard soil HA was provided by the IHSS.

For the atomic force microscopy (AFM) analysis of the precipitate, 5 mg of the solid product and the standard soil HA were dispersed in 12 mL deionized distilled water after ultrasonification (Sonifier Model 350) at 150 watts for 2 min in an ice bath. The HA suspension was adjusted to pH 5.0 with 0.01 M NaOH and HNO_3 and diluted to 15 mL. One drop of the suspension was deposited on a watch glass and air-dried overnight at room temperature (296.5 ± 0.5 K). For the AFM analysis of the solution product, one drop of the supernatant was deposited on a watch glass and air-dried overnight at room temperature. The watch glass then was fastened to a magnetized stainless steel disk (diameter 12 mm) with double-sided tape. The 3-dimensional AFM images were obtained under ambient condition with a NanoScopeTMIII atomic force microscope. The imaging areas were 5 x 5 μm^2. The scanner type was 1881E and the scanner size was 15 μm. A silicon nitride cantilever with a spring constant of 0.12 N/m was used in the contact mode. The scanning rate was 22 Hz. The AFM cantilever was changed frequently to prevent experimental artifacts. Furthermore, the scanning area and scanning angle often were changed by entering different area and angle parameters to detect the artifacts caused by adhesion of HA particles to AFM tips. The diameter and thickness of particles were measured with *Section* analysis of the AFM based on 30 particles.

The concentrations of Al and Si as well as impurities such as Fe, Mn, Zn, and Cu in the catechol, silicic acid, and hydroxy-Al solutions and in the supernatants were determined by atomic absorption spectrometry (AAS, Model 3100 spectrophotometer). The C, H, and N contents of the solid products were measured with a Perkin Elmer Model 2400 CHN Elemental Analyzer. The contents of Al, Si, Fe, Mn, Zn and Cu in the solid product were determined by AAS after the sample was digested in $HF-HClO_4$. The ash contents of the solid product were determined by heating the sample in a muffle furnace at 600 °C for 4 h.[19]

3 RESULTS AND DISCUSSION

The absorption spectra in the wavelength range 350 to 600 nm of the supernatants of the silicic acid-catechol and hydroxy-Al-catechol systems at the end of different reaction periods (Figure 1) indicate that the formation of solution products of catechol humification was enhanced with increasing reaction period. The formation of highly colored solution products in the hydroxy-Al-catechol system was faster compared with the silicic acid-catechol system, as revealed by the absorption spectra of the supernatants at the end of different reaction periods (Figure 1). The absorption spectra of the supernatants of all the treatments at the end of 10-d and 60-d reaction periods revealed that the presence of both hydroxy-Al and silicic acid enhanced catechol humification (Figure 2). However, the power of hydroxy-Al ions to catalyze darkening of the supernatant solutions was much greater than that of silicic acid. Further, black precipitates started to form in the hydroxy-Al-catechol system at the end of a 16-d reaction period, whereas the presence of silicic acid only enhanced the darkening of the catechol solution compared with the pure catechol solution and no precipitates were observed even at the end of a 60-d reaction period. The lower absorbance of the supernatant of the hydroxy-Al-catechol system at > 450 nm at the end of a 60-d reaction period than that of the silicic acid system (Figure 2B) apparently was due to the formation of black precipitates. However, the absorbance of the supernatant of the hydroxy-Al-catechol solution at < 450 nm at the end of a 60-d reaction period was still much higher than that of the silicic acid-catechol solution, indicating that the black precipitates in the hydroxy-Al-catechol system resulted from polymerized catechol materials with a higher molecular weight.

Five absorption bands at 1267, 1386, 1482-1483, 1636 and 1674-1692 cm^{-1} were observed in the difference FTIR spectra between the catechol solution and the supernatant of the hydroxy-Al-catechol or the silicic acid-catechol solution (Figure 3).

The band at 1267 cm^{-1} was attributed to the vibration of aromatic ethers ($-O-CH_2$).[20] The bands at 1386 and 1482-1483 cm^{-1} were from the bending vibration of aliphatic C-H groups. The absorption bands at 1636-1692 cm^{-1} were due to the vibration of carboxylic groups - COO⁻ and aromatic C=C double bond conjugated with C=O groups.[20] These absorption bands revealed that the solution reaction products in the hydroxy-Al-catechol and silicic acid-catechol systems were significantly different from the catechol solution alone.

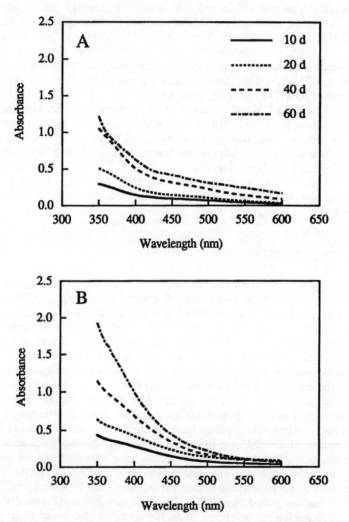

Figure 1 *Absorption spectra of the supernatants of (A) the silicic acid-catechol and (B) hydroxy-Al-catechol systems at the end of reaction periods*

The 60-d aged catechol solution after air-drying was predominated by massive materials with a few spheroidal particles (Figure 4A). These spheroidal particles were attributed to polymerized catechol molecules formed by slow autoxidation of catechol in the presence of dissolved O_2. Film-shaped materials with various diameters of pits were observed in the air-dried silicic acid solution (Figure 4B). The octahedral particles, which are hydroxy-Al polymers, were imaged in the air-dried hydroxy-Al solution (Figure 4C).

By contrast, a large amount of spheroidal particles and their aggregates were observed in the silicic acid-catechol solution and hydroxy-Al-catechol supernatant at the end of a 60 d reaction period after air-drying (Figure 5). These very different surface features of the reaction products in the silicic acid-catechol solution and hydroxy-Al-catechol supernatant compared with catechol, silicic acid, or hydroxy-Al solution alone further revealed that hydroxy-Al and silicic acid substantially catalyzed catechol humification.

The diameter and thickness of the spheroidal particles in the reaction systems after air-drying, respectively, ranged from ca. 80 to 600 nm and ca. 30 to 160 nm (Figure 5), which were much larger than the diameter (ca. 50-150 nm) and thickness (ca. 10-30 nm) of the spheroidal particles of the IHSS standard soil HA.[21] These spheroids larger than the

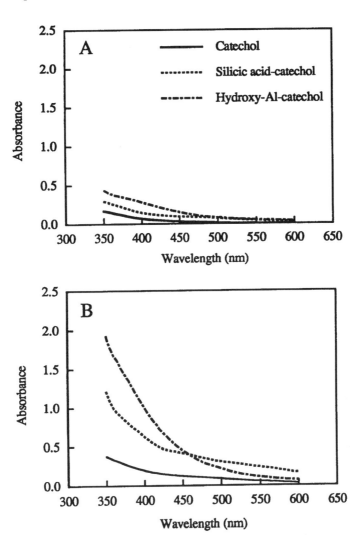

Figure 2 *Absorption spectra of the supernatants of the various systems at the end of (A) 10-d and (B) 60-d reaction periods*

standard soil HA apparently resulted from the aggregation, during the air-drying process, of the polymerized catechol molecules and their complexes with hydroxy-Al and silicic acid.

It has been reported that aerobic oxidation of hydroquinone is catalyzed by silica gel in dilute solutions in an acidic medium and the accelerating effect on the formation of humic acid was attributed to metal impurities such as Fe rather than to silica.[13] In the present study, the impurities in the catechol, silicic acid and hydroxy-Al solutions were analyzed. The data show that 3.7×10^{-7} M Al and 7.1×10^{-8} M Fe were present in the silicic acid solution, whereas Mn, Cu and Zn were not detectable (Table 1). The concentrations of these impurities were about four to five orders of magnitude lower compared with 10^{-3} M Al in the hydroxy-Al-catechol solution (Table 1). However, the absorbance of the silicic acid-catechol solution was about half of the absorbance of the supernatant of the hydroxy-Al-catechol system (Figure 2). Further, the absorbance values of the hydroxy-Al-catechol solution at 3.7×10^{-7} M Al and the hydroxy-Fe-catechol solution at 7.3×10^{-8} M Fe were virtually zero. Therefore, metal impurities at such low concentrations in the silicic acid-catechol solution cannot account for the extent of the darkening of catechol.

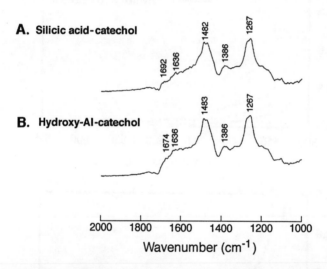

Figure 3 *Difference FTIR spectra between the catechol solution and the solution of silicic acid-catechol or hydroxy-Al-catechol at the end of a 60-d reaction period*

Table 1 *Elemental concentrations of the supernatants of various systems*[a]

	C	Al	Si	Fe	Mn	Cu	Zn
Catechol	0.1	n.d.[b]	1.1×10^{-7}	n.d.	n.d.	n.d.	n.d.
Silicic acid	n.d.	3.7×10^{-7}	1.0×10^{-3}	7.1×10^{-8}	n.d.	n.d.	n.d.
hydroxy-Al	n.d.	1.0×10^{-3}	1.2×10^{-7}	8.0×10^{-8}	n.d.	n.d.	n.d.
Silicic acid-catechol	0.1	3.7×10^{-7}	1.0×10^{-3}	7.3×10^{-8}	n.d.	n.d.	n.d.
Hydroxy-Al-catechol	0.1	7.6×10^{-4}	1.1×10^{-7}	8.2×10^{-8}	n.d.	n.d.	n.d.

[a] The standard error of the concentrations was less than 5 %; [b] Not detectable

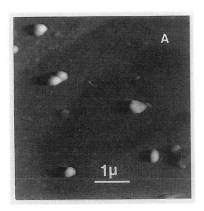
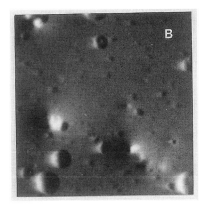
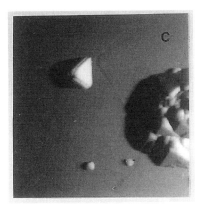

Figure 4 *Atomic force micrographs of the air-dried reaction products of (A) 0.1 M catechol, (B) plus 1 mM silicic acid, and (C) plus 1 mM hydroxy-Al solutions after 60-d aging. The image scale is 1 μ in (A), (B) and (C)*

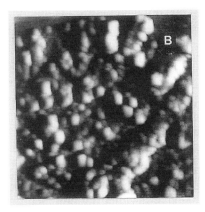

Figure 5 *Atomic force micrographs of the air-dried reaction products of (A) the silicic acid-catechol solution and (B) the supernatant of the hydroxy-Al-catechol system after 60-d aging. The image scale is 1 μ in (A) and (B)*

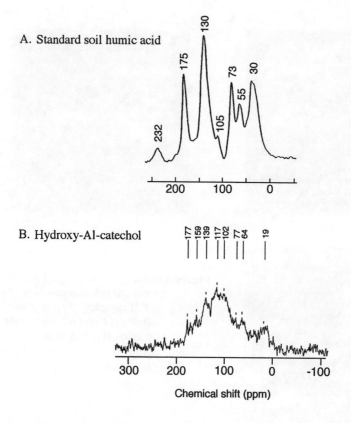

Figure 6 ^{13}C CP-MAS NMR spectra of (A) IHSS standard soil humic acid and (B) the precipitation product of the hydroxy-Al-catechol reaction system at the end of a 60-d reaction period

The precipitate formed in the hydroxy-Al-catechol system was analyzed by solid state ^{13}C CP-MAS NMR spectroscopy (Figure 6B). The chemical shifts at 19 ppm (from alkyl carbons), 64 and 77 ppm (from aliphatic containing C bonded to OH in alcohols, carbohydrates, esters and ethers), 102 ppm (usually attributed to carbohydrates),[22] 117 and 139 ppm (from aromatic carbons), and 177 ppm (from carboxylic carbons and ester groups)[22,23] of the NMR spectrum of the precipitates formed in the hydroxy-Al-catechol system are consistent with the chemical shifts observed in the standard soil HA obtained from the IHSS, although the intensity of the peaks of the *in-situ* NMR spectrum of the precipitate formed in the hydroxy-Al-catechol system is very low compared with the standard purified soil HA (Figure 6). The chemical shift at 159 ppm of the NMR spectrum of the product precipitate in the hydroxy-Al-catechol system (Figure 6B) was attributed to phenolic carbons. The chemical shift at 55 ppm of the NMR spectrum of the IHSS standard soil HA (Figure 6A) is from methoxyl carbons. The chemical shift at 232 ppm of the NMR spectrum of the IHSS standard soil HA (Figure 6A) may be attributable to aldehyde and ketone carbons or due to a side band. The presence of the chemical shift at 19 ppm due to the alkyl carbons in the product precipitate of the hydroxy-Al-catechol system (Figure 6B) revealed cleavage of the ring structure of catechol in the hydroxy-Al-catechol system. The presence of the chemical shift at 177 ppm (Figure 6B) caused by the carboxylic carbons provides evidence for formation of carboxyl groups through oxidation of alkyl fragments in the hydroxy-Al-catechol system. Preston et al.[24] reported that soil humins have the same ^{13}C NMR spectra as soil HAs if the humins are sufficiently de-ashed. The ^{13}C NMR spectra of the standard soil HA and FA provided by the IHSS also show that soil HA and FA have very similar chemical shifts, but the proportion of some chemical shifts is different. The ^{13}C CP-MAS NMR spectra of the standard soil HA and FA provided by the IHSS show that the FA has a relative high intensity of the chemical shift at 175 ppm compared with the HA.

The solid products formed in the hydroxy-Al-catechol system also had the similar FTIR spectra and X-ray diffraction pattern to those of the IHSS standard soil HA (Figures 7 and 8). The absorption bands at 3220-3253 cm^{-1} (Figure 7) were attributed to the OH stretching from adsorbed water and phenolic and alcoholic OH groups in the solid product. The weak absorption bands at 2922-2963 cm^{-1} were from stretching vibrations of aliphatic C-H bonds in methyl and/or methylene units.[20] The absorption band at 1721 cm^{-1} in the standard soil HA (Figure 7A) was attributed to the C=O stretching vibration, due mainly to

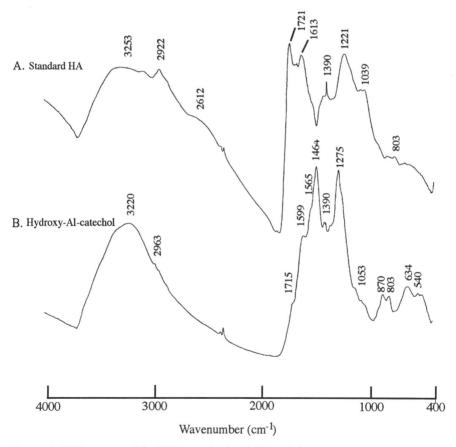

Figure 7 *FTIR spectra of the IHSS standard soil HA and the precipitation product of the reaction of hydroxy-Al ions with catechol at the end of a 60-d reaction period*

carboxyl groups. The shoulder band at 1715 cm^{-1} of the precipitation product of the hydroxy-Al-catechol system (Figure 7B) indicates the formation of -COOH groups through the oxidation of catechol catalyzed by hydroxy-Al ions. The bands at 1613 cm^{-1} (Figure 7A) and 1599 cm^{-1} (Figure 7B) were assigned to aromatic C=C double bonds conjugated with C=O and/or COO$^-$.[20] The band at 1390 cm^{-1} (Figure 7A and B) was attributed to the O-H bending vibrations of alcohols or carboxylate groups.[20] The absorption bands at 1221-1275 cm^{-1} may be due to the C-O stretching vibration and OH bending deformations from carboxyl groups. The absence of the absorption band at 2612 cm^{-1} (due to COOH vibration) and much low intensity of the band at 1715 cm^{-1} in the FTIR spectra of the precipitation product of the hydroxy-Al-catechol system (Figure 7A and B) indicate that less COOH functional groups existed in the precipitate of the hydroxy-Al-catechol system compared with the IHSS standard soil HA. This is accord with the NMR data (Figure 6). However, the ^{13}C CP-MAS NMR spectra of natural humic substances substantially vary with the source and nature of the formation environment.[25]

The absorption band at 1039 cm^{-1} of the standard soil humic acid was assigned to C-O stretching of polysaccharide-like units and Si-O of the silicate impurities.[4] Since very little Si impurity was present in the precipitate of the hydroxy-Al-catechol system, the band at 1053 cm^{-1} in its spectrum (Figure 7B) was apparently from the C-O stretching of polysaccharide-like substances formed. The weak band at 803 cm^{-1} of the standard soil HA (Figure 7A) was probably due to the metal-oxygen vibrations of some metal impurities. The more pronounced bands at 540-870 cm^{-1} of the precipitation product of the hydroxy-Al- catechol system were attributable to the Al-O[26] of Al coprecipitated with humic macromolecules as revealed by the substantial Al content of the solid product of the hydroxy-Al-catechol system (Table 2). Octahedral particles of hydroxy-Al polymers were observed in the solid product of the hydroxy-Al-catechol system by atomic force microscopy (Figure 9C and D). This also provides evidence for the presence of a discrete precipitation product of hydroxy-Al polymers associated with humic macromolecules.

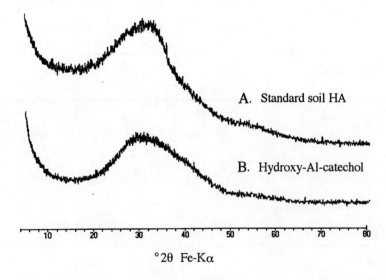

Figure 8 *X-ray powder diffractograms of standard soil HA and the precipitation products of the reaction of hydroxy-Al ions with catechol at the end of a 60-d reaction period*

Table 2 *Elemental contents of the solid product of the hydroxy-Al-catechol system*

C	H	O	Al	Si	Fe	Mn	Cu	Zn	Ash
			g/kg						
489.0(577.5)[a]	27.4(32.4)	361.1	122.4	0.1	n.d.[b]	n.d.	n.d.	n.d.	153.2

[a] The number in the parenthesis is the elemental concentration after de-ashing; [b] not detected.

The AFM images show that the precipitation product of the hydroxy-Al-catechol system consisted of octahedral particles and spheroids with different arrangements (Figure 9C and D). The diameter and thickness of the spheroids in the precipitation product of the hydroxy-Al-catechol system, respectively, ranged from 90 to 120 nm and from 10 to 40 nm (Figure 9B, C and D), which were very similar to the standard soil HA observed under the same condition (Figure 9A). The different arrangements of the spheroids around the octahedral hydroxy-Al polymers observed at different areas of the precipitate were apparently due to electrostatic attraction or repulsion between the hydroxy-Al polymers and humic macromolecules. The AFM images were obtained at pH 5.0. The dissociation constant of carboxyl groups of humic acids is about $10^{-4.5}$.[27] Therefore, over 50 % of the carboxyl groups of humic macromolecules would be deprotonated at pH 5.0. Some negative charges may also be present on the surface of hydroxy-Al polymers due to deprotonation of aluminol groups, although the net pH-dependent charge of the hydroxy-Al polymers should be positive at pH 5.0. When negatively charged sites of hydroxy-Al polymers were close to deprotonated functional groups of the humic macromolecules, the electrostatic repulsion between the hydroxy-Al polymers and macromolecules occurred, resulting in the ring-shape arrangement of humic macromolecules around the precipitated hydroxy-Al polymers (Figure 9D). On the other hand, humic macromolecules can be attracted to positively charged sites of hydroxy-Al polymers, resulting in the formation of hydroxy-Al-humate complexes (Figure 9C). In addition, H-bonding between humic macromolecules and hydroxy-Al polymers should contribute to the interaction.

Complexation of silicic acid with catechol has been investigated by chemists in a few studies,[28-29] although it is not widely recognized that silicon can enter into chelate-type bond formation with some oxygen- and nitrogen-containing organic compounds.[18] A 6-coordinated structure is suggested for the complex anion formed between catechol and silicic acid. To form a hexacoordinated complex of silicon with catechol, the oxygen-oxygen distance of the adjacent hydroxyl groups of catechol should be exactly equal to that required between oxygen atoms in octahedral coordination with a silicon atom.[18] The data of the present study indicate that the darkening and polymerization of catechol was promoted by complexation of silicic acid with catechol.

The role of soluble Al in catalyzing the abiotic oxidation of catechol in the pH range 5 to 7 at Al/catechol molar ratios of 1:1 to 1:3 in reaction periods less than 2 wk was investigated by McBride et al.[15] Although no precipitates were observed in their Al-catechol system, the presence of soluble Al favored the formation of green-colored polymeric products when the reaction system was exposed to air. The data of the present study show that hydroxy-Al ions greatly promoted the oxidation and polymerization of catechol even at pH 4.4 and at a much lower Al/catechol molar ratio (1:100) in a sealed container. Further, black precipitates were formed in the hydroxy-Al-catechol system.

The catalytic effectiveness of a metal ion depends upon its ability to complex with ligands and shift electron density and molecular conformation in ways favorable for the reaction.[30,31] The mechanism for the catalytic effect of Al on the oxidative polymerization of catechol has been proposed. Al^{3+} cations tend to stabilize semiquinone radicals at low pH and direct the manner in which these radicals polymerize;[6,15] aluminum may also promote the formation of semiquinone free radicals by complexing with catechol[6] through delocalization of the electron cloud around the Al-O bond as indicated by the dashed arrow in the Scheme 1. Complexation of Si with catechol has been proposed by Iler,[18] although the reaction mechanism is still obscure. When Al or Si replaces H in catechol, the electron cloud delocalizes from phenolic oxygen into the π-orbital bonding formed from the overlaps between the 2p orbitals of the C atoms of the aromatic ring, thus accelerating the formation of semiquinone free radicals and their coupling to form polycondensates. The semiquinone free radicals formed apparently were partially transformed, through ring cleavage, to aliphatic fragments, resulting in the development of carboxyl groups and the subsequent decarboxylation and CO_2 release.

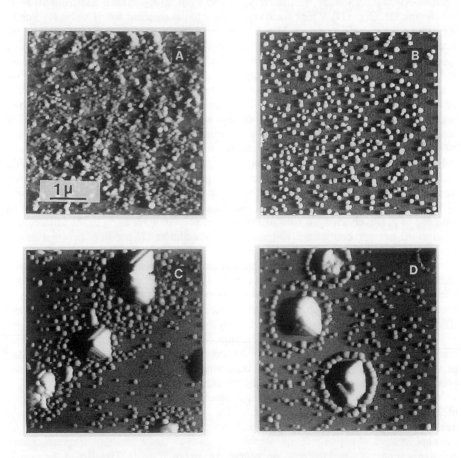

Figure 9 *Atomic force micrographs of (A) the standard soil HA and (B, C and D) the precipitation product of the hydroxy-Al-catechol system. Images b, c, and d were obtained from different areas of the precipitate. The image scale is 1 μ in (A) to (D)*

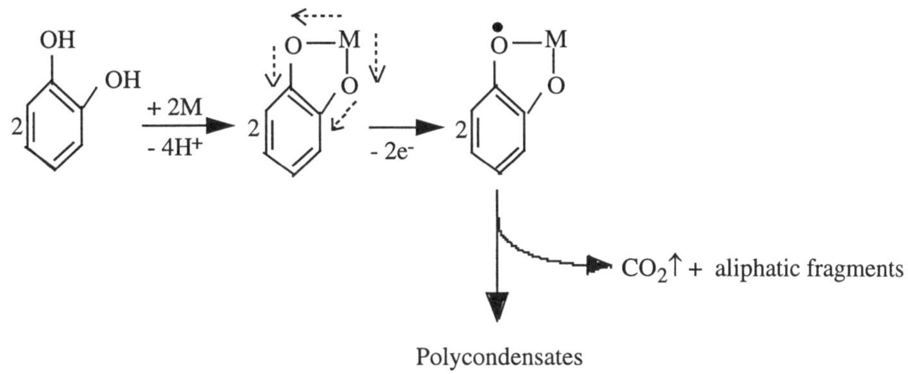

Scheme 1 *Proposed mechanism for the catalysis of hydroxy-Al and silicic acid in the polymerization of catechol (M stands for Al or Si)*

Catechol acts as a hard Lewis base and Al and Si are hard Lewis acids. The stability constant of the Al(III)-catechol complex ($\log K_{\text{Al-catechol}}$ = -6.0 at 25°C and I = 0.1) is over six orders of magnitude higher than that of the Si(IV)-catechol complex ($\log K_{\text{Si-catechol}}$ = -12.42 at 25 °C and I = 0.1),[32] indicating that, compared with silicic acid, hydroxy-Al ions much more easily form complexes with catechol. Compared with the Si-O bond, the Al-O bond formed is more ionic since the electronegativity values of Al, Si, H, and O are, respectively, 1.61, 1.90, 2.20, and 3.44. Therefore, the electron cloud around Al-O should be more delocalized than that around Si-O due to the lower electronegativity of Al than Si. Consequently, the accelerating effect of hydroxy-Al ions on the humification of catechol was much stronger than silicic acid, as observed in the present study.

4 CONCLUSIONS

The results of this study show that the presence of silicic acid and especially hydroxy-Al in solutions substantially promotes the oxidative polymerization and ring cleavage of catechol, resulting in the formation of brown to dark brown humic macromolecules. This accelerating effect on the formation of humic macromolecules significantly increased with increase of the residence time (chemical aging). This finding is very significant in understanding the abiotic formation of humic substances in acidic environments.

ACKNOWLEDGMENT

This research was supported by Research Grant GP2383- and EQP156628-Huang of the Natural Sciences and Engineering Research Council of Canada.

References

1. C. B. Coulson, R. I. Davies and D. A. Lewis, *J. Soil Sci.*, 1960, **11**, 20.
2. F. A. Einhellig, 'Allelopathy: Organisms, Processes, and Applications', K. M. Inderjit, M. Dakshini and F. A. Einhellig, (eds.), ACS Symposium Series 582, Am. Chem. Soc., Washington, D.C., 1995, p. 1.
3. T. S. C. Wang, P. M. Huang, C. -H. Chou and J. -H. Chen, 'Interactions of Soil

Minerals with Natural Organics and Microbes', P. M. Huang and M. Schnitzer (eds.), SSSA Spec. Pub. no. 17., SSSA, Madison, WI, 1986, p. 251.
4. F. J. Stevenson, 'Humus Chemistry', 2nd Edn., Wiley, New York, 1994.
5. P. M. Huang, 'Environmental Impact of Soil Component Interactions. Vol.1. Natural and Anthropogenic Organics', P. M. Huang, J. Berthelin, J. -M. Bollag, W. B. McGill and A. L. Page, (eds.), CRC Lewis Publishers, Boca Raton, 1995, p. 135.
6. P. M. Huang, M. C. Wang and M. K.Wang, 'Principles and Practices in Plant Ecology. Allelochemical Interactions', K. M. Inderjit, M. Dakshini and C. L. Foy, (eds.), CRC Press, Boca Raton, FL, 1999, p. 287.
7. P. M. Huang, 'Handbook of Soil Science', M. E. Sumner (ed.), CRC Press, Boca Raton, FL, 2000, p. B303.
8. H. Shindo and P. M. Huang, *Nature,* 1982, **298,** 363.
9. H. Shindo and P. M. Huang, *Soil Sci. Soc. Am. J.,* 1984, **48,** 927.
10. M. B. McBride, *Soil Sci. Soc. Am. J.,* 1987, **51,** 1466.
11. M. C. Wang and P. M. Huang, *Sci. Total Environ.,* 1992, **113,** 147.
12. J. S. K. Lee and P. M. Huang, 'Environmental Impact of Soil Component Interactions. Vol.1. Natural and Anthropogenic Organics', P. M. Huang, J. Berthelin, J. -M. Bollag, W. B. McGill and A. L. Page, (eds.), CRC Lewis Publishers, Boca Raton, FL, 1995, p. 177.
13. W. Ziechmann, *Z. Pflanzenernaehr. Dueng.,* 1959, **84,** 155.
14. T. S. C. Wang, M. C. Wang and P. M. Huang, *Soil Sci.,* 1983, **136,** 226.
15. M. B. McBride, F. J. Sikora and L. G. Wesselink, *Soil Sci. Soc. Am. J.,* 1988, **52,** 985.
16. S. M. Griffith and M. Schnitzer, *Soil Sci. Soc. Am. J.,* 1975, **39,** 861.
17. S. M. Griffith and M. Schnitzer, *Environ. Biogeochem.,* 1976, **7,** 117.
18. R. Iler, 'The Chemistry of Silica. Solubility, Polymerization, Colloid and Surface Properties, and Biochemistry', Wiley, New York, 1979.
19. C. M. Preston and R. H. Newman, *Can. J. Soil Sci.,* 1992, **72,** 13.
20. P. MacCarthy and J. A. Rice, in 'Humic Substances in Soil, Sediment, and Water', G. R. Aiken, D. M. McKnight, R. L. Wershaw and P. MacCarthy, (eds.), Wiley, New York, 1985, p. 527.
21. C. Liu and P. M. Huang, in 'Understanding Humic Substances: Advanced Methods, Properties, and Applications', E. A. Ghabbour and G. Davies, (eds.), Royal Society of Chemistry, Cambridge, 1999, p. 87.
22. J. M. Bortiaynski, P. G. Hatcher and H. Knicker, in 'Humic and Fulvic Acids. Isolation, Structure, and Environmental Role', J. S. Gaffney, N. A. Marley and S. B. Clark, (eds.), American Chemical Society, Washington, DC, 1996, p. 57.
23. R. L. Wershaw, in 'Humic Substances in Soil, Sediment, and Water', G. R. Aiken, D. M. McKnight, R. L. Wershaw and P. MacCarthy, (eds.), Wiley, New York, 1985, p. 561.
24. C. M. Preston, M. Schnitzer and J. A. Ripmeester, *Soil, Sci. Soc. Am. J.,* 1989, **53,** 1442.
25. R. L. Wershaw, 'Abstracts, Humic Substances Seminar IV', Boston, March, 2000, p. 13.
26. H. W. van der Marel and H. Beutelspacher, 'Atlas of Infrared Spectroscopy of Clay Minerals and Their Admixtures', Elsevier, New York, 1976.
27. B. Manunza, S. Deiana, V. Maddau, C. Gessa and R. Seeber, *Soil Sci. Soc. Am. J.,* 1995, **59,** 1570.
28. V. A. Rosenheim, B. Raibmann and G. Schendel, *Z. Anorg. All. Chem.,* 1931, **196,**

160.
29. C. L. Frye, *J. Am. Chem. Soc.*, 1964, **86,** 3170.
30. M. R. Hoffmann, *Environ. Sci. Technol.*, 1980, **14,** 1061.
31. A. T. Stone and A. Torrents, 'Environmental Impact of Soil Component Interactions. Vol.1. Natural and Anthropogenic Organics', P. M. Huang, J. Berthelin, J. -M. Bollag, W. B. McGill and A. L. Page, (eds.), CRC Lewis Publishers, Boca Raton, FL, 1995, p. 275.
32. A. E. Martell, R. M. Smith and R. J. Motekaitis, 'NIST Critically Selected Stability Constants of Metal Complexes Database, Version 3.0', Texas A & M University, College Station, TX, 1997.

EFFECT OF COVER CROP SYSTEMS ON THE CHARACTERISTICS OF SOIL HUMIC SUBSTANCES

Guangwei Ding,[1] Dula Amarasiriwardena,[2] Stephen Herbert,[1] Jeffrey Novak[3] and Baoshan Xing[1]

[1] Department of Plant and Soil Sciences, Stockbridge Hall, University of Massachusetts, Amherst, MA 01003, USA
[2] School of Natural Science, Hampshire College, Amherst, MA 01002, USA
[3] USDA-ARS-Coastal Plains Soil, Water and Plant Research Center, Florence, SC 29501, USA

1 INTRODUCTION

Soil organic matter (SOM) is an important attribute of soil quality, influencing the productivity and physical well being of soils. Thus, it is important from economic and environmental standpoints to determine how changes in cover crop management will affect SOM and soil quality.

SOM contents and properties are functions of agricultural practices and the amounts and kinds of plant residues returned to the soil.[1-9] It is well established that the labile components of SOM change and reach a new steady state more quickly in response to various management practices than does total organic matter.[10-13] Wander and Traina[14] showed that SOM in crop rotation with cover crops was significantly higher than those rotations without cover crops. However, Lal et al.[15] reported no or minimal change of SOM content. The reason for not detecting any SOM change could be due to natural soil heterogeneity.[14] It is well known that continuous cultivation of cereal crops generally results in substantial losses of soil C and N.[16-18] However, studies of organic matter using ^{13}C nuclear magnetic resonance (NMR) spectroscopy have indicated that the chemical nature of the remaining C shows little change as a result of cultivation and that the stability of SOM appears to depend more on physical protection mechanisms than any inherent recalcitrance of the organic structures.[19-22]

In addition to SOM quantity, the quality (e.g., structure and composition) and distribution of individual fractions (e.g., humic acids, polysaccharides) are essential to the maintenance of soil productivity. Monreal et al.[23] observed a higher lignin dimer to lignin monomer ratio in continuous wheat rotation and this ratio decreased from large to small aggregate sizes, indicating the change in SOM quality. Wander and Traina[24] used diffuse reflectance Fourier transform infrared spectroscopy (DRIFTS) to examine functional groups of SOM fractions and reported that the ratios of reactive to recalcitrant fractions in humic acids best reflected overall SOM bioavailability.

DRIFTS detects molecular vibrations and is useful for functional group analysis and for identification of molecular structures of SOM.[25] But it cannot be used to quantify carbon contents of structural groups. By contrast, ^{13}C NMR spectroscopy provides quantitative data for structural components. NMR has been successfully used to characterize SOM by many scientists.[26-30] Thus, it would be advantageous to use both NMR and DRIFTS to characterize SOM and compare quantitative data of SOM under different cover crop systems. The objective of this study is to determine quantitative, structural and compositional changes of humic substances (humic acid and fulvic acid fractions) caused by cover-crop systems using both NMR and DRIFTS.

2 MATERIALS AND METHODS

2.1 Site Description and Sampling

Since 1990, cover crop experiments have been conducted in the Connecticut River Valley at the Massachusetts Agricultural Experiment Station Farm in South Deerfield, Massachusetts. The soil at the University of Massachusetts at Amherst farm is a fine sandy loam (coarse, mixed, mesic Fluventic Dystrudept) and low in SOM (~2%). Its upper 0.6 m is homogeneous, overlaying inclined layers of coarse and fine material to a great depth. It is a typical soil in the intensively cropped Connecticut River Valley in Massachusetts. Three cover crop treatments with four nitrogen rates (applied to the corn crop after cover crop incorporation) were laid out in a complete factorial design in bordered 3 m x 7.5 m plots in four randomized blocks. Cover crop treatments and seeding rates were 1) check (no cover crop); 2) rye (125 kg/ha); 3) hairy vetch + rye (46+65 kg/ha). Nitrogen fertilizer rates were 0, 67, 135, 202 kg N/ha using NH_4NO_3. Detailed soil sample information is listed in Table1.

Table 1 *Soil samples (UMass South Deerfield Farm)*

Sample number	Depth (cm)	Cover crops	Nitrogen rates (kg N ha^{-1})
VR1	0-25	Vetch/Rye	0
VR4	0-25	Vetch/Rye	202
RA1	0-25	Rye alone	0
RA4	0-25	Rye alone	202
C1	0-25	No cover crop	0
C4	0-25	No cover crop	202

1 = No nitrogen fertilizer treatment, 4 = Nitrogen fertilizer treatment

2.2 Extraction, Fractionation and Purification of Humic Substances

2.2.1 Extraction. Soil was air-dried and passed through a 2 mm sieve. Air-dry soil (50 g) was weighed into a 1000 mL plastic flask, then 500 mL 0.1 M $Na_4P_2O_7$ was added. The air in the flask and solution was displaced with nitrogen (N_2) and the system was shaken for 24 hr at room temperature.

2.2.2 Fractionation. After separation from the $Na_4P_2O_7$ insoluble organic residues by centrifugation at 3000 rpm, the dark-colored supernatant solution was acidified to pH 1 with 6 M HCl and allowed to stand for 24 hr at room temperature for the coagulation of the HA fraction. The soluble material (FA) was separated from the HA by centrifuging at 10000 rpm.

2.2.3 Purification. HA was shaken 3 times for 24 hr at room temperature with 0.1 M HCl/0.3 M HF solution. The insoluble residue (HA) was separated from the supernatant by centrifuging at 10000 rpm, washed with deionized water until free of chloride ions and then freeze-dried. The FA solution was passed through an XAD-8 resin column. The effluent was discarded and the XAD-8 column containing sorbed FA was washed 3 times with deionized water, then the sorbed FA was eluted with 0.1 M NaOH. The solution was immediately acidified with 6 M HCl to pH 1. The resulting solution volume was sufficient to maintain the FA in solution, which was then freeze-dried.

2.3 Diffuse Reflectance Fourier Transform Infrared (DRIFTS) Analysis

DRIFTS was performed in a Midac series M 2010 infrared spectrophotometer with a DRIFTS accessory (Spectros Instruments). All SOM fractions were ground with a sapphire mortar and pestle and stored over P_2O_5 in a dry box. Humic and fulvic acid concentrations for this determination ranged from 2 to 4 mg and were supplemented with KBr to a total weight per sample of 100 mg, then ground with an agate mortar and pestle. The milled sample was immediately transferred to a sample holder and its surface was smoothed with a glass microscope slide. Before analysis, the diffuse-reflectance cell containing the samples was flushed with N_2 gas to eliminate interference from carbon dioxide and moisture. A small jar (20 mL) containing anhydrous $Mg(ClO_4)_2$ was placed inside the sample compartment to further reduce atmospheric moisture.

To obtain DRIFTS spectra, 100 scans were collected at a resolution of 16 cm^{-1} and the spectra with numerical values for major peaks wave-numbers and intensities were recorded. The blank consisted of the powdered KBr stored under the same environmental conditions as the sample-KBr mixtures. Absorption spectra were converted to Kubelka-Munk functions using the Grams/32 software package (Galactic Corporation). Peak assignments and intensity (by height) ratio calculations were made with the methods of Wander and Traina,[24] Baes and Bloom[31] and Niemeyer et al.[32]

2.4 NMR Spectroscopy

CP-TOSS (Cross-Polarization and Total Sideband Suppression) was used.[30] Samples were run at 75 MHz (^{13}C) in a Bruker MSL-300 spectrometer. HA samples were packed in a 7-mm-diameter zirconia rotor with a Kel-F cap. The spinning speed was 4.5 kHz. A 90° ^{1}H pulse was followed by a contact time (t_{cp}) of 500 µs, and then a TOSS sequence was used to remove sidebands.[30,33] The 90° pulse length was 3.4 µs and the 180° pulse was 6.4 µs. The recycle delay was 1 s with the number of scans about 4096.

3 RESULTS AND DISCUSSION

3.1 Carbon-13 NMR Characteristics

For quantitative NMR measurements the conditions that must be met are $t_{cp} \ll T_{1\rho}$ (^{1}H). That is, the contact time must be much shorter than the time constant for proton spin lattice relaxation in the rotating frame.[34] In addition, the delay time between cross-polarization sequences must be long enough to allow for complete ^{1}H spin relaxation (i.e., at least five times $T_1(^1H)$). The ideal situation of $t_{cp} \ll T_{1\rho}$ (^{1}H) may be difficult to satisfy in practice, especially where $T_{1\rho}$ (^{1}H) is very short. Optimum acquisition parameters were therefore chosen after examination of the spin dynamics of the samples to avoid signal suppression by incomplete relaxation.

Although CP-TOSS spectra could not be used for absolute quantitation, they could be compared because all the HAs samples were run under the same conditions and were from the same type of soil. Functional groups were assigned as follows: 0-50 ppm for aliphatic-C (C in straight-chain, branched and cyclic alkanes and alkanoic acids); 50-60 ppm methoxy-C including C from OCH_3 groups as well as C from amino acids; 60-96 ppm carbohydrate-C (aliphatic C bonded to OH groups, or ether oxygens, or occurring in saturated five- or six-membered rings bonded to oxygens); 96-108 ppm O-C-O; 108-145 ppm aromatic-C; 145-162 ppm phenolic-C (i.e., aromatic C bonded to OH groups); 162-190 ppm carboxylic-C; and 190-220 ppm carbonyl-C.[25,34,35]

The ^{13}C NMR data are summarized in Table 2. The HA from the rye alone system differs from other HAs in the aliphatic region (0-50 ppm) and the HA signal in this region is the smallest with no nitrogen fertilizer treatment. The most intense signal in this region appears under the vetch/rye system with no nitrogen fertilizer treatment. When compared to the total aliphatic region (0-108 ppm), the aliphatic-C of the vetch/rye HA without nitrogen fertilizer is 55.3%, and this figure decreases to 48.9% for rye alone without nitrogen fertilizer.

Table 2 *The ^{13}C NMR characteristics of HA samples*

Type of Materials	Distribution of C Chemical shift (ppm) %							
	0-50	50-60	60-96	96-108	108-145	145-162	162-190	190-220
VR1	26.8	8.4	15.9	4.2	22.6	6.5	14.4	3.2
VR4	26.4	8.4	16.9	4.2	21.6	6.6	14.4	1.5
RA1	20.8	6.7	17.1	4.2	25.4	7.5	16.4	1.9
RA4	21.2	7.1	18.2	4.3	23.5	7.2	16.3	2.3
C1	26.1	8.0	17.5	4.5	21.3	6.2	14.4	2.0

Types of Materials	Aliphatic-C %	Aromatic-C %	(Aliphatic-C)/(Aromatic-C)
VR1	55.3	29.1	1.9
VR4	55.9	28.2	2.0
RA1	48.9	32.8	1.5
RA4	50.7	30.7	1.7
C1	56.1	27.6	2.0

The phenolic-C (145-162 ppm) content is higher in the HA extracted from rye alone systems than the other two systems. Similarly, the HA extracted from the rye cover system is also relatively enriched in aromatic-C. An inspection of the data for HA with no cover crop shows that aromatic-C content (including 108-145 and 145-162 ppm regions) is the lowest (Table 2). The nonprotonated C signals (130 ppm) likely are from aromatic carbons, including C in polynuclear aromatic rings. Polymerization and polycondensation of HA macromolecules may have occurred on contact with soil minerals.[34,35] For both sites of rye systems (with or without nitrogen fertilizer), the relative increase in aromatic C and decrease in aliphatic C (Table 2) indicate that humification may be greater than in the other two treatments. These results may indicate a greater stability of HA from rye alone systems because 1) the HA appears to contain fewer biopolymers (proteinaceous materials and carbohydrates) and is therefore less biodegradable and 2) the greater aromaticity implies a more stable chemical structure.

3.2 DRIFTS Spectroscopy of HA and FA

Wavenumbers and assignments for peaks in DRIFTS data are the same as in IR and FTIR spectroscopy.[25,31,32] The main absorbance bands and corresponding assignments obtained for the various HA and FA samples are listed in Table 3. All peaks observed in the HA and FA spectra are typical, as reported in the literature. However, more detailed information on the reactivity of HA and FA is provided by calculating O/R ratios.

The O/R ratios are computed by dividing the peak heights of oxygen-containing functional groups by those of aliphatic and aromatic (referred to as recalcitrant) groups.[24] The impact of cover crop systems on spectral composition is summarized by these ratios (Figures 1 and 2). The HA fraction isolated from vetch/rye plots with nitrogen fertilizer has the highest ratio

$$R_1 = (1727+1650+1160+1127+1050)/(2950+2924+2850+1530+1509+1457+1420+779))$$

and the lowest R_1 ratio appears at the vetch/rye without nitrogen fertilizer (Figure 1). The relatively high O/R ratios of the HA support the notion that SOM in the vetch/rye with nitrogen fertilizer is more biologically active. When compared with the R_1 values of the HA fractions with nitrogen fertilizer, the highest R_1 is for the vetch/rye system, followed by no cover crop and rye alone systems. For the treatments without fertilizer, R_1 values follow an order of no cover crop > rye along > vetch/rye.

R_2 is the ratio of the peak heights of ketonic and carboxyl (1720 cm^{-1}) groups divided by those of CH_2 and aromatic (1457+1420+779 cm^{-1}) groups. Its value is the highest in the HA spectrum of rye alone without fertilizer. This ratio is the lowest in the HA spectrum of rye alone with nitrogen fertilizer. We expect that since HA makes up the largest single SOM pool in mineral soils, HA O/R ratios would be a good indicator of overall SOM characteristics. However, the O/R ratio may not reflect the bioavailability of SOM.[24]

Fulvic acids have relatively low molecular weights and high oxygen contents, and as a result are more polar and therefore mobile than HAs. Thus, FA may be more representative of the available organic matter pool.[24,25] In contrast to the HA O/R ratios, the FA O/R ratio ((1850+1650+1400+1080+560)/(3340+2924+1535+1457)) is greater in the vetch/rye without fertilizer than with fertilizer treatment (Figure 2). Meanwhile, the O/R ratios of FA from both the vetch/rye treatments are significantly higher than rye treatments. This means that FA fractions from vetch/rye systems are more active than from the rye alone systems. The O/R ratios of FA do not differ much between the rye alone systems and no cover crop system with nitrogen fertilizer. This indicates that FA fractions may undergo a similar change either by chemical oxidation or as a microbial carbon source for those systems. When we compare the R_1 ratios of HA with O/R ratios of FA in different cover crop systems, the HA R_1 from both the vetch/rye and rye alone systems is higher with nitrogen fertilizer treatment than without nitrogen fertilizer. The reverse is true for the O/R ratios of FA. Both HA R_1 and FA O/R ratios of no cover crop system are higher without nitrogen fertilizer treatment than with nitrogen fertilizer. This suggests that cover crop systems (with or without nitrogen fertilizer) affect the HA and FA composition. However, even though FA O/R ratios reflect chemical reactivity, they are not positively correlated with SOM bioavailability (based on indirect measures of N and C mineralization).[8] It is generally assumed that the composition of humic substances, including HA and FA, remains relatively constant in a given soil.

Table 3 *Peak assignments of diffuse reflectance Fourier transform infrared spectra (DRIFTS) for HA and FA samples*[24,25,31]

Wavenumber (cm^{-1})	Functional groups	HA	FA
3279-3340	Phenol OH, amide N-H	X	X
2962-2950	CH$_2$, symmetric stretch	X	
2924-2930	CH$_2$, asymmetric stretch	X	X
2850	CH$_2$, symmetric stretch	X	
2500	CO-OH H bonded	X	X
1850	C=O stretch		
1735-1713	C=O ketonic, COOH	X	
1650	C=O, C=O-H bonded, amide H	X	X
1630-1608	C=C aromatic		
1550	Aromatic ring, amide		
1535-1520	C=C aromatic ring, amide	X	X
1509	Aromatic ring, amide	X	
1457	CH$_3$ asymmetric stretch, CH bond	X	X
1420	Aromatic ring stretch		X
1400	COO$^-$ salt, COOH		X
1379-1327	COO$^-$, CH$_3$, symmetric stretch		
1260-1240	CO, COOH, COC, phenol OH	X	
1190-1127	Aliphatic, alcoholic OH	X	
1080-1050	CO aliphatic alcohol	X	X
1030	Aliphatic COC, aromatic ether, SiO		
918-912	OH, COOH, Al-OH		
850-830	CH aromatic bend, Al-O-Si	X	
799	Fe-O-Si		
779	CH aromatic out-of-plane bend	X	
750	Unknown mineral peak		
694	Unknown mineral peak		
560	COO salt, Mg/Si-O aliphatic	X	
530-520	Si-O		
480	Aromatic ring bending	X	

4 CONCLUSIONS

In this study, the chemically isolated HA and FA fractions display the quantitative and qualitative differences resulting from various cover crop treatments. These fractions are affected by agronomic and environmental factors. Although DRIFTS cannot be used to quantify the absolute C contents of structural groups, it can be used to generate peak ratios from which we can assess the relative enrichment or depletion of specific functional groups. The HA O/R ratio is a good indicator of overall SOM characteristics, and cover crop systems and nitrogen fertilizer rates can change the O/R ratios of HA and FA. From CP-TOSS ^{13}C NMR data, the HA extracted from rye alone systems is more aromatic and less aliphatic than the HA from vetch/rye, indicative of the impact of cover crop systems on the structure and composition of humic substances. Future research needs to address how those changes of SOM affect soil productivity and sustainability.

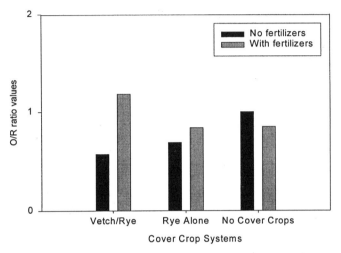

Figure 1 *Ratios of selected peak heights from DRIFTS spectra of HA samples under different cover crop systems*

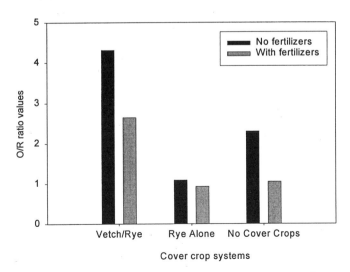

Figure 2 *Ratios of selected peak heights from DRIFTS spectra of FA samples under different cover crop systems*

ACKNOWLEDGEMENTS

This work was supported in part by the U.S. Department of Agriculture, National Research Initiative Competitive Grants Program (97-35102-4201 and 98-35107-6319), the Federal Hatch Program (Project No. MAS00773), and a Faculty Research Grant from University of Massachusetts, Amherst. We would also like to thank Dr. L. C. Dickinson for his technical support.

References

1. C. A. Campbell and W. Souster, *Can. J. Soil Sci.*, 1982, **62**, 651.
2. L. K. Mann, *Geoderma*, 1985, **36**, 241.
3. J. W. Doran, D. G. Fraser, M. N. Culick and W. C. Liebhardt, *Am. J. Altern. Agric.*, 1987, **2**, 99.
4. M. Y. Cheshire, B. T. Christensen and L. H. Sorensen, *J. Soil Sci.*, 1990, **41**, 29.
5. C. A. Cambardella and E. T. Elliott, *Soil Sci. Soc. Am. J.*, 1992, **56**, 777.
6. K. Paustian, W. J. Parton and J. Persson, *Soil Sci. Soc. Am. J.*, 1992, **56**, 476.
7. B. Eghball, L. N. Mielke, D. L. McCallister and J. W. Doran, *J. Soil Water Conserv.*, 1994, **49**, 201.
8. M. M. Wander, S. J. Traina, B. R. Stinner and S. E. Peters, *Soil Sci. Soc. Am. J.*, 1994, **58**, 1130.
9. C. A. Campbell, V. O. Biederbeck, B. G. McConkey, D. Curtin and R. P. Zentner, *Soil Biol. and Biochem.*, 1999, **31**, 1.
10. W. B. McGill, J. F. Dormaar and E. Reinl-Dwyer, in 'Land Degradation and Conservation Tillage', Proceedings of the 34th Annual Canadian Society of Soil Science Meeting, Calgary, 1988, p. 30.
11. E. Gregorich, in 'Soil Quality Indicators for Sustainable Agriculture in New Zealand', Proceedings of Workshop held for MAF Policy at Lincoln University, Christchurch, New Zealand. Feb., 1996, p. 40.
12. J. W. Doran, *Soil Sci. Soc. Am. J.*, 1980, **44**, 765.
13. D. A. Angers, M. A. Bolinder, M. R. Carter, E. G. Gregorich, C. Drury, B. C. Liang, R. P. Voroney, R. R. Simard, R. Bayaert and J. Martel, *Soil and Tillage Res.*, 1997, **41**, 191.
14. M. M. Wander and S. J. Traina, *Soil Sci. Soc. Am. J.*, 1996, **60**, 1081.
15. R. E. Lal, D. J. Regnier, W. M. Eckert and R. Hammond, in 'Soil and Water Conserv. Soc. Am', Proc. Int. Conf., Jackson TN, Apr. 1991, Ankeny, IA.
16. J. F. Dormmar, *Can. J. Soil Sci.*, 1979, **59**, 349.
17. R. C. Dalal and R. J. Henry, *Soil Sci. Soc. Am. J.*, 1988, **52**, 1361.
18. C. A. Campbell, K. E. Bowren, M. Schnitzer, R. P. Zentner and L. Townley-Smith, *Can. J. Soil Sci.*, 1991, **71**, 377.
19. J. Q. Skjemstad, R. C. Dalal and P. F. Barron, *Soil Sci. Soc. Am. J.*, 1986, **50**, 354.
20. J. Q. Skjemstad and R. C. Dalal, *Aust. J. Soil Res.*, 1987, **25**, 323.
21. J. M. Oades, A. G. Waters, A. M. Vassalo, M. A. Wilson and G. P. Jones, *Aust. J. Soil Res.*, 1988, **26**, 289.
22. P. Capriel, P. Harter and D. Stephenson, *Soil Sci.*, 1992, **153**, 122.
23. C. M. Monreal, M. Schnitzer, H. R. Shulten, C. A. Campbell and D. W. Anderson, *Soil Biol. Biochem.*, 1995, **27**, 845.
24. M. M. Wander and S. J. Traina, *Soil Sci. Soc. Am. J.*, 1996, **60**, 1087.
25. F. J. Stevenson, 'Humus Chemistry: Genesis, Composition, Reactions', 2nd Edn,

Wiley, New York, 1994.
26. M. A. Wilson, 'NMR Techniques and Application in Geochemistry and Soil Chemistry', Pergamon, Oxford, 1987.
27. B. Xing, W. B. McGill and M. J. Dudas, *Environ. Sci. Technol.*, 1994, **28**, 1929.
28. B. Xing, *Chemosphere,* 1997, **35**, 633.
29. M. Schnitzer, H. Kodama and J. A. Ripmeester, *Soil Sci. Soc. Am. J.*, 1991, **55**, 745.
30. B. Xing, J. Mao, W. -G. Hu, K. Schmidt-Rohr, G. Davies and E. A. Ghabbour, in 'Understanding Humic Substances: Advanced Methods, Properties and Applications', E. A. Ghabbour and G. Davies, (eds.), Royal Society of Chemistry, Cambridge, 1999, p. 49.
31. A. U. Baes and P. R. Bloom, *Soil Sci. Soc. Am. J.*, 1989, **53**, 695.
32. J. Niemeyer, Y. Chen and J. -M. Bollag, *Soil Sci. Soc. Am. J.*, 1992, **56**, 135.
33. K. Schmidt-Rohr and H. W. Spiess, 'Multidimensional Solid-State NMR and Polymers', Academic Press, London, 1994.
34. M. Schnitzer and C. M. Preston, *Soil Sci. Soc. Am. J.*, 1986, **50**, 326.
35. B. Xing and Z. Chen, *Soil Sci.*, 1999, **164**, 40.

STRUCTURAL-GROUP QUANTITATION BY CP/MAS ^{13}C NMR MEASUREMENTS OF DISSOLVED ORGANIC MATTER FROM NATURAL SURFACE WATERS

R. L. Wershaw, G. R. Aiken, J. A. Leenheer and J. R. Tregellas

U.S. Geological Survey, Denver Federal Center, Denver, CO 80225

1 INTRODUCTION

The reactivity of dissolved organic matter (DOM) components of natural water systems depends on the chemical structures of the components.[1-8] In order to be able to predict the concentrations of products of a given DOM reaction such as chlorination, it is necessary to have precise quantitation of the reactive groups in the DOM. A number of different degradation techniques have been developed for chemical structural-group identification in humic substances.[9-11] However, these techniques have not been used for structural-group quantitation because of low yields, uncertainties in yields and the potential alteration of the component chemical structures.

Nuclear magnetic resonance spectroscopy (NMR) provides a solution to the problem of quantitation of carbon-containing structural groups because, in principle, it gives a quantitative measure of the number of atoms of any isotope (such as ^{13}C) that possesses spin. Liquid-state ^{13}C NMR spectrometry has been employed routinely by organic chemists and biochemists for structural elucidation and structural-group quantitation because, for most low-molecular-weight organic compounds that are soluble in a suitable solvent, sharp, well-resolved lines are obtained for each type of carbon atom in the compound.[12] However, the liquid-state ^{13}C NMR spectra of humic substances (HSs) generally consist of broad bands because of the heterogeneity of humic isolates. As expected, fractionation of the isolates improves the resolution of the spectra.[13]

Solid-state cross polarization magic-angle spinning (CP/MAS) ^{13}C NMR spectrometry has been used more extensively than liquid-state ^{13}C NMR spectrometry for the measurement of relative concentrations of structural groups of HSs.[14] The main reasons are that the analyses can be carried out much more rapidly than liquid-state NMR spectrometry, require minimal preparation and are nondestructive. Unfortunately, the quantitative reliability of solid-state ^{13}C NMR measurements of complex macromolecular systems is subject to a number of uncertainties.[15,16] Wind et al.[16] have pointed out that the uncertainties arise from the factors listed in Table 1. We shall attempt to evaluate the importance of these factors on the quantitative reliability of ^{13}C CP/MAS NMR spectra of DOM and to develop a procedure for calculating structural-group concentrations from ^{13}C CP/MAS NMR data.

Table 1 *Factors that affect the quantitation of ^{13}C CP/MAS NMR measurements (adapted from Wind et al.[16])*

Limiting Factors	Possible Remedies
General MAS Factors	
Sample heterogeneity	Isolate different phases
Unpaired electrons in organic free radicals and paramagnetic metal ions	Remove paramagnetic species chemically
Interference with proton decoupling from molecular motion	Measure spectra at different proton decoupling powers
Spinning sidebands arising from MAS	Increase spinning rate or use lower magnetic field
Broadening arising from interference between MAS and the motion of sample molecules	Measure spectra at different spinning rates
Recycle delay	Use delay sufficient to allow for relaxation
Factors specific to CP	
Hartmann-Hahn matching conditions	Precisely adjust power levels for match
MAS modulation of cross polarization time constant (T_{CH})	Make measurement at low field with low spinning rate
Possible dephasing of proton magnetization	Proton rf field should be as high as possible (short as possible proton 90° pulse)
Cross-polarization spin dynamics	Measure $T_{1\rho H}$ to determine if $T_{1\rho C} > T_{1\rho H} > T_{CH}$ is met

Jurkiewicz and Maciel[15] used direct polarization magic-angle spinning (DP/MAS) ^{13}C NMR spectroscopy to measure structural-group concentrations in a suite of coals. An internal ^{13}C-enriched intensity standard was incorporated into the coal samples during the measurements. They found that repetition delays as long as 335 s were necessary to obtain quantitative results from some of their samples because of the long spin-lattice relaxation time constants (T_{1S}) of the most slowly-relaxing aromatic spins in these samples.

Kinchesh et al.[17,18] performed a careful study of the use of CP/MAS ^{13}C NMR spectroscopy for structural-group quantitation of natural organic matter in whole soils. They pointed out that "three distinct quantitation regimes" are possible: (1) quantitation is obtainable from a single CP/MAS spectrum; (2) quantitation is obtainable from a series of CP/MAS spectra; and (3) quantitation is not possible using CP/MAS. Regime 3 generally results from high concentrations of paramagnetic centers in the sample. Therefore, in the absence of such paramagnetic centers the first two regimes generally will obtain. If the carbon atoms in all of the functional groups have similar CP cross-relaxation time constants (T_{CH}) and if the expression $T_{1\rho H} > T_{CH}$, where $T_{1\rho H}$ is the proton spin-lattice relaxation time constant in the rotating frame, is satisfied for all of the carbon atoms, then a single contact time (CT) may be used to obtain a quantitative spectrum. The CT should be chosen so that $T_{CH} \ll CT \ll T_{1\rho H}$.

Two types of paramagnetic centers are found in soils and natural organics samples: those arising from paramagnetic metal ions in the sample and those arising from unpaired electrons in organic free radicals. Paramagnetic metal ions pose the main problem in whole soil samples. Kinchesh et al.[18] found that high paramagnetic metal-ion concentrations (Fe:C ratios greater than 1) made it impossible to obtain a satisfactory spectrum of a sample. Jurkiewicz and Maciel[15] showed that the percentage of carbon atoms observed in CP/MAS ^{13}C NMR measurements of coals is inversely related to the concentrations of unpaired electrons in the coals. For the concentration range between about 1×10^{19} and 3×10^{19} spins/gram in coal samples, they found that 81 to 97% of the carbon nuclei were observed. The unpaired electrons in free radicals interact with nearby ^{13}C nuclei to produce very broad NMR lines that are essentially invisible in most NMR experiments.

Paramagnetic metal ions can readily be eliminated from DOM isolates using the methods indicated in Section 2 of this paper. Analyses of three of the samples (Coal Creek FA, Everglades F1 HPOA and Everglades U3 HPOA) indicate that the iron concentrations are reduced to less than 0.01%. However, stable free radicals cannot readily be removed and will remain in the isolates.[19] These free radicals appear to be associated with phenolic species. The concentration of free radicals in solutions of humic substances generally increases with increasing pH above 5.[19] The reported concentration of unpaired electrons in humic isolates is generally between 1×10^{17} and 1×10^{18} spins/gram.[19] These concentrations are at least an order of magnitude less than the spin concentration reported for a coal sample where 97% of the carbon atoms were detected.[15] From these results we propose that the free radical concentrations in most DOM samples probably are too low to reduce the number of carbon atoms detected in NMR measurements appreciably. Further spin-counting experiments, however, should be conducted with actual DOM samples to verify this proposal because the spin-lattice relaxation times of the free radicals in DOM may be different from those observed in coal samples.

Direct polarization (DP) liquid-state ^{13}C NMR measurements are susceptible to fewer quantitative uncertainties than DP/MAS or CP/MAS solid-state ^{13}C NMR measurements. In the absence of high concentrations of unpaired electrons, liquid-state ^{13}C NMR spectra measured with pulse delays long enough to allow all the structural groups to relax and gated decoupling to suppress nuclear Overhauser enhancement (NOE) should provide accurate quantitative data.[20,21] Martin et al.[22] have reviewed the conditions necessary for quantitative liquid-state measurements. Fründ and Lüdemann[23] and Schnitzer and Preston[24] compared the CP/MAS quantitation with liquid state quantitation and found that aromatic contents were generally underestimated and aliphatic contents overestimated in the CP/MAS data. Cook et al.[25] and Cook and Langford[26] compared NMR spectra obtained using a standard ^{13}C CP/MAS pulse sequence and a ramp CP/MAS pulse sequence to liquid-state spectra of a soil humic acid and a soil fulvic acid. They found that the ramp CP/MAS sequence provides better agreement with liquid-state spectra measured under quantitative conditions than did the standard CP/MAS sequence. However, inspection of their published spectra indicates that there are still significant differences between the solid-state and liquid-state spectra such that the concentrations derived from the solid-state spectra of aromatic and carboxyl groups are underestimated and aliphatic groups are overestimated.

We shall demonstrate below that comparison of the integral areas of CP/MAS spectra of a suite of DOM samples with those of the corresponding quantitative liquid-state spectra allows one to derive general relationships that may be used to correct structural-unit concentrations calculated from solid-state CP/MAS spectra.

2 MATERIALS AND METHODS

2.1 Sample Isolation and Characterization

Aquatic humic substances were isolated from surface waters collected from locations listed in Table 2 using Amberlite XAD-8* resin according to established methods.[27] In brief, filtered water samples were acidified to pH 2 and passed through appropriately sized columns of XAD-8 resin. The hydrophobic acid fraction (HPOA) of DOM was recovered by back elution with 0.1 N NaOH. The Coal Creek, Ogeechee River and Williams Fork HPOA isolates were further fractionated into fulvic acids (FAs) and humic acids (HAs) by lowering the pH of a solution of each HPOA to 1.0 with HCl and then separating the HAs precipitate by centrifugation.[28] All fractions were desalted on XAD-8 resin, H-saturated using AG-MP 50 cation exchange resin and lyophilized. An isolation and desalting procedure very similar to the one used here was found to reduce metal ion concentrations to very low levels in Suwannee River fulvic and humic acid isolates.[29] Fe(III) concentrations in the Suwannee River samples were at least two orders of magnitude lower than those found to distort the NMR spectra of compost and sewage sludge humic and fulvic acids.[30] NMR analyses were performed on Na-saturated samples obtained by dissolving isolates in distilled water, passing solutions through Na-saturated AGMP-50 cation exchange resin and lyophilization.

Table 2 *Sample site locations and descriptions*

Sample	Date collected	Site description
2BS	April 1997	Oligotrophic wetland, Water Conservation Area 2A, Florida Everglades. Vegetation: Saw-grass periphyton.
3A-33	April 1997	Oligotrophic wetland, Water Conservation Area 3A, Florida Everglades. Vegetation: Saw-grass periphyton.
Coal Creek	June 1982	Small stream draining Flat Tops Wilderness Area, Colorado. Vegetation: Spruce-Fir forest.
Pacific Ocean	February 1986	Pacific Ocean near Hawaii at a depth of 800 feet.
F-1	April 1997	Eutrophic wetland, Water Conservation Area 2A, Florida Everglades. Vegetation: Cattails.
Ogeechee River	May 1982	Small river draining Appalachian piedmont near Grange, Georgia. Vegetation: Oak-Hickory-Pine forest.
S10E	March 1995	Canal in Florida Everglades between Loxahatchee National Wildlife Refuge and Water Conservation Area 2A.
Williams Fork	June 1983	Williams Fork Reservoir, Colorado
U3	April 1997	Wetland, Water Conservation Area 2A, Florida Everglades. Vegetation: Sawgrass periphyton

Elemental compositions of the organic matter isolates were determined by Huffman Laboratories by the methods described in Huffman and Stuber.[31] Specific UV absorbance

---

* The use of trade names in this report is for identification purposes only and does not constitute endorsement by the US Geological Survey.

data at 254 nm (SUVA)$_{254}$ were obtained on the isolates by measuring the UV spectra of solutions containing approximately 5mgC/L organic matter in 0.001 N NaHCO$_3$. The UV absorbances were then divided by the dissolved organic carbon (DOC) concentration of the sample to yield absorbances per mg C. The UV absorbance measurements were made on a Hewlett-Packard 8453 UV/VIS spectrophotometer with a 1 cm cell. DOC concentration data were measured with an Oceanography International Model 700 Carbon Analyzer. The elemental composition and (SUVA)$_{254}$ data for the hydrogen saturated samples are given in Table 3.

Table 3 *Ash free elemental compositions and specific UV absorbance (in units of L/mg C cm) at 254 nm (SUVA)$_{254}$ data for hydrogen saturated samples*

Sample	Elemental composition(%)						SUVA
	C	H	O	N	S	Ash	
Coal Creek FA	53.1	4.5	39.9	1.0	0.6	2.2	0.039
Ogeechee River FA	53.0	4.8	38.8	1.1	1.6	1.6	0.038
Ogeechee River HA	54.6	4.9	36.8	1.6	1.8	3.9	0.053
Williams Fork FA	49.9	4.5	42.6	1.1	1.9	11.8	0.030
Pacific Ocean FA	56.2	6.0	36.3	1.1	0.4	0.4	0.006
Everglades 3A-33 HPOA	53.7	4.7	38.9	1.8	0.8	5.5	0.032
Everglades F1 HPOA	52.7	4.5	39.3	1.8	1.7	4.5	0.040
Everglades U3 HPOA	53.9	4.5	38.4	1.8	1.3	2.5	0.035
Everglades S10E HPOA	54.5	4.9	37.4	1.9	1.3	1.8	0.035

2.2 NMR Spectrometry

Solid-state, cross-polarization, magic-angle-spinning (CP/MAS) ^{13}C spectra were measured on a 200 MHz Chemagnetics CMX spectrometer with a 7.5 mm-diameter probe. The spinning rate was 5000 Hz. All of the experiments were performed with a pulse delay of 1 second and a pulse width of 4.5 microsec for the 90° pulse. Spectra were measured for different contact times between 1 and 10 msec. Adequate signal to noise could generally be obtained from samples that consisted of at least 200 mg of material with 1000 to 2000 transients. Smaller samples required that additional transients be collected in order to obtain satisfactory signal to noise ratios. A line broadening of 100 Hz was applied in the Fourier transformation of the free induction decay data.

Direct polarization magic-angle spinning (DP/MAS) measurements were also made on the Chemagnetics CMX spectrometer using the same probe and spinning rate as in the CP/MAS experiments. An excitation pulse of 2.00 microsec duration (corresponding to a 40° tip angle) was applied to the samples. Pulse delays of 8 to 48 sec were used in order to determine when complete relaxation was attained. It was found that the receiver delay had to be increased from 18 to 20 microsec to obtain flat baselines in the DP/MAS spectra.

Liquid-state ^{13}C NMR spectra were measured on approximately 200 mg/mL of sample dissolved in D$_2$O in 10-mm diameter tubes on a Varian XL 300 spectrometer at 75.429 MHz. Quantitative spectra were obtained using inverse gated decoupling in which the proton decoupler was on only during the acquisition of the free induction decay (FID) curve; an 8-sec delay time and a 45° tip angle were used. The sweep width was 30 kHz.[32] The inverse gated decoupling procedure eliminates the effect of different nuclear Overhauser enhancement (NOE) factors for different functional groups in the same spectrum.

The NMR spectra were integrated by scanning each spectrum into a computer at 200 dpi with Hewlett Packard Deskscan II software. The resulting TIFF file of the spectrum was imported into Adobe PhotoShop and converted to grayscale. An appropriately scaled grid containing the desired regions was overlaid on the spectrum, and the number of pixels between the baseline and the spectral curve within each spectral region was displayed and divided by the number of pixels under the entire spectral curve. The resulting number was multiplied by 100.

2.3 Iron Analyses

DOC solutions containing 10 mg carbon/L were prepared by dissolving the samples in deionized water. The solutions were analyzed for iron by inductively coupled plasma-emission spectrometry.[33]

2.4 Model Compound Preparation

A model plant flavonol condensation product was prepared by reacting 1 g of 2,4,6-trihydroxybenzoic acid (Aldrich; according to the manufacturer it contained approximately 10% phloroglucinol) with 2 g of propionaldehyde (Chemservice) in 300 mL of water buffered at pH 8 with sodium bicarbonate. The molar concentration of propionaldehyde was approximately fivefold that of the 2,4,6-trihydroxybenzoic acid in order to promote maximum condensation. The reactants were stirred in a stoppered 500 mL erlenmeyer flask overnight and the solution was then acidified to pH 1 with HCl to give a white precipitate. This precipitate was separated by filtration and redissolved in butanol. The butanol was back-extracted with water several times to remove salts and low-molecular weight reactants. The butanol was then removed by vacuum evaporation. The reaction product changed color from white to light orange as oxidation occurred. Approximately 600 mg of reaction product was obtained.

3 RESULTS AND DISCUSSION

3.1 DOM Samples

Solid-state CP/MAS ^{13}C NMR spectra at different contact times and liquid-state ^{13}C NMR spectra of the DOM samples listed in Table 3 were measured. These samples were chosen to represent a variety of different surface-water environments. The variations in the elemental compositions and SUVA values indicate that these samples are markedly different in chemical structure. Spectral regions representative of carbon atoms in aliphatic hydrocarbons (0-62 ppm); carbons atoms in carbohydrates and other aliphatic alcohols (62-90 ppm); anomeric carbons (90-110 ppm); aromatic carbon atoms (110-160 ppm); carbonyl carbons in carboxylic acids, amino acids, peptides and quinones (160-190); and carbon atoms in ketones and aldehydes (190-230 ppm) were integrated (Figure 1). There is significant overlap of the chemical shift ranges of the various structural units listed above.

Therefore, the choice of boundaries for the various spectral regions is somewhat arbitrary. For this reason the boundaries of the spectral regions used in this study are slightly different than those used by Cook et al.[25]

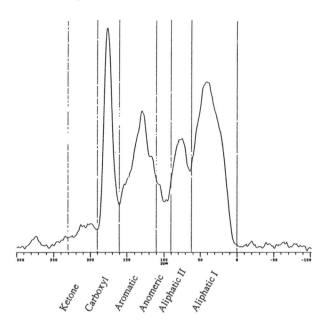

Figure 1 *Spectral regions representative of carbon atoms in different structural groups*

DP/MAS spectra of the Pacific Ocean fulvic acid and the S10E HPOA were measured using pulse delays of 8 to 48 sec. Very little change in the spectra was observed for delays greater than 32 sec, indicating that a 32 sec delay is adequate for complete relaxation of these samples. The spectral regions representative of the various structural components indicated above were integrated for the spectra and the integrals were compared with those of the corresponding liquid-state spectra. Figure 2 shows that there is very close agreement between the liquid-state and DP/MAS spectra of each sample.

The close agreement of corresponding integrated areas of the spectra measured by the two different direct polarization methods provides additional support for the presumption that direct polarization provides an accurate means of measuring the relative concentrations of the various structural groups in the two samples. The only two of the factors listed in Table 1 that would distort both the solid-state and liquid-state measurements are use of a pulse delay that is too short to allow for complete relaxation and the presence of high concentrations of free radicals. The close agreement of integrated areas of the DP/MAS spectra to those of liquid-state spectra indicates that the other factors are not distorting the DP/MAS spectra of the fulvic acid samples. The DP/MAS variable pulse delay experiments indicated that a 48 second delay was more than adequate for complete relaxation of the solid samples and previous studies have shown that 8 seconds is adequate for the liquid samples. Thus, the only factor that might alter the quantitation is the presence of free radicals in the samples. As pointed out in the Introduction, humic

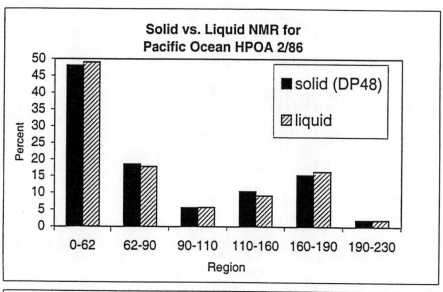
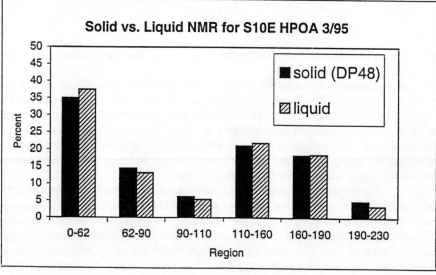

Figure 2 *Comparison of integrated areas of the spectral regions of the liquid-state and DP/MAS spectra of Pacific Ocean fulvic acid and S10E HPOA*

isolates generally have free radical concentrations that are too low to appreciably reduce the number of carbon atoms detected in NMR measurements. Support for this assumption is provided by the fact that relatively long pulse delays are required for complete relaxation in the DP/MAS measurements. If high concentrations of free radicals were present one would expect much more rapid relaxation.

Comparison of the CP/MAS spectra of the samples used in this study to the corresponding quantitative liquid-state spectra indicate that, in general, CP/MAS does not provide accurate structural-group quantitation. Longer contact times improve the quantitation, but the results still diverge from those obtained from the liquid-state spectra. These results are illustrated in Figure 3 for three very different DOM isolates. Aromatic and carbonyl carbons are consistently underestimated and aliphatic carbons overestimated in this very diverse suite of samples. These results indicate that it may be possible to develop a general method for correcting the CP/MAS quantitation which will be applicable to most if not all DOM isolates.

To this end, the integrated areas of the solid-state CP/MAS spectra at three different contact times versus those of the liquid-state DP spectra were plotted for each spectral region (Figure 4). These plots indicate that there is a linear relationship between the liquid-state DP and solid-state CP/MAS integrated areas in each of the spectral regions shown for each contact time measured. The ketone-aldehyde plots have been excluded because there was a great deal of scatter in the data in this region. This scatter was probably the result of two factors: (1) areas in this region were small and (2) keto-enol tautomerism is likely in the samples, giving rise to different tautomeric forms in the liquid and solid states.

The observed linear relationships between the liquid-state and solid-state integrated areas are most likely indicative of the presence of similar organic structural units in all of the DOM samples used in this study. DOM in natural waters arises to a large extent from the oxidative degradation of plant litter.[32] Wershaw et al.[34] isolated DOM from the senescent leaves of different species of trees. The DOM of the leachate from each species was fractionated by polarity in order to obtain more homogeneous isolates. The NMR spectra of the fractions indicated that they are composed of identifiable structure units derived from carbohydrates, lignin, hydrolyzable and nonhydrolyzable tannins and lipids. The NMR spectral bands of natural DOM occur in the same regions as those observed in the spectra of the leaf leachate DOM isolates. However, the natural DOM spectral bands are broader and less well-resolved than those of the leaf leachate isolates indicating, not surprisingly, that natural DOM is more heterogeneous than the fractions isolated from the leaves of single species of trees.

The integrated areas of the 5 msec contact-time spectra are closer to those of the liquid-state spectra than those of the shorter contact times. However, the data in Figure 4 indicate that the areas of the 5 msec spectra are still not the same as those of the liquid state spectra. Alemany et al.[35,36] have studied the cross polarization dynamics of organic compounds with nearby protons and those with remote protons. They found that the relative rates of cross polarization of carbon atoms by protons in different structural groups fit the order CH_3(static) > CH_2 > CH $\approx$ CH_3(rotating) > C(nonprotonated) predicted from the model of Demco et al.[37] and that carbon atoms remote from protons cross polarize more slowly than those nearby to protons. Thus, remote carbon atoms may not fully polarize because of decay of proton magnetization before transfer of the magnetization to the carbon atoms is completed.

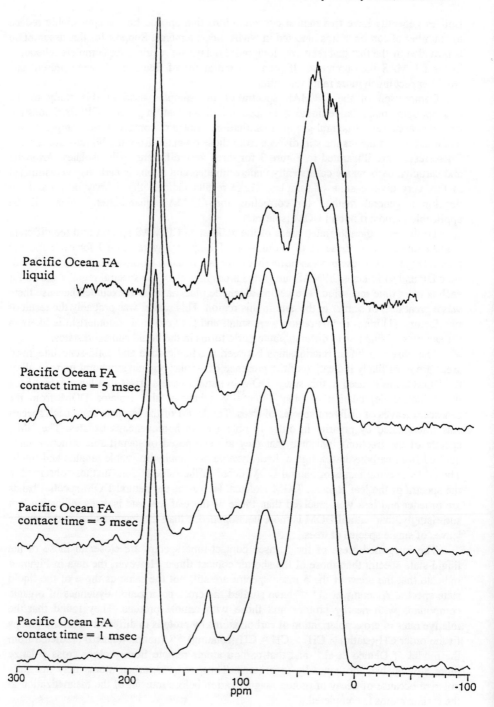

Figure 3a *CP/MAS at three different contact times and the corresponding quantitative liquid-state spectra for Pacific Ocean Fulvic Acid*

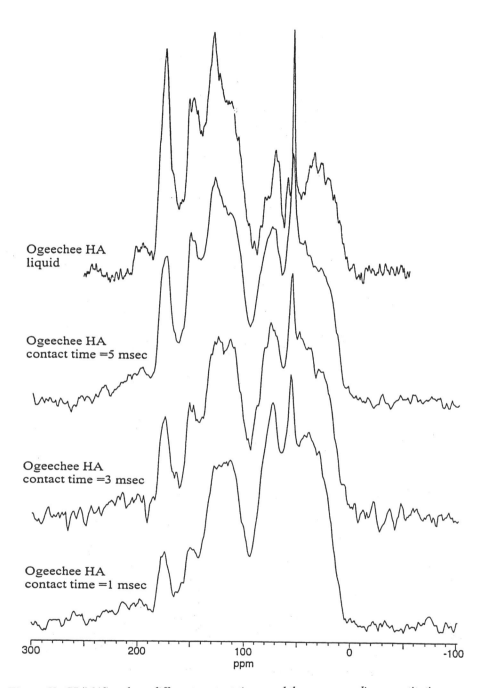

Figure 3b *CP/MAS at three different contact times and the corresponding quantitative liquid-state spectra for Ogeechee Humic Acid*

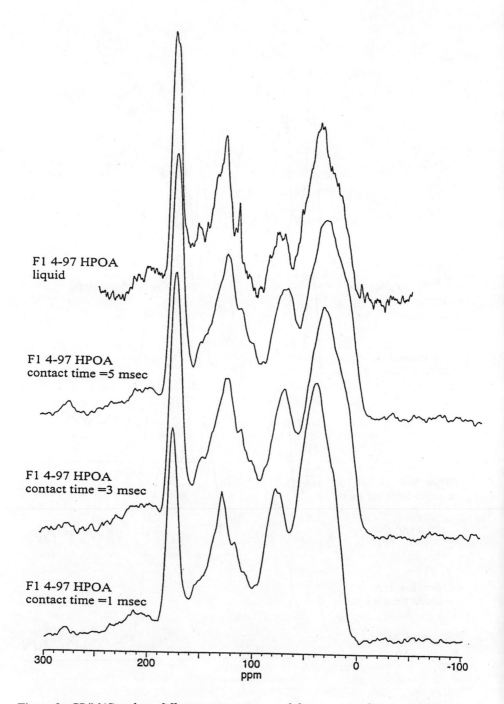

Figure 3c *CP/MAS at three different contact times and the corresponding quantitative liquid-state spectra for F-1 HPOA*

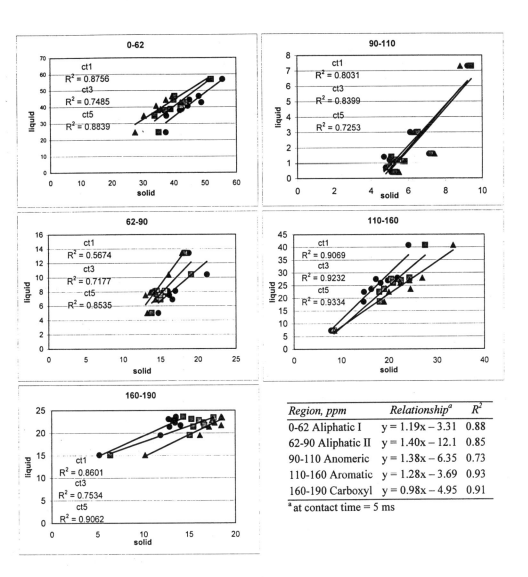

Figure 4 *Plots of the integrated areas of each spectral region of the solid-state CP/MAS spectra at three different contact times versus the regional areas of the corresponding liquid-state DP spectra. Contact times are ●, 1ms; ■, 3ms; ▲, 5ms. The area for each region of a spectrum is plotted as a percentage of the total area under the spectrum*

3.2 Model Polymer

Natural DOM consists mainly of fragments derived from the partial degradation of plant polymers; however, some DOM may arise from the condensation of some of these fragments.[32] An attempt was made to synthesize a model DOM component with carbon atoms remote from protons that would result from the degradation and condensation of a plant flavonol pigment. One of the most abundant plant flavonols, quercetin, has been shown to undergo autoxidation at neutral pH in aqueous solution according to reaction 1.[38]

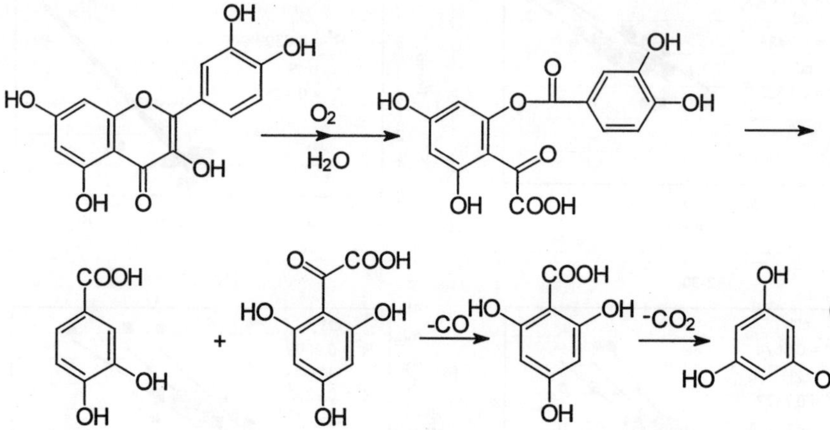

Phloroglucinol, the end product of reaction 1, is known to rapidly polymerize with aldehydes to give products of the type shown below:[39]

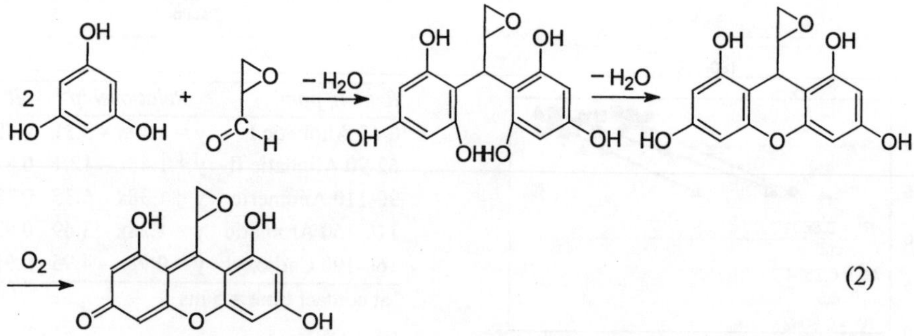

A variation of reaction 2 was attempted in which propionaldehyde was reacted with 2,4,6-trihydroxybenzoic acid to form a model DOM component that might be used for NMR quantitation studies. 2,4,6-trihydroxybenzoic acid was chosen rather than pholoroglucinol in order to eliminate one proton on the ring. It was hoped that the polymerization would be rapid enough to prevent extensive decarboxylation of the acid. Propionaldehyde was chosen as the aldehyde in this reaction in order to avoid overlap of chemical shifts of the sugar anomeric carbons in the region between 95 and 105 ppm with some of the phloroglucinol ring carbons even though the aldehydes with which phloroglucinol is most likely to condense during leaf senescence are reducing sugars.

Condensation of propionaldehyde with 2,4,6-trihydroxybenzoic acid should not be limited to one position on the ring; we expected that both protons would react, thereby eliminating all of the ring protons. CP/MAS spectra of the product at 1 and 5 msec contact times are given in Figure 5; the spectrum of 2,4,6-trihydroxybenzoic acid is also given in the Figure. The assignments of the chemical shifts for 2,4,6-trihydroxybenzoic acid are shown in Figure 6. No evidence of residual 2,4,6-trihydroxybenzoic acid is present in the spectrum of the reaction product. The band at 95 ppm in the reaction product most likely represents carbons attached to carboxylic acid groups. The band at 110 ppm is indicative of substitution on the phloroglucinol ring.[40] If we assume that the band at about 76 ppm is due to an aliphatic alcohol impurity, then the NMR spectrum is consistent with a structure in which all of the ring protons have been replaced by substituent groups.

Variable contact time experiments show that most of the changes in the integrated areas take place between contact times of 1 and 5 msec. Longer contact times provide only relatively small increases in the integrated areas of the aromatic bands and concomitant small decreases in the integrated areas of the aliphatic bands. Comparison of the 1 and 5 msec contact time spectra shows that the relative intensity of the terminal methyl band at 13 ppm is greater in the 5 msec spectrum than in the 1 msec spectrum. This is consistent with the finding of Alemany et al.[36] that freely rotating CH_3 groups cross polarize more slowly than CH_2 groups. In addition to the increase in intensity of the CH_3 band relative to the CH_2 band at 25 ppm, there is an increase in intensity relative to the band at 35 ppm that probably represents CH groups. Alemany et al.[36] found that for their samples the rates of cross polarization of CH_3 and CH groups are approximately equal; that does not appear to be the case here. It also is apparent from these spectra that for this particular sample it would have been better to have chosen 165 ppm for the boundary between aromatic and carboxylate carbons.

4 CONCLUSIONS

Apparent linear relationships have been demonstrated between the integrated areas of the spectral regions representative of carbon atoms in aliphatic hydrocarbons, carbon atoms in carbohydrates and other aliphatic alcohols, anomeric carbons, aromatic carbon atoms, carbonyl carbons in carboxylic acids, amino acids, peptides, and quinones in the liquid-state NMR spectra and the corresponding regions in the solid-state spectra of a suite of DOM samples from different environments. We propose that the suite of DOM samples measured here is representative of DOM samples in general, and therefore that the apparent linear relationships measured in this study may be used to correct structural-unit concentrations calculated from solid-state CP/MAS spectra of any given DOM sample. The integrated areas of the 5 ms contact-time spectra are closer to those of the liquid-state spectra than those with shorter contact times, and therefore it is preferable to use a contact time of 5 ms for the CP/MAS analysis of DOM.

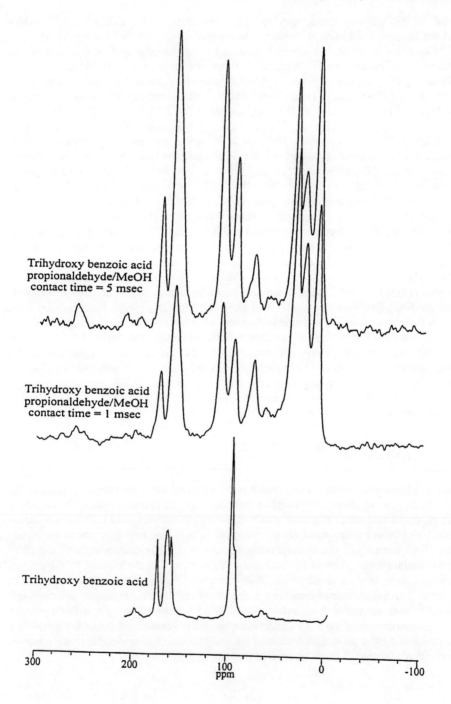

Figure 5 *CP/MAS spectra at 1 and 5 ms contact times of the condensation product of propionaldehyde with 2,4,6-trihydroxybenzoic acid*

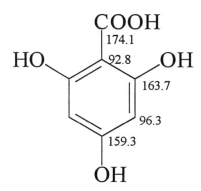

Figure 6 ^{13}C NMR chemical shift assignments for 2,4,6-trihydroxybenzoic acid

References

1. R. A. Larson and E. J. Weber, 'Reaction Mechanisms in Environmental Organic Chemistry', Lewis Publishers, Boca Raton, FL, 1994.
2. R. A. Larson and A. L. Rockwell, *Environ. Sci. Technol.*, 1979, **13**, 325.
3. M. Dore, N. Merlet, J. De Laat and J. Goichon, *Jour. Amer. Water Works Assoc.*, 1982, **74**, 103.
4. S. Lin, R. J. Liukkonen, R. E. Thom, J. G. Bastian, M. T. Lukasewycz and R. M. Carlson, *Environ. Sci. Technol.*, 1984, **18**, 932.
5. J. V. Hanna, W. D. Johnson, R. A. Quezada, M. A. Wilson and L. Xiao-Qiao, *Environ. Sci. Technol.*, 1991, **25**, 1160.
6. S. D. Boyce and J. F. Hornig, *Environ. Sci. Technol.*, 1983, **17**, 202.
7. S. Fam and M. K. Stenstrom, *Jour. WPCF*, 1987, **59**, 969.
8. C. Flodin, E. Johansson, H. Borén, A. Grimvall, O. Dahlman and R. Mörck, *Environ. Sci. Technol.*, 1997, **31**, 2464.
9. R. F. Christman, D. L. Norwood, Y. Seo and F. H. Frimmel, in 'Humic Substances II-In Search of Structure', M. H. B.Hayes, P. MacCarthy, R. L. Malcolm and R. S. Swift, (eds.), Wiley, Chichester, 1989, p. 33.
10. M. H. B. Hayes and M. R. O'Callaghan, in 'Humic Substances II-In Search of Structure', M. H. B.Hayes, P. MacCarthy, R. L. Malcolm and R. S. Swift, (eds.), Wiley, Chichester, 1989, p. 143.
11. J. M. Bracewell, K. Haider, S. R. Larter and H. -R. Schulten, in 'Humic Substances II-In Search of Structure', M. H. B.Hayes, P. MacCarthy, R. L. Malcolm and R. S. Swift, (eds.), Wiley, Chichester, 1989, p. 181.
12. W. S. Warren, *Science*, 1998, **280**, 398.
13. R. L. Wershaw, 'Membrane-micelle model for humus in soils and sediments and its relation to humification', United States Geological Survey Water-Supply Paper 2410, 1994.
14. C. M. Preston, *Soil Sci.*, 1996, **161**, 144.
15. A. Jurkiewicz and G. E. Maciel, *Anal. Chem.*, 1995, **67**, 2188.

16. R. A. Wind, G. E. Maciel and R. E. Botto, in 'Magnetic Resonance of Carbonaceous Solids', R. E. Botto and Y. Sanada, (eds.), Amer. Chem. Soc., Washington, D. C., 1993, p. 3.
17. P. Kinchesh, D. S. Powlson and E. W. Randall, *Eur. Jour. Soil Sci.*, 1995, **46**, 125.
18. P. Kinchesh, D. S. Powlson and E. W. Randall, *Eur. Jour. Soil Sci.*, 1995, **46**, 146.
19. N. Senesi and C. Steelink, in 'Humic Substances II-In Search of Structure', M. H. B. Hayes, P. MacCarthy, R. L. Malcolm and R. S. Swift, (eds.), Wiley, Chichester, 1989, p. 373.
20. J. K. M. Sanders and B. K. Hunter, 'Modern NMR Spectroscopy-A Guide to Chemists', Oxford University Press, Oxford, 1993.
21. C. M. Preston, in 'NMR of Humic Substances and Coal', R. L. Wershaw and M. A. Mikita, (eds.), Lewis Publishers, Chelsea, MI, 1987, p. 3.
22. M. L. Martin, G. J. Martin and J. -J. Delpuech, 'Practical NMR Spectroscopy', Heyden, London, 1980.
23. R. Fründ and H. -D. Lüdemann, *Sci. Total Environ.*, 1989, **81/82**, 157.
24. M. Schnitzer and C. M. Preston, *Soil Sci. Soc. Amer. J.*, 1986, **50**, 326.
25. R. L. Cook, C. H. Langford, R. Yamdagni and C. M. Preston, *Anal. Chem.*, 1996, **68**, 3979.
26. R. L. Cook and C. H. Langford, *Environ. Sci. Technol.*, 1998, **32**, 719.
27. G. R. Aiken, D. M. McKnight, K. A. Thorn and E. M. Thurman, *Org. Geochem.*, 1992, **18**, 567.
28. G. R. Aiken, in 'Humic Substances in Soil, Sediment, and Water: Geochemistry, Isolation, and Characterization', G. R. Aiken, D. M. McKnight, R. L. Wershaw and P. MacCarthy, (eds.), Wiley, New York, 1985, p. 363.
29. H. E. Taylor and J. R. Garbarino, in 'Humic Substances in the Suwannee River, Georgia: Interactions, Properties, and Proposed Structures', R. C. Averett, J. A. Leenheer, D. M. McKnight and K. A. Thorn, (eds.), U.S. Geological Survey Open-File Report 87-557, 1989, p. 85.
30. P. E. Pfeffer, W. V. Gerasimowicz and E. G. Piotrowski, *Anal. Chem.*, 1984, **56**, 734.
31. E. W. D. Huffman Jr. and H. A. Stuber, in 'Humic Substances in Soil, Sediment, and Water: Geochemistry, Isolation, and Characterization', G. R. Aiken, D. M. McKnight, R. L. Wershaw and P. MacCarthy, (eds.), Wiley, New York, 1985, p. 433.
32. R. L. Wershaw, J. A. Leenheer, K. R. Kennedy and T. I. Noyes, *Soil Sci.*, 1996, **161**, 667.
33. J. R. Garbarino and H. E. Taylor, *Appl. Spectrosc.*, 1979, **33**, 220.
34. R. L. Wershaw, J. A. Leenheer and K. R. Kennedy, in 'Humic Substances: Structures, Properties and Uses', G. Davies and E. A. Ghabbour, (eds.), Royal Society of Chemistry, Cambridge, 1998, p. 47.
35. L. B. Alemany, D. M. Grant, R. J. Pugmire, T. D. Alger and K. W. Zilm, *J. Amer. Chem. Soc.*, 1983, **105**, 2133.
36. L. B. Alemany, D. M. Grant, R. J. Pugmire, T. D. Alger and K. W. Zilm, *J. Amer. Chem. Soc.*, 1983, **105**, 2142.
37. D. E. Demco, J. Tegenfeldt and J. S. Waugh, *Phys. Rev. B,* 1975, **11**, 4133.
38. C. G. Nordstrom, *Suomen Kem.*, 1968, **B41**, 351.
39. J. Pritzker and R. Jungkunz, *Z. Untersuch. Lebensm.*, 1927, **54**, 247.

40. Z. Czochanska, L. Y. Foo, R. H. Newman and L. J. Porter, *J. Chem. Soc. Perkin I Trans.*, 1980, 2278.

STRUCTURAL INVESTIGATION OF HUMIC SUBSTANCES USING 2D SOLID-STATE NUCLEAR MAGNETIC RESONANCE

Jingdong Mao,[1] Klaus Schmidt-Rohr[2] and Baoshan Xing[1]

[1]Department of Plant and Soil Sciences, Stockbridge Hall, University of Massachusetts, Amherst, MA 01003, USA
[2]Department of Chemistry, Iowa State University, Ames, IA 50011, USA

1 INTRODUCTION

Humic substances (HSs) play important roles in many environmental and agricultural reactions and processes. Their structural information is critical for understanding these roles.[1-3] Although a great deal of structural information has been obtained from different kinds of methods such as classical chemical analysis and modern spectroscopic techniques, it is almost impossible to conclude a HS structure due to its complexity and heterogeneity.[4] Classical chemical methods are based on elemental compositions, which requires the cleavage of HS and can only show the average of molecular agglomeration. The best way to study HSs is with non-destructive spectroscopic techniques. Among them, solid-state nuclear magnetic resonance (NMR) has proven to be one of the most powerful methods.

There are many solid-state NMR techniques. The predominant ones used in studying HSs are one-dimensional (1D) techniques such as CP/MAS (cross-polarization with magic angle spinning). Undoubtedly, 1D techniques have provided a large quantity of useful structural data, but the typical broad HS 1D spectra cannot be used for detailed structural investigation.[5-9] However, the development of advanced 2D NMR techniques overcame this problem.[10] In 2D NMR experiments, based on the second parameter the complex spectra are separated into simpler, more resolved subspectra, which are easier to interpret. Furthermore, one of the 2D techniques, 2D heteronuclear ^{13}C-^{1}H NMR (HETCOR) can even tell how different carbon units are connected. This technique was first used in solution-state NMR experiments. It was then successfully applied to the solid-state and used to investigate simple polymers and later complicated coals.[11-14] But this technique has not been used for the study of HSs despite its usefulness. The reasons that limit the application of this technique are: 1) paramagnetic materials in HSs can shorten the ^{1}H T_1 (spin-lattice relaxation times) and lead to broad and unacceptable ^{1}H spectra; and 2) for ideal HETCOR experiments, the sample should be confined to the center of a small radiofrequency coil so that strong and uniform radiofrequency pulses will be allowed. Actually, this situation cannot be achieved for HSs because large HSs samples must be used to overcome reduced sensitivity due to the broadness of the overall ^{13}C spectrum and of the individual lines.

Compared with the solution HETCOR technique, the solid spectrum can give some information that cannot be obtained in the solution state. Generally, the difference between solution and solid HETCOR arises from the different character of heteronuclear interaction.[10] The heteronuclear J coupling in liquids is a through-bond interaction and the dipolar interaction in solids is through space. Thus, the transfer of magnetization in solids can be of a less local nature. Although this character may complicate spectral analysis, it allows the ^{1}H-^{13}C correlation analysis between unprotonated carbons and protons, showing the spatial proximity of those nuclei. By adjusting the transfer distance of magnetization through the contact time and insertion of dipolar dephasing, new information on HSs can be obtained. The objective of this study is to structurally examine a humic acid using 2D HETCOR solid-state NMR.

2 MATERIALS AND METHODS

2.1 Samples

A peat humic acid (Amherst HA) was used. The details of extraction and purification were described elsewhere.[1] The elemental composition of this sample is %C, 52.9; %H, 5.32; %O, 42.5; and %ash, < 0.1.

2.2 NMR Measurements

The HETCOR pulse sequence employed in this study is shown in Figure 1. It has the classic four-part structure: preparation, evolution (t_1), mixing and detection (t_2).[10] This experiment requires proton evolution with BR-24 multiple-pulse homonuclear dipolar decoupling and TOSS (total sideband suppression)[15,16] sequences before detection. BR-24 is one of the multipulse sequences that are used mainly for the homonuclear decoupling of protons and consists of specifically designed cycles of several pulses, possibly separated by windows without irradiation.[10,17] Through homonuclear decoupling of BR-24 the homonuclear dipolar interaction is averaged to zero.

The sample was packed in a 4-mm-diameter zirconia rotor with a Kel-F cap and run at a ^{13}C frequency of 75.48 MHz in a Bruker DSX-300 spectrometer at a spinning speed of 3.7 kHz. The proton 90° pulse length was 3.5 µs and carbon pulse length 6 µs.

Several 2D HETCOR experiments were performed: 1) short CP (0.1 ms); 2) short CP (0.1 ms) with 40 µs dipolar dephasing; and 3) long CP (1 ms) with 10 ms spin diffusion and 40 µs dipolar dephasing.

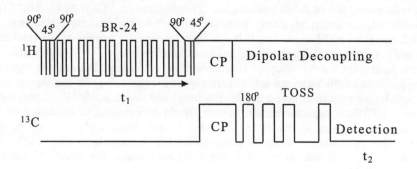

Figure 1 *HETCOR pulse sequence using BR-24 as the multipulse*

3 RESULTS AND DISCUSSION

Figure 2 shows the 2D HETCOR spectrum with a contact time of 0.1 ms. The 2D HETCOR technique provides well-resolved proton spectra because it separates the proton resonance over a much larger ^{13}C chemical-shift range. With 0.1 ms contact time, the magnetization can only transfer one to two bonds away. That is, it can transfer only within a short distance.

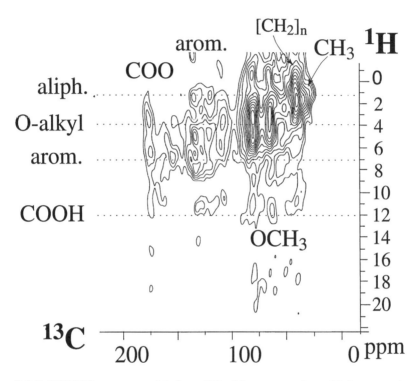

Figure 2 *2 D HETCOR spectrum of Amherst HA with a contact time of 0.1 msec and without dipolar dephasing*

Figure 3 is the 2D HETCOR spectrum with short CP (0.1 ms) and 40 μs dipolar dephasing. With dipolar dephasing, only unprotonated carbons and mobile groups like CH_3 can be detected whereas with short CP, only carbons with directly bonded or adjacent protons (two or three bonds away) can be detected. Thus, this experiment can only show unprotonated carbons with adjacent protons and mobile groups like CH_3.

Figure 4 shows the 2D HETCOR spectrum with long CP (1 ms) and 10 ms spin diffusion and 40 μs dipolar dephasing. Spin diffusion is the transfer of magnetization through space. With long CP and spin diffusion, the magnetization can transfer over a long distance.

We now discuss the information from each peak from low to high chemical shifts (ppm). The proton spectrum can be approximately assigned as follows: 0.8-3 ppm, aliphatic protons; 3-5.5 ppm, protons associated with oxygen-containing functional groups; and 6-8.5 ppm, aromatic/amide protons.

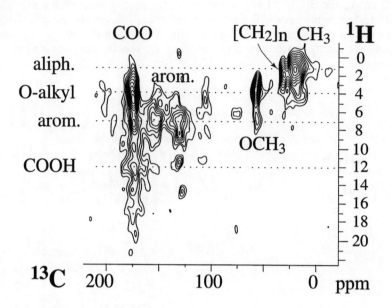

Figure 3 *2D HETCOR spectrum of Amherst HA with a contact time of 0.1 ms and 40 µs dipolar dephasing*

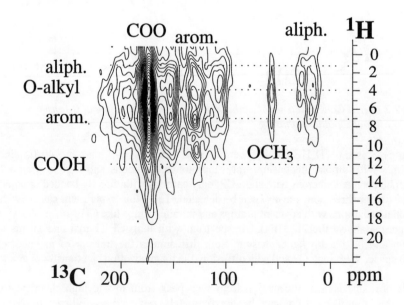

Figure 4 *2D HETCOR spectrum of Amherst HA with a contact time of 1 ms, 10 ms spin diffusion and 40 µs dipolar dephasing*

CH₃ groups. The signal in the 0-24 ppm region is due to CH_3 groups. The evidence for the assignment to CH_3 group is: 1) their ^{13}C and ^1H chemical shifts correspond to CH_3 groups; and 2) the signals of this range remain after dipolar dephasing due to the high mobility of CH_3 (Figures 3 and 4). The CH_3 groups at the end of the aliphatic chains appear at 15 ppm and the ones attached to aryl rings are at 20 ppm.[18] For Amherst HA, CH_3 groups are attached to aliphatic and O-alkyl carbons but not aromatic carbons (Figure 3) because the CH_3 proton chemical shifts did not extend as far as the chemical shifts of CH_3 attached to aromatic rings.

(CH₂)ₙ groups. Both amorphous and crystalline $(CH_2)_n$ chains were discovered in humic substances by Hu et al.[19] The amorphous peak is at 31 ppm and crystalline peak at 32.9 ppm in the ^{13}C spectra. In Figure 3, there is only the signal from amorphous $(CH_2)_n$ chains left because it survives dipolar dephasing due to its high mobility.

OCH₃ groups. The peaks at around 55 ppm of ^{13}C and 3.9 ppm of ^1H are due to OCH_3. This peak exists in both dipolar dephasing HETCOR spectra (Figures 3 and 4) due to the high mobility of the CH_3. In Figures 3 and 4 there are cross peaks between OCH_3 and aromatic carbons, showing that OCH_3 is connected directly to aromatic rings, that is in a lignin-like structure. The ratio of the OCH_3 to the aromatic C-O is 1:3 from the quantitative DP/MAS spectrum of Amherst HA.[20] Thus we surmise that about 1 out of 3 aromatic C-O groups are not phenolic groups but $C-OCH_3$ groups.

Carbohydrate groups. In both dipolar dephasing spectra, the carbohydrate groups except the anomeric groups disappear due to the abundant protons (Figures 3 and 4). But we observe all of their signals in the normal 2D HETCOR spectrum (Figure 2). The close correlation between the carbohydrate groups and the anomeric groups confirms that the anomeric group is part of the carbohydrate (Figure 2).

Anomeric groups. There are two kinds of anomeric groups: ketal and acetal.[18] The ^1H-^{13}C 2D HETCOR spectra permit unambiguous identification of otherwise unresolved anomeric carbons (O-CHR-O), which are the characteristic of polysaccharides (Figures 2-4). The anomeric groups are not resolved in a typical 1D ^{13}C NMR spectrum,.

Aromatic groups. There are two kinds of aromatic groups: protonated and unprotonated. The protonated aromatic groups disappeared in the dipolar dephasing spectrum (Figure 3). Not much information on this group is obtained from this study.

Aromatic C-O groups. The aromatic C-O group is well resolved in the 2D HETCOR experiments (Figures 2-4), which cannot be expected from 1D experiments. The aromatic C-O groups are closely connected with the aromatic protons (Figure 3).

COO groups. The 2D HETCOR experiment is valuable in characterizing the environment of COO groups and other unprotonated carbons. The COO carbons are found prominently near O-alkyl protons, but some are also close to aromatics as well as aliphatics (Figure 3). The ^1H spectrum of COO groups can extend as high as 20 ppm due to COOH protons participating in hydrogen bonds.[21]

Ketonic groups. Ketonic groups are not observed in the 2D spectra with short CP (Figure 3), but are detected in the 2D spectra with long CP (Figure 4). This indicates the nonprotonated characteristic of the carbonyl group. That is, this group is mainly the keto group.

4 CONCLUSIONS

New structural information on HSs can be obtained from a series of two-dimensional ^1H-^{13}C heteronuclear solid-state NMR (HETCOR) spectra with different contact times and different spectral-editing techniques (e.g., dipolar dephasing and spin-spin relaxation time filters). For Amherst HA, the CH_3 groups are found to be connected with O-alkyl and aliphatic components. The OCH_3 groups are connected directly with the aromatic rings. The COO carbons are found prominently in O-alkyl environments, but some are also bonded to aromatics as well as aliphatics. The hydrogen bonding of COOH causes its ^1H chemical shift to extend up to 20 ppm. The carbonyl groups are primarily composed of keto groups. 2D HETCOR experiments are especially valuable for characterizing the environments of COO groups and other unprotonated carbons. With better resolution using BR-24, the ^1H chemical shift differences are also useful in separating the signals of sp^3 sites that are poorly resolved in the 1D ^{13}C spectra.

ACKNOWLEDGEMENTS

This work was supported in part by the U.S. Department of Agriculture, National Research Initiative Competitive Grants Program (97-35102-4201 and 98-35107-6319), the Federal Hatch Program (Project No. MAS00773), and a Faculty Research Grant from University of Massachusetts, Amherst. We would also like to thank Dr. L. C. Dickinson for his technical support.

References

1. F. J. Stevenson, 'Humus Chemistry: Genesis, Composition, Reactions', 2nd Edn., Wiley, New York, 1994.
2. B. Xing, W. B. McGill and M. J. Dudas, *Environ. Sci. Technol.*, 1994, **28**, 1929.
3. B. Xing and J. J. Pignatello, *Environ. Sci. Technol.*, 1997, **31**, 792.
4. M. H. B. Hayes, in 'Humic Substances, Peats and Sludges. Health and Environmental Aspects', M. H. B. Hayes and W. S. Wilson, (eds.), Royal Society of Chemistry, Cambridge, U.K., 1997, p. 3.
5. C. M. Preston, *Soil Sci.*, 1996, **161**, 144.
6. R. L. Wershaw, and M. A. Mikita (eds.), 'NMR of Humic Substances and Coal: Techniques, Problems and Solutions', Lewis Publishers, Chelsea, MI, 1987.
7. M. A. Wilson, 'NMR Techniques and Applications in Geochemistry and Soil Chemistry'; Pergamon Press, Oxford, 1987.
8. P. Kinchesh, D. S. Powlson and E. W. Randall, *Eur. J. Soil Sci.*, 1995, **46**, 125.
9. D. E. Axelson, 'Solid State Nuclear Magnetic Resonance of Fossil Fuel', Multiscience Publications Ltd., Canadian Government Publishing Center, Supply and Services, Canada, 1985.
10. K. Schmidt-Rohr and H. W. Spiess, 'Multidimensional Solid-State NMR and Polymers', Academic Press, London, 1994.
11. A. Bielecki, D. P. Burum, D. M. Rice and F. E. Karasz, *Macromolecules,* 1991, **24**, 4820.
12. D. P. Burum and A. Bielecki, *J. Magn. Reson.,* 1991, **94**, 645.
13. C. E. Bronnimann, C. F. Ridenour, D. R. Kinney and G. E. Maciel, *J. Magn. Reson.,* 1992, **97**, 522.
14. M. A. Wilson, J. V. Hanna, K. B. Anderson and R. E. Botto, *Org. Geochem.*, 1993, **20**, 985.
15. W. T. Dixon, *J. Chem. Phys.*, 1982, **77**, 1800.
16. W. T. Dixon, J. Schaefer, M. D. Sefcik, E. O. Stejskal and R. A. McKay, *J. Magn. Reson.,* 1982, **49**, 341

17. H. Rosenberger, G. Scheler and E. Künstner, *Fuel,* 1988, **67**, 508.
18. M. A. Wilson, R. J. Pugmire and D. M. Grant, *Org. Geochem.,* 1983, **5**, 121.
19. W. -G. Hu, J. -D. Mao, B. Xing and K. Schmidt-Rohr, *Environ. Sci. Technol.,* 2000, **34**, 530.
20. J. -D. Mao, W. -G. Hu, K. Schmidt-Rohr, G. Davies, E. A. Ghabbour and B. Xing, *Soil Sci. Soc Am. J.,* 2000 (in press).
21. R. K. Harris, P. Jackson, L. H. Merwin and B. J. Say, *J. Chem. Soc. Faraday Trans.,* 1988, **84**, 3649.

PROCEDURES FOR THE ISOLATION AND FRACTIONATION OF HUMIC SUBSTANCES

Michael H. B. Hayes[1] and Colin L. Graham[2]

[1] Chemical and Environmental Sciences, University of Limerick, Ireland
[2] School of Chemistry, University of Birmingham, Edgbaston, Birmingham B15 2TT, England

1 INTRODUCTION

It is appropriate in considerations of procedures for the isolation and fractionation of humic substances (HSs) to take account of the abundances of these substances, of the ways in which they are formed and of aspects of their chemical compositions and properties.

HSs are found in all soils and waters that contain organic matter (OM)[1] and result from the biological and chemical transformations of organic debris. The amounts in soils are several times greater than those in waters. Estimates of the global abundances of organic carbon (OC) in soil organic matter (SOM) range from 11-30 x 10^{17} g,[2] but the most widely accepted values for this OC are in the range of 14 to 15 x 10^{17} g.[2-4] The OC in mineral soils can be composed of 70-80% HSs. Recognizable plant remains constitute a small percentage of the SOM of mineral soils. Thus the abundance of OC in HSs is of the order of 2 to 3 times greater than the terrestrial biomass estimated to be of the order of 5.6 x 10^{17} g.[3] There are significant reserves of HSs in deep sea environments and these are almost entirely *autochthonous* (i.e., produced within the aquatic system),[5] as are those in soils, but the HSs in streams and in fast flowing waterways are largely *allochthonous* (i.e., produced outside the system, usually in the soils of the watersheds).[6] HSs in lakes, reservoirs, estuaries and coastal waterways are both autochthonous and allochthonous.[1,5]

Labile plant materials decompose rapidly on entering aerobic soil environments with adequate water supplies, but more resistant components transform slowly in the same environment. Because of the compositional diversities and differences in the transformation modes of the components, it is impossible to accurately define the gross mixtures that compose SOM or the dissolved organic matter (DOM) or the particulate organic matter (POM) of waters.

Hayes and Swift[7] have distinguished between recognizable plant/animal debris and the highly transformed materials that contain no recognizable plant, animal or microbial structures. Recognizable plant debris is composed of identifiable classes of organic macromolecules such as carbohydrates, peptides, lignins, suberins, hydrocarbons, fatty acids and esters and nucleic acids, which are non-HSs, and the transformed, dark coloured amorphous materials, or *humus*, composed largely of HSs.

It is important to realise that the classical definitions of HSs are operational and based on aqueous solubility properties. Aiken et al.[1] state that humic substances are *"a general category of naturally occurring heterogeneous organic substances that can generally be characterized as being yellow to black in color, of high molecular weight, and refractory.*" This definition is still broadly relevant but as discussed in a number of the communications in *Soil Sci.*, 1999, **164**, the classical interpretations suggesting that HSs have high molecular weight (HMW) values may not be as acceptable now (see Section 4.2.2). The term refractory can be considered to apply only when HSs are in protected environments.

The degree of resistance that HSs have to further microbial degradation may well be attributable to self associations of the molecules, through associations with difficult to degrade substances such as long chain hydrocarbons in fatty acids and esters, to associations with the soil mineral colloids and to entrapment in soil aggregates. Accessibility to transforming microorganisms and to their degradative enzymes is implicit for resistance. Thus because of protection some labile molecules can survive for a time, but all are eventually biodegraded. Without biodegradation, the surface of the earth would be deeply covered in HS[8] and the waters would be dark coloured 'soups.'

The generalized terms *humic acids* (HAs), *fulvic acids* (FAs), and *humins* cover the major fractions still used to describe HSs components, but the boundaries between these fractions are not clear. HAs and FAs can be considered as a continuum.

Humic acids (HAs) as defined by Aiken et al.[1] are *"the fraction of HSs that is not soluble in water under acid conditions but becomes soluble at greater pH."* A pH of 1 is the standard used by soil scientists whereas pH 2 is more generally accepted by water scientists. The term *'humic fraction'* can include biomolecules such as peptides, sugars, nucleic acid residues and fats that are not covalently linked to the humic molecules. Such molecules can be sorbed to or co-precipitated (at pH 1 or 2) with the HAs and such water soluble components invariably will be present in solution with the FAs after the HA fraction is precipitated. By dissolving HAs in dimethyl sulfoxide (DMSO) containing 1% HCl and passing the solution onto an XAD-8 [(poly)methylmethacrylate] resin column, Häusler and Hayes[9] significantly decreased the amino acid and neutral sugar contents of the HAs. This indicated that saccharide and peptide materials were associated, possibly by sorption or/and co-precipitation to and with the HAs.

Fulvic acids (FAs) are defined[1] as the *"fraction of humic substances that is soluble under all pH conditions."* Soil scientists take FAs to be the fraction that stays in solution when basic soil extracts are adjusted to pH 1. Supernatants from the acidified base will, of course, contain non-humic materials and are best defined as the *"fulvic acid fraction."* In the IHSS (International Humic Substances Society) procedure for isolating FAs,[10] the acidic FA fraction is passed onto XAD-8 resin. The true FAs are retained on the resin and the polar non-HSs elute from the column.

Humin is defined[1] as *"that fraction of humic substances that is not soluble in water at any pH value."* On the basis of that definition, humin can include any humic-type material that dissolves in non-aqueous solvents, which in the cases of soils have been exhaustively pre-extracted with basic aqueous solvents.

Significant advances were made during the last half century in the isolation and fractionation of HSs, although the technique first described by Achard[11] in 1786 is the basis of the major solvent systems still used. In this paper we describe the principles behind the procedures for the isolation and fractionation of HSs and emphasize the advances made in this generation. It will be seen on the basis of these advances that it will be appropriate to consider new criteria for defining the different fractions of HSs.

2 ISOLATION OF HUMIC SUBSTANCES FROM SOILS

2.1 Some Relevant Compositional Properties of Humic Substances

Hayes[12,13] has discussed some aspects of HSs composition and structure that are relevant to the present topic. Composition refers to the content of the elements, the functional groups and the component molecules or 'building blocks' of HSs molecules. Analytical data are most meaningful when the samples analysed and compared have been subjected to the same isolation and fractionation procedures. A set isolation and fractionation procedure was followed when preparing the IHSS Standard HSs;[10] thus useful comparisons can be made with these Standards when samples are isolated and fractionated in the same way.

Acidic functionalities have the greatest influences in the solubilization of HSs, and cation-exchange capacity (CEC) data at different pH values (obtainable from titration experiments) can give good indications of the distributions of the acidic functionalities with different pK_a values. These functionalities range from strong (activated carboxyls) to weak (phenol and enol) acid groups. Leenheer et al.[14] have shown that activating substituents to carboxyl groups give rise to strong acids and in his comprehensive review Perdue[15] has stressed how the same functional groups can have different pK_a values in different local molecular environments. Weakly dissociable groups (phenols and enols) ionize in the alkaline range and such functionalities contribute significantly to the CEC values at the higher pH values.

When the acidic groups are undissociated (at low pH values), HSs can be considered to have properties similar to neutral molecules with significant inter-and intramolecular hydrogen bonding. As dissociation takes place the conjugated bases solvate in aqueous media and that is why aqueous basic media have been the traditional extractants for HSs since the time of Achard.[11]

Swift[10] has updated the wet chemical and some spectroscopy procedures for determinations of functional groups in HSs. There are comprehensive reviews of acidic functionalities[15] and of various spectroscopy procedures such as infrared[16] and of applications of both proton and ^{13}C nuclear magnetic resonance (NMR)[17-21] to studies of the acidic and other functionalities of HSs.

Elemental analyses data (see Huffman and Stuber[22] for the methods used for the analyses of the IHSS Standard and Reference samples) are useful indicators of the extents of polar functionalities in HSs. Much of the oxygen contents can be attributed to carboxyl (especially) and to phenolic hydroxyl groups.

Recently attention has focused on the non-polar components in humic fractions. It is now clear that non-polar hydrocarbons and long chain fatty acids and esters are present to varying extents in humic fractions. It is reasonable to conclude that where such components are not covalently linked to the humic molecules they are co-extracted because of their associations with the humic molecules.

2.2 Some Relevant Properties of Solvents

Hayes[23] has given detailed accounts of some of the properties of different solvent types that might be considered for the isolation of HSs from soils and has outlined the most important of these properties in recent publications.[12,13] Some of the relevant properties are summarized in Table 1.

Boiling point is important should it be necessary to recover the solute by evaporation of the solvent. Secondary reactions, including decarboxylation, will take place when elevated temperatures are used to recover HSs from solvent systems. Also, it is important that the solvents be in the liquid state at ambient temperatures. Density is important only if the solutes cannot be dissolved in the solvent system and in considerations of separations involving gravimetry. Use can be made of mixtures to regulate density since the density of a mixture is close to the arithmetic average of the densities of the components. The densities of the solvents in Table 1 will not prevent dissolution when the solvents have the abilities to solvate the humic molecules. The relative permittivity (K_r) or dielectric constant refers to the ability of a solvent to decrease the coulombic field of an ion. The value for water is ca. 78.5, which indicates the extent to which the field is decreased in the liquid compared with a vacuum at a distance r from the solute species.

Table 1 *Boiling point, density (ρ), relative permittivity (K_r), dipole moment (μ), electrostatic factor (EF) and base parameter (pK_{HB}) values for selected solvents*

Solvent	Boiling Point (°C)	Density (ρ)	Relative Permittivity (K_r)	Dipole Moment (μ)	EF	pK_{HB}
Diethylether	35	0.61	4.34	1.36	5.90	0.98
Methanoic acid	101	1.22	58.0	-	-	-
Ethanoic acid	118	1.04	6.13	0.83	5.09	-
Pyridine	115	0.98	12.4	-	-	1.88
Methyl cyanide	82	0.78	37.5	3.84	144.0	1.05
Acetone	56	0.78	20.7	2.88	59.62	1.18
N-Methyl-2-pyrrolidinone	202	1.03	32.0	-	-	2.37
Formamide	210	1.13	109.5	3.37	369.0	-
N,N-Dimethylformamide	153	-	36.7	3.82	140.2	2.06
Dimethyl sulfoxide	189	1.10	46.6	4.49	209.2	2.53
Water	100	1.00	78.5	1.84	144.4	-

In general, less polar solutes dissolve best in solvents with low K_r values and higher values favour the dissolution of polar molecules. However, definite trends that relate to the extent to which a solute is dissolved cannot be established because of the involvements of specific interactions and especially hydrogen bonding.[24]

Dipole moment (μ) values are important for predicting the interactions of solvents with polar or charged molecules because the ability of a solvent to disrupt molecular associations depends on the extent to which it can solvate the component molecules or ions and decrease the interactions that hold them together. Where specific solvent-solute interactions are not important then the dipole moment of the solvent has the major role in determining the orientation of the solvent around the solute. This orientation is essential for the electrostatic solvation process. Self-association is inhibited when solvent shells are formed around the molecules.

Electrostatic factor (EF) values are the product of K_r and μ that take account of the influence of both factors on the electrostatic solvation of solutes.[25] It can be predicted that solvents with EF values <50 will be poor solvents for HSs.[23]

The base parameter (pK_{HB})[26] measures the relative strength of the acceptor when a hydrogen-bonded complex is formed using any suitable hydroxyl reference acid. The higher the pK_{HB} value the better the compound is an acceptor in hydrogen bonding.

Other relevant properties of solvents include refractive index and viscosity values. The overall tendency of compounds to interact through dispersion forces is related to the refractive index values and the greater the refractive index the stronger are the dispersive interactions. Where refractive index is used to measure the concentration of solute it is important to maximise the differences in values between the solvent and the solute.

Low viscosity is important for considerations of handling and mixing. In general, for the solvents listed in Table 1, there is with the exception of N,N-dimethylformamide (which has a relatively low viscosity) a degree of matching of viscosity and boiling point and the viscosities of mixtures are intermediate between the values for the components.

Some *dipolar aprotic solvents* are useful for the solubilization of HSs. Parker[27] has considered these solvents to have K_r values >15 and to be incapable of donating hydrogen atoms to form strong hydrogen bonds. Among the compounds listed in Table 1 acetone, methyl cyanide, N-methyl-2-pyrrolidinone, N,N-dimethylformamide and dimethylsulfoxide (DMSO) can be classified as dipolar aprotic solvents. These compounds, with the exception of methyl cyanide, have exposed electronegative O atoms, as have the dipolar but protic formamide and N-methylformamide (which have sites for solute to solvent hydrogen bonding). Anions are less well solvated in dipolar aprotic solvents than in water. Anions solvate in water and in protic solvents by ion-dipole interactions on which strong hydrogen bonding is superimposed. The anions also solvate by ion-dipole interactions, but without the influence of hydrogen bonding in the cases of the dipolar aprotic solvents. Anion solvation is aided by less energetic interactions arising from the polarizability of the anions and solvent molecules.[27]

2.3 Considerations of Solubility Parameters

Solution takes place when the self-attractive forces in the pure solute and pure solvent molecules are of the same order of magnitude. However, often it is found that a mixture of two solvents, one with a solubility parameter (δ) above and the other with a δ value below that of the solute will provide a better solvent for the solute than either solvent alone.

Hildebrand and coworkers[28-30] introduced the concept of the one-component solubility parameter (δ). It is defined as the square root of the cohesive energy density (CED), Eq. (1), where V_l is the molar volume of the liquid and $-E$ is the molar cohesive

$$\delta = (-E/V_l)^{0.5} = (CED)^2 \qquad (1)$$

energy, or the molar energy of vaporization, expressed as $(cal/cm^3)^{0.5}$, or 2.046 $(J/cm^3)^{0.5}$. Detailed discussions of the theory and derivations is given in the references cited.[28-30]

The one δ component is appropriate for solutions lacking in polarity and in specific interactions. To some extent this parameter is now replaced by multicomponent solubility parameters that give values for each of the different interaction forces. These include dispersion forces, dipole-dipole and dipole-induced dipole interactions, and specific interactions, especially hydrogen bonding. The total solubility parameter is expressed in Eq. (2), where δ_i represent the empirical estimates of the contributions from dispersion

$$\delta_o = (\sum_i \delta_i^2)^{0.5} \qquad (2)$$

forces (δ_d), polar forces (δ_p) and hydrogen bonding (δ_h). Values of δ and δ_o should not necessarily be equal. The hydrogen-bonding parameter can be further subdivided[31] into the acid or proton donor (δ_a) and the base or proton acceptor (δ_b) parameters.[24,31-33]

Hayes[23] has provided, from a compilation by Barton,[33] the Hilderbrand (δ), total (δ_o), dispersive (δ_d), polar (δ_p) and hydrogen-bonding (δ_h) parameters and from those by Karger et al.[35] the data for proton donor and proton acceptor parameters for some of the solvents listed in Table 1. Reference is made in Section 2.5 to predictions based on solubility parameter data that can be made about the solubilization of HSs.

2.4 Solubilization of Soil Humic Substances: General Considerations

The dominant exchangeable cations in agricultural soils are di- and trivalent (especially Ca^{2+}, Mg^{2+}, Fe^{3+} and Al^{3+}). These cations are strongly held by the anionic functionalities of HSs. Such cations form inter- and intramolecular bridges between anionic sites, suppressing repulsion and inhibiting solvation. Cation bridging does not take place when the charge balancing metal cations are monovalent. Thus, when repulsion between the charged species takes place the molecules (or the molecular associations) assume expanded conformations. Solvation can occur readily provided the ratio of charged to neutral moieties is adequate.

Cation bridging, hydrogen bonding, hydrophobic interactions and van der Waals association effects (which also apply for the H^+-exchanged species) cause HSs molecules to assume shrunken or condensed conformations and water is partially excluded from the matrix. These effects can be overcome by replacing the divalent/polyvalent cations with monovalent species (other than H^+).

Stevenson[36] has outlined four criteria for the ideal method of extracting soil HSs: 1) the method should lead to the isolation of unaltered materials; 2) the extracted HSs should be free of inorganic contaminants such as clay and polyvalent cations; 3) extraction should be complete, thereby ensuring representative fractions from the entire molecular weight range; and 4) the method should be universally applicable to all soils.

An effective extraction method requires the use of solvents that will not degrade or alter the HSs but be capable of releasing all of the HSs from the other soil components.

Whitehead and Tinsley[37] have listed four criteria for solvents for HSs. They proposed that an effective solvent should have 1) a high polarity and a high dielectric constant (or permittivity) to assist the dispersion of the charged molcules; 2) a small molecular size to penetrate humic structures; 3) the ability to disrupt the existing hydrogen bonds and to provide alternative groups to form humic-solvent bonds; and 4) the ability to immobilise metallic cations.

2.5 Isolation of Humic Substances in Aqueous Media

The behaviour and properties of H^+-exchanged HSs are similar to those of neutral polar macromolecules that are hydrogen bonded to each other. In the same way they can be compared with negatively charged polyelectrolytes whose charges are neutralized with counterions. The relative sizes of humic molecules are a matter of debate at this time, but that need not influence considerations of approaches to their isolation. Hydrogen bonding together with associations of the non-polar moieties through hydrophobic bonding and van der Waals forces and through charge transfer processes can confer properties on HSs molecules that are similar to those of HMW materials, even if the associations actually are molecules that are not macromolecular.

Only the most polar and the least associated HSs molecules will dissolve in water. As the pH is raised from 2, at which all of the HAs can be considered to be precipitated, dissociation of the acid groups commences and the conjugate bases solvate in water. The strong acids dissociate at the lower pH values and as the pH continues to rise more of the weaker acid groups dissociate until in the alkaline range phenols and eventually enols and very weak acidic functionalities dissociate. Dissociation of the acid functionalities weakens the hydrogen bonding and eventually the solvation energy causes the humic associations to break up and the humic molecules to dissolve. The most acidic groups will be first to be solvated. Use can be made of these principles to fractionate HSs on the basis of charge density differences.

Mention was made in Section 2.4 of the solvation inhibition effects that arise from bridging between humic molecules attributable to charge-neutralizing divalent and polyvalent cations. Thus, in order to isolate HSs in aqueous media from the soil environment using the classical isolation procedures it is necessary to replace the divalent and polyvalent charge neutralizing cations with H^+ or with monovalent metal cations.

In this way, two techniques are applied in the classical procedures. In the first the divalent and polyvalent cations are replaced by H^+-ions. 1 M HCl generally is used for this. The soil is then dialyzed against distilled water or washed with water until the chloride is removed (i.e., no AgCl precipitate is obtained on addition of $AgNO_3$). Peptization of the HSs can take place when the chloride and excess acid is removed and some FAs can be solubilized. However, the bulk of the H^+-exchanged HSs are strongly associated through hydrogen bonding and only the most polar and least associated of the molecules are dissolved in water. Then 0.1 M or 0.5 M NaOH is applied. Even the weak acids are dissociated in the alkaline media and solvation of the polydisperse (with respect to charge) polyelectrolytes takes place. However, because oxidation of SOM takes place in alkaline media[38] it is desirable to carry out the isolations under reducing conditions when alkaline conditions are used for the extraction process.[39]

In the second procedure the divalent and polyvalent cations are replaced with monovalent species. The bridging effect then is lost, the conjugated bases of the acidic functionalities are solvated and dissolution of humic molecules takes place. Bremner and Lees[40] were aware of this when they introduced applications of 0.1 M sodium pyrophosphate ($Na_4P_2O_7$) adjusted with acid (usually phosphoric) to pH 7 for the extraction of OM from soils. The carboxyl functionalities are ionized at pH 7 and their Na^+-salts are solvated. Amounts extracted with pyrophosphate are significantly less than with 0.1 M NaOH, which suggests that dissociation of the more weakly acidic functionalities has a considerable influence on the extents to which HSs are solvated. Molecular associations through van der Waals forces are strong enough to inhibit dissolution of some fractions of HSs (especially humin) molecules with low charge densities, regardless of the pH of the medium.

Based on these considerations, Clapp and Hayes,[41] Hayes[42] and Hayes et al.[43] used a sequential extraction procedure for the isolation of HSs from soils. Exhaustive extraction (i.e., repeated extractions with one solvent until the colour released is minimal before proceeding to the next solvent in the series) with water isolates the components that might be expected to be contained in drainage waters. This was followed by extraction with neutral 0.1 M sodium pyrophosphate (Pyro), then with Pyro (0.1 M) at pH 10.6 and then with a solution 0.1 M with respect to NaOH and 0.1 M with respect to Pyro (pH 12.6).

The HSs that are not isolated in that series of aqueous solvent systems are classed as humin. Consideration might be given to the use of concentrated sulphuric acid (under cooling conditions) to dissolve the humin. This approach would require careful control of temperature as the acid is diluted. Another approach is to follow the NaOH/Pyro solvent system with organic solvents, such as DMSO and methyl isobutylketone (see Section 2.6).

2.6 Applications of Organic Solvents for the Solubilization of Soil Humic Substances

Hayes et al.[44] (see also Hayes[23]) repeatedly extracted (till the colour of the extracts was negligible) an air dried H^+-exchanged Sapric Histosol soil with 2.5 M ethylenediamine (EDA), with anhydrous EDA, with the sodium salt of 1 M ethylenediaminetetraacetic acid (EDTA), with pyridine, with sulfolane, with N,N-dimethylformamide (DMF) and with dimethylsulfoxide (DMSO) and they compared the yields with the amounts extracted with 0.5 M NaOH. Only the 2.5 M EDA solution could match the amounts of HSs isolated with 0.5 M NaOH and that was attributed to the basicity of the EDA solution. Before extraction with anhydrous EDA, the H^+-exchanged Histosol was dried in vacuo at 75°C over phosphorus pentoxide and then thoroughly mixed with EDA that had been dried by distillation from solid sodium hydroxide. The extraction process was repeated twice.

In the absence of water, EDA was a very poor solvent for the HAs. This suggests that the solute-solvent interactions were not sufficiently strong to overcome the strong hydrogen bonding forces operating between humic molecules. It also is probable that the drying process caused hydrophobic functionalities (either components of or associated with the humic molecules) to be exposed to the exterior and these were not solvated by the EDA. This effect would shield the more polar and more readily solvated functionalities in the humic molecules.

Where EDA made contact with the acidic functionalities protonation of the solvent was inevitable and the conjugate acid was then held by ion exchange with the conjugate base. However, the size of the EDA molecule and of its protonated derivative would inhibit penetration into the tightly hydrogen bonded matrix. Thus, in the absence of water HAs would not swell readily in EDA or in amine solvents. Yields of FAs in anhydrous EDA were significant and that could signify the greater availability of polar functionalities in FAs compared to HAs. Pyridine followed by water extracted more HSs than did similar DMF, DMSO and sulfolane systems. The greater solubilization by the pyridine system is attributable to pH effects. However, because of the low buffering effect of the solvent the pH of extraction with pyridine was only 4.2. Theory suggests that substantially more HSs would be extracted should the pH be maintained at pH >9 (the pK_b of pyridine is 8.96). At pH > 9 pyridine molecules that dominate in the medium are involved in the humic solvation processes. As the pH falls there is a logarithmic increase in pyridinium ions and the cationic species are held by ion exchange to the conjugate bases of the acidic humic functionalities. Even though humic pyridinium salts are less dissociated than humic monovalent inorganic cation species, they are significantly more dissociated than the H^+-exchanged and divalent and polyvalent cation-exchanged humates. However, the relatively high yields (34% compared with 58% for 0.5 M NaOH) of HAs from the pyridine system can be explained by the fact that the extraction with pyridine of the air-dried Histosol was followed by extraction with water. The water solvated the conjugated bases in the HSs.

Elemental analyses of the humates extracted with 2.5 M EDA and with pyridine clearly indicate that these solvents alter the compositions of the HSs. The N contents of the HAs and especially of the FAs were significantly raised. EDA would form Schiff base derivatives and would react with the carbon α to the keto structures of quinones. Pyridine could be held by van der Waals forces and by charge transfer interactions. Retention by ion exchange would be a major binding process should the extracts not be H^+-ion exchanged. The free radical contents of the EDA extracts were significantly raised, while those of the pyridine extracts were not.

DMSO was the best of the dipolar aprotic solvents (DMSO, DMF, sulfolane) as an extractant for HSs in the system used by Hayes et al.[44] Some coloured materials precipitated during the dialysis of the FA-type substances isolated in the aprotic solvent systems and it was considered that the FAs in the FA fraction were contaminated with HA materials solvated by the residual solvents in the acidified mixtures.

Hayes[23] used a 'level playing field' test to compare the solvating capabilities of a number of organic solvents with 0.5 M NaOH. H^+-exchanged HAs isolated from a Sapric Histosol (Glade soil, Florida) by the IHSS procedure[10] were extracted in the solvents (0.2% w/v) listed in Table 2. The mixtures were left to swell overnight. Then each system was centrifuged and the supernatants were diluted with the solvent until an absorbance reading (against the solvent blank) could be obtained at 400 nm. In each case the absorbance values in Table 2 represent the product of the reading obtained and the dilution factor used.

From the data in Table 2 it is evident that acetonitrile, dioxane, ethanol, pyridine, methyl cyanide, methanoic acid (90%) and water are poor solvents for H^+-exchanged HAs. This could be predicted, as discussed below, on the basis of the pK_{HB} values (Table 1). Pyridine was a poor solvent, as predictable from its low K_r value and because of the low moisture content of the freeze-dried (and subsequently air-dried) sample. This effect is similar to that described for anhydrous EDA.

Table 2 *A comparison of the absorbance values for solutions isolated from H^+-exchanged humic acids by various organic solvents and by 0.5 M NaOH*[23]

Solvent	Absorbance Value
Water	1.0
Dioxane	0.0
Methyl cyanide	0.0
Ethanol	1.0
Methanoic acid (90%)	4.0
Pyridine	5.0
N,N-Dimethylformamide (DMF)	18.0
Formamide	19.0
Dimethyl sulfoxide (DMSO)	21.0
0.5 M NaOH, pH 9.2	23.0
0.5 M NaOH	24.0

The results for 90% methanoic acid might seem disappointing when compared with those of Sinclair and Tinsley.[45] However, these authors used anhydrous methanoic acid, which can be expected to have all the properties of a good solvent. The presence of water induced ionization and the anions would not solvate the humic molecules.

Compared with aqueous NaOH, formamide, DMF and DMSO are good solvents for the Glade H^+-exchanged HAs. Reference to Table 1 shows that each of these 'good solvents' has EF values >140 and pK_{HB} values >2 (we do not have pK_{HB} values for formamide, but it can be predicted from the other data in Table 1 that the value will be >2). Its relatively large d_h and d_p parameters (discussed below and in Section 2.3) indicate that this solvent is a good acceptor in hydrogen bonding systems (a value for pK_{HB} is not listed in Table 1). The low pK_{HB} value for acetonitrile indicates that it is incapable of breaking the hydrogen bonds of the H^+-exchanged HAs. N-methyl-2-pyrrolidinone was not included among the solvents tested but the relevant parameters recorded in Table 1 would suggest that it satisfies the criteria for a 'good solvent' for HSs. However, the hydrogen bonding parameter (d_h) (discussed below and in Section 2.3) is not favourable. Also, the molar volume of the compound (96.5 cm^3/mol) is greater than those for the 'good solvents' (71.0 and 71.3 cm^3/mol, respectively, for DMSO and DMF) and that would hinder penetration of the solvent into the hydrogen bonded humic matrix. However, N-methyl-2-pyrrolidinone was found to be a better solvent for HSs in a Ca^{2+}-exchanged than for a H$^+$- exchanged Histosol.[23] This might be attributed to the more 'open' structure (hence the more ready penetration by the solvent into the Ca^{2+}-soil matrix) and possibly also to an ability to complex the exchangable cations.

The dispersion force (d_d) parameters are similar (7-9) for the organic solvents listed in Table 2 but there are significant differences in the polar (d_p) and hydrogen bonding (d_h) parameters. The values for d_h are the most important. The 'good solvents' for HSs all have d_h values >5 (9.3, 8.1, 5.5 and 5.0 for formamide, methanoic acid, DMF and DMSO, respectively).

On the basis of the logic of this discussion, water should be a very good solvent for H$^+$-HSs. It has K$_r$ and pK$_{HB}$ values of 78.5 and 144.44, respectively, and d_p and d_h values of 7.8 and 20.7, respectively. Also, it has large values for d_a and d_b (see Section 2.3), as does formamide. Solution is greatest when the product d_a (solvent) x d_b (solute), are maximal. The very large values of d_h, d_a and d_b indicate the extents of self association through hydrogen bonding of water molecules. It is clear that the d_a and d_b values of the H$^+$-exchanged HAs are not sufficient to disrupt these attractive forces. Hydrogen bonding is disrupted as the pH is raised and the acid functionalities dissociate. Then water becomes an excellent solvent as it solvates the conjugate bases of the dissociated acid functionalities.

On the basis of d_h, d_a and d_b values (9.5, 6.9, and 6.9, respectively), ethanol might be considered to be a good solvent for H$^+$-exchanged HSs. Its failure may be attributable to its d_p (4.3) and K$_r$ and EF values (Table 1), which are significantly less than those for the 'good solvents'.

The solubility parameter approach has a sound basis in theory. However, we cannot apply the approach on a quantitative basis because we do not have the necessary values for the humic components. To be able to apply this approach it will be necessary to work with humic fractions (or SOM components) that are relatively homogeneous with respect to size and composition. Also, applications of polymer solution theory[23] to studies of the dissolution of HSs and of their isolation from soils are hindered because interactions between each pair of components must be known. HAs, FAs and the other components of SOM are parts of multicomponent systems and the interactions between the different components can only be guessed at.

In the final stage of their sequential extraction process, Clapp and Hayes[41,46] used DMSO (following the exhaustive extraction at pH 12.6 (see Section 2.5)) to isolate OM that in the classical definition might be regarded as humin. For DMSO to be an effective solvent it is essential that the humic materials be in the H$^+$-exchanged form (DMSO is a good solvent for cations but a poor solvent for anions)[47] and behave like hydrogen bonded, polar, neutral molecules. Thus Clapp and Hayes[41] used a mixture of DMSO (94%) and 12 M HCl (6%).

There are strong interactions between the polar face of DMSO and water, carboxyl and phenolic groups, and DMSO-water interactions are stronger than the associations between water molecules. Thus, DMSO will associate with the phenolic and carboxyl groups to break the intra- and intermolecular hydrogen bonds. The non-polar face of DMSO can associate with the hydrophobic moieties in HSs and the combination of hydrogen-bond breaking and the disruption of hydrophobic associations makes DMSO a very effective solvent for HSs. Removal of the solvent is not a problem because the HSs sorb to the XAD-8 resin and the DMSO and acid wash through; see Section 4.2.

An alternative to exhaustive sequential extraction using a series of solvents would be to extract with the NaOH/Pyro system and subsequently to fractionate by eluting at different pH values from XAD resins (see Sections 4.1 and 4.2). HSs extraction with DMSO/HCl would follow the exhaustive extraction with base.

Rice and MacCarthy[48,49] have described the use of methylisobutylketone (MIBK) to isolate a nominal humin fraction from soil after exhaustive extractions with base had been completed. Their observations indicated that the isolates were composed of lipid and HA-type materials. Consideration might be given to the use of MIBK after DMSO/HCl in the sequential extraction system.

2.7 Application of Mixed Solvent Systems for the Isolation of Soil Humic Substances

A review by Suggett[50] suggests that the addition of urea (10%) to DMSO gives an effective breaker of hydrogen bonds in cellulose that leads to dissolution of the polymer. When urea (10%) was added to DMSO, to sodium tetraborate ($Na_2B_4O_7$, 0.025 M; pH 9.2) and to sodium pyrophosphate (0.1 M, adjusted to pH 7 with H_3PO_4) and these solutions were added (separately) to the H^+-exchanged HAs from the Glade soil (a Sapric Histosol), it was found that urea depressed the solubilization by the borate, did not affect solubilization by the neutral pyrophosphate and when water (5% of the total mass) was added to the urea (10%) in DMSO complete dissolution of the H^+-HA took place.[23] Dissolution gradually was depressed as additional increments of water were added to the system. The use of urea in solvents cannot be recommended because it raises the N contents of the HSs, indicating that it interacts with the solute.

Hayes[23] has described experiments in which DMSO was mixed with varying amounts of water and HCl, singly and in combination, for the dissolution of H^+-exchanged HAs and for the extraction of HSs from tropical soils. The results were compared with those for 0.1 M NaOH. Dissolution of the HAs was complete for all systems. Results were similar for the NaOH and DMSO + H_2O systems in the case of the H^+-HAs, but the addition of HCl increased the E_4/E_6 ratios, which might be interpreted in terms of changes in the conformations of the molecules.[51] For tropical soils, addition of acid to DMSO significantly increased the solubilization of the coloured HSs, as would be expected because the acid H^+-exchanged the soils and, as pointed out in Section 2.6, DMSO could be expected to be a poor solvent for ionized humates. Additions of small amounts of water (<5%) did not influence the E_4/E_6 ratios of the humic extracts. However, these ratios were significantly higher for a tropical rainforest soil than for a tropical savanna soil.

3 ISOLATION OF HUMIC SUBSTANCES FROM WATERS

In a comprehensive discussion of the geochemistry of stream fulvic and humic substances, Malcolm[52] has defined aquatic HSs as "that portion of the organic substances in waters which passes through a 0.45 μ membrane filter, and upon acidification to pH 2 with HCl has a column distribution coefficient (k') greater than 50 on XAD-8 resin at 50% breakthrough of the column for visually dark-coloured, non-specific amorphous carbonaceous material." XAD-8 [(poly)methylmethacrylate] resin technology has set new standards for the isolation of aquatic HSs and more recently it has provided a new approach for the fractionation of HSs (see Sections 4.1 and 4.2).

Aiken[53] has listed a variety of procedures for the isolation of HSs from waters. These include filtration and applications of co-precipitation, ultrafiltration, reverse osmosis, solvent extraction, ion exchange and sorption, including uses of alumina, carbon, and non-ionic macroporous and weak anion exchange resins in which the active functionalities are supplied by secondary amines and their salts.

Leenheer[54] has referred to six fractions of dissolved organic carbon (DOC). These are the hydrophobic acids, bases and neutrals and the hydrophilic acids, bases and neutrals. The hydrophobic and hydrophilic fractions are retained by XAD-8 [(poly)methylmethacrylate] and XAD-4 (styrenedivinylbenzene)[55] resins, respectively. XAD-8 resin technology was used to isolate the IHSS Standard aquatic HAs and FAs from the Suwannee River. The procedure is described by Aiken,[53] Thurman and Malcolm[56] and Thurman.[57] It involves first filtering the water through 0.2 μ pore size filters (originally 0.4 μ pore size silver filters were used). As a degree of clogging of the pores takes place, a very efficient removal of solid particles of colloid dimensions is achieved. The pH of the filtrate is adjusted to 2 and the liquid is passed onto an XAD-8 resin equilibrated with water at pH 2 (0.01 M HCl). The XAD-8 resin has a low polarity but the ester functionality provides some sites for hydrogen bonding. The hydrophobic faces of the humic molecules associate through van der Waals forces with the hydrophobic backbone of the resin and some hydrogen bonding can take place between the humic and fulvic carboxyls and the ester functionality. Polar, non-humic components are washed through the resin. When the resin is then back eluted with 0.1 M NaOH the acidic functionalities in the humic structures are ionized, changes in conformations of the sorbed molecules take place and the humic components are desorbed. The pH of the centre cut is immediately adjusted to *ca.* 4 and the dilute fractions on either side of the centre cut are recycled with the subsequent batch of water to be added to the regenerated resin (i.e., neutralized with 0.1 M HCl and then equilibrated with 0.01 M HCl). The fractionation of the humic extracts is described in Sections 4.1 and 4.2. 'Hydrophobic neutrals' are retained by the resin after back elution with base. These are recovered in organic solvents (ethanol and methyl cyanide). Soxhlet extraction is required to remove the most tightly bound components.

Nowadays, XAD-8 and XAD-4 resins are used in tandem and hydrophilic components are retained on XAD-4. Prior to back elution the contents of the resin column are desalted by passing distilled water through the resin until the conductivity of the eluate is close to that of the distilled water. In practice a conductivity of <100 mS cm^{-1} is acceptable. After desalting, the column is back eluted in base (0.1 M NaOH), the eluate is H$^+$-ion exchanged (using a cation exchange resin such as IR 120) and the XAD-4 acids are recovered by freeze drying. The 'hydrophilic neutrals' retained by the XAD-4 resin after back elution with base are recovered as described for the hydrophobic neutrals.

The reverse osmosis (RO) procedure can be used to concentrate OM from large volumes of water under ambient conditions.[58] Serkiz and Perdue[59] and Sun et al.[60] have described instrumentation that processes 150 - 200 L of water hr^{-1} with 90% recovery of organic carbon and without exposing the DOM to harsh chemical reagents. All the solutes, including inorganic salts and non-HSs, are concentrated in the process. Thus it is appropriate to subject the concentrated solutions to the resin procedures described above.

4 FRACTIONATION OF HUMIC SUBSTANCES

Fractionations of HSs on the basis of charge density differences and on the basis of size differences are achievable. This paper focuses on preparative scale fractionations.

4.1 Fractionation of Aquatic Humic Substances

The pH of the material recovered by back elution from XAD-8 (see Section 3) is adjusted to 2, the mixture is refrigerated and the fine HA precipitates are allowed to settle. After centrifugation, the FA supernatants are desalted by passing slowly onto an XAD-8 resin column (equilibrated at pH 2), then washed with distilled water until the conductivity is <100 mS cm^{-1} (ideally desalting might be continued until the conductivity approaches that of the water, but this risks loss of some FA materials from the column). The usual procedure back elutes the FAs in base, passes the eluates through a cation-exchange resin and freeze dries. However, a degree of fractionation on the basis of charge density differences can be achieved by back eluting first with water (the eluate can be recovered directly by freeze drying) and then batchwise at different pH values, culminating in 0.1 M NaOH. It should be borne in mind that back elution with any solvent that generates salt will require that the eluate be desalted.

In general, the HA fraction (precipitated at pH 2) is desalted by dialysis. However, if the HA is dissolved in an aqueous medium that is sufficiently alkaline to ensure complete solvation then diluted to <50 mg L^{-1} (or to a level at which all of the HAs are in solution when the pH is adjusted to 2) and after adjusting the pH to 2 the sample is applied slowly to XAD-8 (equilibrated at pH 2) as described for the FA. Desalting can be carried out as described above for FAs (see Section 4.2.1 for a procedure if the HAs associate at pH 2). Batchwise back elution with water and then with aqueous systems of increasing pH allows fractionation of the HAs on the basis of charge density differences.

4.2 Fractionation of Humic Substances from Soils and Sediments

Isolation procedures that are relevant are discussed in Sections 2.5 and 2.6. In general, the procedures described deal only with the isolation of HSs mixtures and there was no discussion of methods applicable to the recovery of the components of the DOM.

4.2.1 Fractionation Based on Charge Characteristics. In the classical procedures for separating HAs from FAs in aqueous solution extracts, the pH of the solution is adjusted to 1 (using, generally, 6 M HCl), the HAs are precipitated and the FA fraction is contained in the supernatant solution. This fraction contains significant amounts of salts that may be removed by dialysis. However, dialysis leads to the loss of the smaller-sized FAs. The HA fraction will be contaminated by some finely divided inorganic colloids. HCl/HF is used to dissolve these in the IHSS isolation procedure[10] and dialysis leads to their removal with the salts. However, this procedure leads to excessive losses of the HAs.

The IHSS procedure[10] recovers the 'true FAs' from the FA fraction by passing this fraction onto an XAD-8 resin (as described in Section 4.1). The FAs are sorbed by the resin and the polar non-humic components pass through. When XAD-8 and XAD-4 resins are used in tandem, the major part of the polar non-humic fraction is retained by the XAD-4 resins and these 'XAD-4 Acids' are recovered as described in Section 4.1. The FAs are recovered from the XAD-8 as described in Section 4.1.

A procedure developed in the authors' laboratory and based on a method introduced by Malcolm and MacCarthy[61] adapts the techniques used in water chemistry for the fractionation of soil HSs. Aspects of this procedure were first used to isolate HSs from solution in DMSO/HCl.[9] That approach is discussed below.

The aqueous solutions recovered as described in Section 2.6 are treated in a manner similar to that for HSs in waters. For alkaline solutions, the media are diluted and the pH adjusted to 7-8. Then the solutions are filtered through 0.2 μ pore size filters and the filtrates are diluted to <50 mg L^{-1} (ideally to <30 mg L^{-1} or to a concentration in which there is no evidence for coagulation or precipitation when the pH is adjusted to 2; *vide infra* should coagulation take place). This solution is applied to XAD-8 and XAD-4 resins in tandem. The XAD-4 resin is desalted and the 'XAD-4-Acids' are recovered as described in Section 4.1.

There are a number of options for dealing with the HAs and FAs held by the XAD-8 resin. Where these materials were isolated by the exhaustive extraction procedures outlined in Section 2.5, a degree of fractionation on the basis of charge density differences will have been achieved. Thus the retentates (on the resin) may be recovered, as outlined in Section 4.1, by back elution in 0.1 M NaOH and the HAs and FAs isolated as described for the aquatic humic fractions. In cases where extraction has involved only one solvent system (say 0.1 M pyrophosphate + 0.1 M NaOH) the procedure outlined may also be applied. However, a fractionation on the basis of charge density differences may be achieved by back eluting batchwise with aqueous systems of increasing pH values. Recovery of the HAs and FAs from each batch may be carried out as described for the aquatic samples.

Similar procedures may be applied for the fractionation of HSs isolated in organic solvent systems when the solvents are soluble in water. We have experience only with DMSO systems.[9,41] When a solution of HSs in DMSO/HCl is applied to an XAD-8 column at pH 2, the DMSO/HCl solvent passes through the resin and the HSs are sorbed to the resin. Co-extracted saccharides and peptides that are not covalently linked to the HSs molecules move with the DMSO/HCl.[9] XAD-8 and XAD-4 resins in tandem have not been used for this work and so it is not known to what extent the materials that move in the DMSO/HCl system will be sorbed by XAD-4. The column is washed exhaustively with 0.01 M HCl to remove the DMSO from the system.

There are several possibilities for dealing with the HSs sorbed to the XAD-8. One of these could involve desalting, back eluting with aqueous systems at different pH values and recovering the HAs and FAs as described above. Another possibility is to back elute batchwise with aqueous solutions of increasing pH values, as described in the example given above for the hypothetical case of extraction with aqueous pyrophosphate and NaOH.

Recent studies in our laboratory have attempted to use the XAD resin techniques to fractionate HAs isolated with 0.1 M NaOH from a Fibric Peat. These HAs have a relatively low charge density compared with those from a sapric Histosol. We found that in the course of introducing the dilute HA solution onto the XAD-8 resin the solution became faintly cloudy and HSs appeared to concentrate at the top of the XAD-8 resin column. In the course of desalting, this evidently non-dissolved fraction was eluted. The material that stayed sorbed to the column was removed by back elution with base and recovered by application of the procedures described above. The non-solubilized material was redissolved in dilute base, diluted as before and applied to the XAD-8 column at pH 2.5. Again a faint cloudiness was observed as the sample was introduced to the XAD-8 resin and non-soluble material accumulated at the top of the column. This also eluted in the course of the desalting process and the fraction that stayed sorbed to the resin was removed in base and recovered as before. The process was repeated at pH 3 and at pH 3.5. All of the remaining material was retained by the column at pH 3.5 and was recovered as described. In this way four HA fractions were obtained and their compositional differences are being investigated. We obtained two fractions when the HA from a Mollisol soil was fractionated on XAD-8. The NMR spectra indicated that the fractions were compositionally different.

Analytical fractionation techniques based on charge density differences include applications of electrophoresis, which involves the movement of charged species in solution in response to an applied electrical potential. Reference is made by Hayes[13] to applications of the different electrophoretic separation methods, including zone electrophoresis, moving boundary electrophoresis, isotachophoresis, isoelectric focusing (IEF) and polyacrylamide gel electrophoresis (PAGE), which also can give a fractionation based on size differences. Preparative column electrophoresis and continuous flow paper electrophoresis techniques have achieved separation of saccharide and peptide materials from the humic components in soil extracts but have not yet achieved clear cut separations of humic components. These are distributed in a broad diffuse band,[62] which is to be expected because HSs are polydisperse with respect to charge.

4.2.2 Fractionation Based on Size Differences. There is disagreement with regard to the sizes of HSs. One school of thought considers that HSs are polydisperse macromolecules with molecular weight (MW) values ranging from <1000 to more than 10^6 daltons.[63] Another approach considers that HSs are pseudomacromolecular and only consist of aggregations or associations of relatively small molecules[64-66] that may take the forms of micelle-like structures. The different points of view are discussed in a series of articles in *Soil Sci.*, 1999, **164**.

Considerable emphasis has been placed on the uses of gel chromatography or gel permeation chromatography (GPC) for the fractionation of HSs on the basis of molecular size differences (MSD) and reference is made to applications of this technique.[7,10,12,13,63] Part of the differences in the concepts that exist might be attributable to pitfalls in applications of gel chromatography procedures (see *Soil Sci.*, 1999, **164**). Swift[10] has provided a diagrammatic representation of the tedious procedure required to produce fractions that have a degree of molecular size homogeneity. In principle, the procedure involves the reprocessing (through the gel column) of substances eluted between specific volume boundaries from a standardized gel column until all of the material is contained within these volume boundaries. In theory, it should be possible to isolate several homogeneous fractions in that way.

Pressure filtration through membranes of discrete but different pore sizes may offer an alternative to gel chromatography for the fractionation of HSs on the basis of size differences. Using this approach Tombácz[67] fractionated a HA into six components that were compositionally different. Piccolo and his colleagues[65] have used high performance size exclusion chromatography (HPSEC). Separate inclusions of methanol, ethanoic acid and HCl to the mobile phases of constant ionic strength resulted in progressively smaller molecular size fractions. The decrease of HAs sizes as the result of the amendments was attributed to disruptions of supramolecular associations.

There is considerable logic behind the concept of macromolecular associations. However, the associations are likely to include non-humic substances and especially components with low polarities, which include long chain hydrocarbons, long chain fatty acids and esters, waxes and suberins. Such components are readily detectable by pyrolysis/gas chromatography-mass spectrometry techniques.[68]

5 CONCLUSIONS

There has been much emphasis since the time of Achard[11] more than two centuries ago on finding a solvent or solvent system that will isolate quantitatively the HSs from soils. Considerable progress has been made in this generation with regard to the isolation of HSs from waters. It is considerably easier, however, to isolate these substances from water because aquatic HSs, for the most part, are the most polar and the least associated of the HSs in the natural environment. That is why the allochthonous HSs of waters were released from their associations in soils or sediments. Aqueous base (NaOH or KOH) can be expected to quantitatively dissolve HSs that have sufficient ionizable acidic functionalities to allow solvation to take place provided, of course, that there are no materials or conditions that inhibit solvation and the release of the solvated molecules into the bulk solvent. It can be expected that HSs that are not soluble in base (the humin fraction) would be soluble in some organic solvents.

We still do not have a clear concept of the nature of humin materials. It is clear, however, that humins are abundant and important components of SOM. The evidence we have suggests that humin materials expose a hydrophobic face and may therefore have an important role in the binding of non-polar hydrophobic anthropogenic compounds. It is conceivable that humin could consist of altered HAs that have undergone a high degree of decarboxylation and are proceeding towards coalification. That concept, however, has no evidence to sustain it. The evidence we have associates humin materials with the colloidal mineral components of soils. It can be appreciated how, in associating with mineral (hydr)oxides below their point of zero charge when the mineral surface is positively charged, that the hydrophobic faces of bound humic molecules may be exposed to the exterior. In that case, should the acid functionalities be inaccessible the molecules might be difficult to solvate in aqueous media. Cation-bridging between the acidic functionalities of HAs and clay surfaces could generate a similar but less convincing concept. Logic suggests that there is more to the abundance of humin than these binding mechanisms indicate. Using MIBK, the work of Rice and MacCarthy[48,49] has isolated lipid and HA materials from the humin fraction (see Section 2.6). Similarly, Clapp and Hayes[41] isolated with DMSO materials that could be classified as humin, but had the compositions of HAs and FAs. Lipid-type materials would have been retained by the XAD-8 resin used to recover the HSs isolated in the DMSO extract of the Mollisol soil. It is tempting to infer that humin is an association of HAs and FAs with lipids, hydrocarbons and other non-humic and non-polar materials that coat the humic molecules, rendering them insoluble in aqueous solvents.

In the course of our ongoing studies on the isolation, fractionation and characterization of the HSs of a Fibric Peat we have observed that lipid-type materials in the alkaline extracts are sorbed to the 0.2 µ pore-size cellulose acetate membranes used to filter the HSs isolates prior to applications to the XAD-8 resin columns. It remains to be seen to what extents these lipid-type materials were associated with humic components. More attention should be focused on the extraction of humin materials and consideration might be given to the uses of supercritical fluids.

The XAD resin procedures described are new but relatively simple techniques for fractionating HSs on the basis of charge density differences. The discussion and degree of controversy that prevails with regard to the sizes (and arising from that, the shapes) of humic molecules is good for the science. Concepts of macromolecularity and the notion of random coil solution conformations[7,63] have served us well in this generation with regard to explanations of reactions and interactions of HSs in the soil environment. However, it may be necessary to change these concepts to considerations of pseudomacromolecularity and of molecular associations. The end result in so far as explanations of reactivities may be similar. We are but guessing until such matters are resolved.

References

1.   G. R. Aiken, D. M. McKnight, R. L. Wershaw and P. MacCarthy, in 'Humic Substances in Soil, Sediment, and Water', G. R. Aiken, D. M. McKnight, R. L.

Wershaw and P. MacCarthy, (eds.), Wiley, New York, 1985, p. 1.
2. W. H. Schlesinger, in 'The Role of Terrestrial Vegetation in the Global Carbon Cycle: Measurement by Remote Sensing', G. M. Woodwell, (ed.), Wiley, Chichester, 1984, p. 111.
3. W. H. Schlesinger, R. Lal, J. Kimble, E. Levine and B. A. Stewart, (eds.), 'Soils and Global Change', CRC Lewis, Boca Raton, 1995, p. 9.
4. H. Eswaran, E. Van den Berg and P. Reich, *Soil Sci. Soc. Am. J.*, 1993, **57**, 192.
5. G. R. Harvey and D. A. Boran, in 'Humic Substances in Soil, Sediment, and Water', G. R. Aiken, D. M. McKnight, R. L. Wershaw and P. MacCarthy, (eds), Wiley, New York, 1985, p. 233.
6. B. E. Watt, R. L. Malcolm, M. H. B. Hayes, N. W. E. Clark and J. K. Chipman, *Water Res.*, 1996, **6**, 1502.
7. M. H. B. Hayes and R. S. Swift, in 'The Chemistry of Soil Constituents', D. J. Greenland and M. H. B. Hayes, (eds.), Wiley, Chichester, 1978, p. 179.
8. D. S. Jenkinson, in 'The Chemistry of Soil Processes', D. J. Greenland and M. H. B. Hayes, (eds.), Wiley, Chichester, 1981, p. 505.
9. M. J. Häusler and M. H. B. Hayes, in 'Humic Substances and Organic Matter in Soil and Water Environments', C. E. Clapp, M. H. B. Hayes, N. Senesi and S. M. Griffith, (eds.), Proc. 7th Int. Conf. IHSS, Univ. of Minnesota, St. Paul, 1996, p. 25.
10. R. S. Swift, in 'Methods of Soil Analysis. Part 3. Chemical Methods', D. L. Sparks, A. L. Page, P. A. Helmke and R. H. Loeppert, (eds.), Soil Science Society of America and American Society of Agronomy, Madison, WI, 1996, p. 1011.
11. F. K. Achard, *Crell's Chem. Ann.*, 1786, **2**, 391.
12. M. H. B. Hayes, in 'Humic Substances: Structures, Properties and Uses', G. Davies and E. A. Ghabbour, (eds.), Royal Society of Chemistry, Cambridge, 1998, p. 1.
13. M. H. B. Hayes, in 'Humic Substances, Peats and Sludges. Health and Environmental Aspects', M. H. B. Hayes and W. S. Wilson, (eds.), Royal Society of Chemistry, Cambridge, 1997, p. 3.
14. J. A. Leenheer, R. L. Wershaw and M. M. Reddy, *Environ. Sci. Technol.*, 1995, **29**, 393.
15. E. M. Perdue, in 'Humic Substances in Soil, Sediment, and Water', G. R. Aiken, D. M. McKnight, R. L. Wershaw and P. MacCarthy, (eds.), Wiley, New York, 1985, p. 493.
16. P. MacCarthy and J. A. Rice, in G. R. Aiken, D. M. McKnight, R. L. Wershaw and P. MacCarthy, (eds.), 'Humic Substances in Soil, Sediment, and Water', Wiley, New York, 1985, p. 527.
17. C. Steelink, R. L. Wershaw, K. A. Thorn and M. A. Wilson, in 'Humic Substances II: In Search of Structure', M. H. B. Hayes, P. MacCarthy, R. L. Malcolm and R. S. Swift, (eds.), Wiley, Chichester, 1989, p. 282.
18. M. A. Wilson, in 'Humic Substances II: In Search of Structure', M. H. B. Hayes, P. MacCarthy, R. L. Malcolm and R. S. Swift, (eds.), Wiley, Chichester, 1989, p. 309.
19. R. L. Malcolm, in 'Humic Substances II: In Search of Structure', M. H. B. Hayes, P. MacCarthy, R. L. Malcolm and R. S. Swift, (eds.), Wiley, Chichester, 1989, p. 340.
20. A. J. Simpson, R. E. Boersma, W. L. Kingery, R. P. Hicks and M. H. B. Hayes, in 'Humic Substances, Peats and Sludges. Health and Environmental Aspects', M. H. B. Hayes and W. S. Wilson, (eds.), Royal Society of Chemistry, Cambridge, 1997,

p. 3.
21. A. J. Simpson, J. Burdon, C. L. Graham, M. H. B. Hayes and N. Spencer, *Eur. J. Soil Sci.*, 2000, in press.
22. E. W. D. Huffman and H. A. Stuber, in 'Humic Substances in Soil, Sediment, and Water', G. R. Aiken, D. M. McKnight, R. L. Wershaw and P. MacCarthy, (eds.), Wiley, New York, 1985, p. 433.
23. M. H. B. Hayes, in 'Humic Substances in Soil, Sediment, and Water', G. R. Aiken, D. M. McKnight, R. L. Wershaw and P. MacCarthy, (eds.), Wiley, New York, 1985, p. 329.
24. L. R. Snyder, in 'Techniques of Chemistry, Vol. XII. Separation and Purification', E. S. Perry and A. Weissberger, (eds.), Wiley, New York, 1978, p. 25.
25. M. R. J. Dack, in 'Techniques of Organic Chemistry', A. Weissberger, (ed.), Vol. VIII, 1976; and in 'Solutions and Solubilities', M. R. J. Dack, (ed.), part II, Wiley, New York, 1976, p. 95.
26. R. W. Taft, D. Gurka, L. Joris, R. Von Schleyer and J. W. Rakshys, *J. Am. Chem. Soc.*, 1969, **91**, 4801.
27. A. J. Parker, *Quart. Rev. Chem. Soc.*, 1962, **16**, 163.
28. J. H. Hildebrand and R. L. Scott, 'Solubility of Non-Electrolytes', 3rd Edn., Reinhold, New York, 1951.
29. J. H. Hildebrand and R. L. Scott, 'Regular Solutions', Prentice-Hall, Englewood Cliffs, NJ, 1962.
30. J. H. Hildebrand J. N. Prausnitz and R. L. Scott, 'Regular and Related Solutions', Van Nostrand-Reinhold, New York, 1970.
31. C. M. Hansen, 'Three-Dimensional Solubility Parameter and Solvent Diffusion Coefficient', Danish Technology Press, Copenhagen, 1967.
32. B. L. Karger, L. R. Snyder and C. Horvath, 'An Introduction to Separation Science', Wiley, New York, 1973.
33. A. F. M. Barton, *Chem. Rev.*, 1975, **75**, 731.
34. L. R. Snyder, in 'Techniques of Organic Chemistry, Vol. XII. Separation and Purification', E. S. Perry and A. Weissberger, (eds.), 3rd. Edn., 1978, p. 25.
35. B. L. Karger, L. R. Snyder and C. Eon, *J. Chromatog.*, 1976, **125**, 71.
36. F. J. Stevenson, 'Humus Chemistry: Genesis, Composition, Reactions', 2nd. ed., Wiley, New York, 1994.
37. D. C. Whitehead and J. Tinsley, *Soil Sci.*, 1964, **97**, 34.
38. J. M. Bremner, *J. Soil Sci.*, 1950, **1**, 198.
39. M. B. Choudri and F. J. Stevenson, *Soil Sci. Soc. Amer. Proc.*, 1957, **21**, 508.
40. J. M. Bremner and H. Lees, *J. Agric. Sci.*, 1989, **39**, 274.
41. C. E. Clapp and M. H. B. Hayes, in 'Humic Substances and Organic Matter in Soil and Water Environments', C. E. Clapp, M. H. B. Hayes, N. Senesi and S. M. Griffith, (eds.), Proc. 7th Int. Conf. IHSS, Univ. of Minnesota, St. Paul, 1996, p. 3.
42. T. M. Hayes, PhD Thesis, University of Birmingham, 1996.
43. T. M. Hayes, M. H. B. Hayes, J. O. Skjemstad, R. S. Swift and R. L. Malcolm, in 'Humic Substances and Organic Matter in Soil and Water Environments', C. E. Clapp, M. H. B. Hayes, N. Senesi and S. M. Griffith, (eds.), Proc. 7th Intern. Conf. IHSS, University of Minnesota, St. Paul, 1996, p. 13.
44. M. H. B. Hayes, R. S. Swift, R. E. Wardle and J. K. Brown, *Geoderma*, 1975, **13**, 231.
45. A. H. Sinclair and J. Tinsley, *J. Soil Sci.*, 1981, **32**, 103.
46. C. E. Clapp and M. H. B. Hayes, *Soil Sci.*, 1999, **164**, 899.
47. D. Martin and H. G. Hauthal, 'Dimethyl Sulphoxide', (translated by E.S.

Halberstadt), Van Nostrand-Reinhold, New York, 1975.
48. J. A. Rice and P. MacCarthy, *Sci. Total Environ.*, 1989, **81/82**, 61.
49. J. A. Rice and P. MacCarthy, *Environ. Sci. Technol.*, 1990, **24**, 1875.
50. A. Suggett, in 'Water, A Comprehensive Treatise: Vol 4, Aqueous Solutions of Amphiphiles and Macromolecules', F. Franks, (ed.), Plenum, New York, 1975, p. 519.
51. Y. Chen, N. Senesi and M. Schnitzer, *Soil Sci. Soc. Am. J.*, 1977, **41**, 352.
52. R. L. Malcolm, in 'Humic Substances in Soil, Sediment, and Water', G. R. Aiken, D. M. McKnight, R. L. Wershaw and P. MacCarthy, (eds.), Wiley, New York, 1985, p. 181.
53. G. R. Aiken, in 'Humic Substances in Soil, Sediment, and Water', G. R. Aiken, D. M. McKnight, R. L. Wershaw and P. MacCarthy, (eds.), Wiley, New York, 1985, p. 363.
54. J.A. Leenheer, in 'Humic Substances in Soil, Sediment, and Water', G. R. Aiken, D. M. McKnight, R. L. Wershaw and P. MacCarthy, (eds.), Wiley, New York, 1985, p. 409.
55. R. L. Malcolm and P. MacCarthy, *Environ. Int.*, 1992, **18**, 597.
56. E. M. Thurman and R. L. Malcolm, *Environ. Sci. Technol.*, 1981, **15**, 463.
57. E. M. Thurman, 'Organic Geochemistry of Natural Waters', Martinus Nijhoff/Dr W. Junk Publishers, Dordrecht, The Netherlands, 1985.
58. M. Deinzer, R. Melton and D. Mitchell, *Water Res.*, 1975, **9**, 799.
59. S. M. Serkiz and E. M. Perdue, *Water Res.*, 1990, **24**, 911.
60. S. M. Sun, E. M. Perdue and J. F. MacCarthy, *Water Res.*, 1995, **29**, 1471.
61. R. L. Malcolm and P. MacCarthy, *Environ. Int.*, 1992, **18**, 597.
62. M. H. B. Hayes, J. E. Dawson, J. L. Mortensen and C. E. Clapp, in 'Volunteered Papers', M. H. B. Hayes and R. S. Swift, (eds.), 2nd Intern. Conf., International Humic Substances Society, Department of Soils, Water and Climate, Univ. of Minnesota, St. Paul, 1985, p. 31.
63. R. S. Swift, in M. H. B. Hayes, P. MacCarthy, R. L. Malcolm and R. S. Swift (eds.), 'Humic Substances II: In Search of Structure', Wiley, Chichester, 1989, p. 467.
64. R. L. Wershaw, *Soil Sci.*, 1999, **164**, 803.
65. A. Piccolo, P. Conte, A. Cozzolino and R. Spaccini, in 'Humic Substances and Chemical Contaminants', C. E. Clapp, M. H. B. Hayes, N. Senesi, P. Bloom and P. M. Jardine, (eds.), Soil Science Society of America, Madison, WI., 2000, in press
66. R. von Wandruszka, *Soil Sci.*, 1998, **163**, 921.
67. E. Tombácz, *Soil Sci.*, 1999, **164**, 814.
68. Y. Huang, G. Eglinton, E. R. E. van der Hage, J. J. Boon and P. Ineson, *Eur. J. Soil Sci.*, 1998, **49**, 1.

DIFFERENCES IN HIGH PERFOMANCE SIZE EXCLUSION CHROMATOGRAPHY BETWEEN HUMIC SUBSTANCES AND MACROMOLECULAR POLYMERS

A. Piccolo, P. Conte and A. Cozzolino

Dipartimento di Scienze Chimico-Agrarie, Università di Napoli "Federico II", 80055 Portici, Italy

1 INTRODUCTION

Humic substances (HSs) are natural organic materials present in water, soil and sediments that play a fundamental role in supporting crop production in soils[1] and in controlling both the fate of environmental pollutants[2,3] and the biogeochemistry of organic carbon in the global ecosystem.[4] Despite the ecological and environmental importance of HSs and the intensive research conducted in the last 100 years, their chemistry still remains obscure.[5,6]

The variability of sources and the chemical heterogeneity of HSs have prevented definite knowledge of their secondary chemical structure and absolute molecular weight.[7] However, detailed information on molecular weights and sizes is essential in order to compare different humic materials and study their reactivity with other chemical species. A traditionally accepted view is that humic substances consist of coiled conformations made up of long chain molecules which may be slightly cross-linked.[1,8] The polymeric random coil model[9] depicts humic matter as most densely coiled at high concentrations, low pH and high ionic strength, whereas at neutral pH, low ionic strength, and low concentration the humic macromolecules behave like flexible linear colloids. Several reviewers[1,10] have pointed out problems associated with the techniques used to determine molecular weight values (such as freezing point depression, vapor pressure osmometry, light scattering and sedimentation methods) and molecular sizes (by methods such as size exclusion chromatography, ultrafiltration, small-angle X-ray scattering) of HSs. The main problem is the lack of model compounds of known composition and molecular weight (MW) that can be used as calibration standards, thereby resulting in considerable uncertainty with regard to sizes and real MW values. Data in the literature suggest that these values may vary from 500 Da for some aquatic HS to more than 10^6 Da for soil humic acids[1] and there is not agreement between the different methods used to evaluate MW values.

High Performance Size Exclusion Chromatography (HPSEC) has been the most widely used method[11-18] to assess molecular sizes of both aquatic and terrestrial humic material, but interpretation of the often erratic results never departed from the polymeric model of humic substances. Nevertheless, efforts to correlate humic MW values obtained by HPSEC to those calculated by vapor-pressure osmometry[17] or to obtain viscosity data to derive the "universal calibration equation" as is customary in HPSEC of polymers[18] were regularly unsuccessful. Moreover, if the theory of random coiled polymers implies that it should be possible to isolate several homogeneous humic fractions by size exclusion chromatography, reprocessing of fractions showed that smaller sized components are released and isolated fractions would appear to have similar molecular sizes.[19] These contradictions appeared to be resolved by a number of chromatographic results[20-23] that indicated that the apparently high molecular size of humic substances can be reversibly disrupted into smaller size associations by the actions of organic acids. These findings suggest that dissolved humic substances, rather than being polymeric coils as previously believed mostly by analogy to biological macromolecules[1] appear to reflect the structure of randomly self-associating small heterogeneous molecules that are loosely held together by weak hydrophobic forces. These results appeared to be confirmed in another approach[24] where protection by HSs of pyrene fluorescence from bromide quenching was found to be lost when acetic acid, followed by base, was added to the medium.

While these findings suggest that HSs are loosely bound associations of heterogeneous molecules of relatively small molecular weight instead of being polyelectrolyte polymers, it is well known that non-size exclusion effects such as ionic exclusion or specific adsorption may lead to erroneous HPSEC evaluation of MW in macromolecules.[12-14,16,25-27]

The objective of this study was to verify that HPSEC results really are due to conformational changes of humic substances rather than to changes in column performance upon addition of acids. We thus compared the size-exclusion chromatographic behavior of different humic substances with that of undisputed macromolecular polymers such as polysaccharides and polystyrenesulfonates of known molecular weight.

2 MATERIALS AND METHODS

2.1 Humic Substances. Four humic acids (HAs) were isolated from different raw materials: HA-A from an agricultural soil (Typic Eutrochrepts) near Roskilde (Denmark), HA-B from a volcanic soil (Typic Xerofluvent) near Rome (Italy), HA-C from an oxidized coal provided by Eniricerche SpA (Italy), and HA-D from a North Dakota Leonardite (Mammoth International Chemical Company). HAs were extracted and purified by common procedures.[1] The original materials were shaken overnight in a solution of 0.5 M NaOH and 0.1 M $Na_4P_2O_7$ under a N_2 atmosphere. HAs were precipitated from the alkaline extracts by lowering the pH to 1 with 6 M HCl. The HAs were extensively purified by dissolution in 0.1 M NaOH solution followed by centrifugation and separation of supernatant from any contaminating solid. Subsequent acidification of supernatant precipitated the HAs, which were then recovered by centrifugation and the supernatant was separated. This procedure was repeated three times. The HAs were then treated with a 0.5% (v/v) HCl-HF solution for 36 h, dialyzed (Spectrapore 3 dialysis tubes, 3,500 MW cut-off) against distilled water until chloride-free and freeze-dried. HAs were then redissolved in 0.5 M NaOH and passed through a strong cation-exchange resin (Dowex 50) to further eliminate divalent and trivalent metals and freeze-dried again.

2.2 Characterization of Humic Samples. The elemental contents of HAs (Table 1) were determined with a Fisons EA 1108 Elemental Analyzer and the ash contents, obtained by burning 50-100 mg of the HAs in an oven at 750°C for 8 hours were less than

Table 1 *Elemental analyses (on ash-free and moisture-free bases) of humic acids*

HA	C/%	H/%	N/%	C/H	C/N
HA-A	47.4	5.0	4.4	4.4	10.8
HA-B	53.7	4.9	4.3	11.0	12.5
HA-C	48.0	3.0	1.0	16.0	48.0
HA-D	45.9	3.7	1.0	12.4	45.9

5% (w/w) for all humic materials.

Cross-polarization Magic Angle Spinning Carbon-13 Nuclear Magnetic Resonance Spectroscopy (CPMAS ^{13}C-NMR) experiments were carried out on a Bruker AMX400 instrument operating at 100.625 MHz. A recycle time of 1 sec and an acquisition time of 13 msec were used. All the experiments were conducted with a Variable Contact Time (VCT) pulse sequence in order to find the Optimum Contact Time (OCT) for each sample and to minimize the error on the evaluation of the peak areas.[28] OCT ranged between 0.8 and 1.0 msec. A line broadening of 50 Hz was used to transform all the FIDs. The area in the 110-140 ppm region was corrected for the side band of carboxylic group signals by subtracting the side band area in the 190-230 ppm region from that of the 110-140 ppm region. The areas of each region of the spectra in Table 2 were attributed to non polar carbons such as the aliphatic (0-45 ppm), and aromatic (110-160 ppm) ones, to polar carbons such as the C-O, C-N groups and anomeric carbons (45-110 ppm) and to carboxylic carbons (160-190 ppm). The areas of the 0-45 and 110-160 ppm regions were used to calculate hydrophobicity (HB), whereas those of the 45-60, 60-110, and 160-190 ppm regions were used to obtain hydrophilicity (HI) of HAs. The HI/HB ratios also are given in Table 2.

2.3 HPSEC System. The HPSEC system consisted of a high pressure Perkin-Elmer LC200 solvent pump and of two detectors in series: a UV/Vis variable wavelength detector (Perkin-Elmer LC295) set at 280 nm and a refractive index (RI) detector (Fisons Instruments, Refractomonitor IV). A Rheodyne rotary injector equipped with a 100 µL sample loop was used to load the calibration standard and humic solutions. Size exclusion separation occurred through a G3000SW (600 mm x 7.5 mm i.d.) TSK column (Toso Haas). The stationary phase is a rigid spherical silica gel chemically bonded with hydrophilic compounds[29] with an alleged low residual hydrophobicity and minimal ion exchange capacity. The column was preceded by a 7.5 cm TSK Guard-Column (7.5 mm i.d.) packed with G3000SW and by a 0.2 µm stainless-steel inlet frit. The column system was thermostated at 25°C by a water bath. The column manufacturer only reports a calibration for globular proteins of 5 to 300 kDa as the nominal separation range. Experimental results have shown that the column separation range varied with calibration standards and their molecular size.[27] The flow rate was set to 0.6 mL min^{-1} and the HPSEC eluent was a 0.05 M $NaNO_3$ and 4.0 x 10^{-3} M NaN_3 solution (the latter as a bacteriostatic agent). This neutral mobile phase was suitable for rapid and efficient HPSEC of humic substances and for avoiding phenomena of solute-stationary phase interactions.[13,16,27] Moreover, the 0.05 M concentration was considered the best compromise between sample solubility and ability of humic molecules to form a fully coiled conformation in solution.[9] The mobile phase was made with MilliQ water and HPLC-grade reagents filtered through a Millipore 0.45 µm filter and He degassed. The void volume (V_0 = 11.18 mL) and total permeation volume (V_t = 20.57 mL) of the columns were determined using Blue Dextran

Table 2 *Distribution (%) of ^{13}C in resonance intervals (ppm) of CPMAS-NMR spectra and HI/HB ratios of HAs*

HA	0-45	45-60	60-110	110-160	160-190	HI/HB^a
HA-A	34.7	16.7	18.8	13.9	18.1	1.11
HA-B	35.3	12.8	30.8	16.0	10.3	1.05
HA-C	22.1	9.8	16.4	36.9	23.4	0.84
HA-D	25.3	8.4	18.0	39.4	16.6	0.66

^{a} HI/HB = (45-60) + (60-110) + (160-190)/ (0-45) + (110-160)

2000 and water, respectively.

2.4 Size Exclusion of Polymeric Standards. Polysaccharides (Polymer Sciences Laboratories, UK) of known MW (100, 48, 23.7, and 12.2 kDa) and polystyrenesulfonates (Polymer Standard Services, Germany) of known MW (130, 32, 16.8, 6.78 and 4.3 kDa) were used as calibration standards. While only polystyrenesulfonates were measurable with the UV detector (260 nm), both standards were revealed by the RI detector. Solutions of both aqueous polymers were prepared by dissolving 2 mg of each polymer of known MW into 10 mL of the HPSEC mobile phase and eluted into the chomatographic system with and without previous addition to pH 3.5 of either HCl or acetic acid (AcOH) prior to HPSEC analysis as was done for humic solutions (see below). All calibration curves obtained with these two series of standards (untreated and treated with acids) were semi-log linear over the range defined by standards of known MW and were used to determine the molecular weight of a humic analyte, M_i, at some eluted volume *i*. Relative standard deviations of triplicate analyses were lower than 2% for all neutral and polyelectrolyte standards.

2.5 Humic Solutions. Purified HA samples (50 mg) were first suspended in distilled water (50 mL) and titrated to pH 7 with a CO_2-free solution of 0.5 M NaOH in an automatic titrator (VIT 90 Videotitrator, Radiometer, Copenhagen) under a N_2 atmosphere and with stirring. After having reached constant pH 7, the solution containing sodium humates was left after titration for two more hours, filtered through a Millipore 0.45 μm filter and freeze-dried. Sodium humates pre-titrated to pH 7 were used to exclude the random occurrence of negative charges on solute molecules when dissolved into the HPSEC mobile phase and to depress ionic exclusion phenomena. Uncontrolled formation of negative charges on the solute is believed to change ionic strength of humic solutions thereby affecting sample exclusion.[30] Humic solutions for HPSEC analysis were then prepared by dissolving 2.0 mg of each sodium humate sample in 10 mL of the HPSEC eluent to obtain a final HS concentration of 0.2 mg mL^{-1} in the control solution. Prior to injection in the HPSEC system, each control solution was titrated to pH 3.5 with HCl or acetic acid. Addition of HCl and acetic acid did not significantly affect the ionic strength of humic solutions, which remained constant at 0.05 M. Humic solutions were freshly prepared before each injection.

2.6 Molecular Weight Determination. Size exclusion chromatograms for both the UV and RI detectors were evaluated by using Perkin-Elmer-Nelson Turbochrom 4-SEC peak integration and molecular weight software, a SEC noise threshold of 5, and a filter size of 5 for the Savitzky-Golay smoothing. Calculations of weight (M_w) and number-averaged (M_n) molecular weights and polydispersity (M_w/M_n) were done by the method of Yau et al.[25] using the following equations:

$$M_w = \sum_{i=1}^{N} h_i(M_i) / \sum_{i=1}^{N} h_i \qquad \text{and} \qquad M_n = \sum_{i=1}^{N} h_i / \sum_{i=1}^{N} h_i / M_i$$

where M_i and h_i are the molecular weight and the height of the i-th chromatographic slice in the chromatogram of each sample eluted at volume i, respectively. M_w and M_n values from chromatograms of control HAs as well as from chromatograms of HAs treated with either HCl or acetic acid were obtained by using calibration curves of both neutral and polyelectrolytic polymers treated in the same manner as the HAs. Based on described methods,[25] the system dispersion was found to be less than the error (<5%) and the chromatograms were evaluated without additional correction factors. The relative standard deviation of calculated values among triplicates of each chromatogram varied only to a maximum of 7%, thereby confirming the good reproducibility of the HPSEC system similarly reported elsewhere.[23, 27]

3 RESULTS AND DISCUSSION

Absolute measurements of molecular weights of HSs by column calibration are not possible because their precise chemical structures is not known. For any possible calibration standard, the hydrodynamic radius and the interaction with the stationary phase are necessarily different from those of HSs.[8] A number of investigators[12,14,31] have used globular protein standards despite their recognized overprediction of humic substances molecular weights. Polystyrenesulfonates (PSS) are also popular standards in exclusion studies of humic substances.[12,13,31] However, a similarity in charge density between PSS and humic substances cannot be generally accepted.[21] Though uncharged (unlike HSs), nonionic hydrophilic polymeric biomolecules such as polysaccharides (PYR) or nonionic polyethyleneglycol (PEG) have been also used to evaluate molecular size distributions of dissolved humic samples from different sources.[16-18,21,27]

Despite the noted differences from humic material, both the nonionic PYR and polyelectrolytes PSS were used in this study for the very reason of their undisputed polymeric nature to compare their chromatographic behavior to that of HSs upon changes of mobile phase composition. Figures 1 and 2 show the calibration curves obtained from the responses of UV and RI detectors, respectively, when PSS standards of different MW were dissolved in the mobile phase at pH 7 or when their control solution was brought to pH 3.5 with either HCl or acetic acid before HPSEC injection. While high-molecular weight (130 and 32 kDa) PSS standards eluted at elution volumes not significantly different from those of the control solution (pH 7), a progressive increase in elution volumes was noted for the lower molecular-weight PSS standards (16.8, 6.78 and 4.3 kDa) when treated with acids prior to injection. Changes in elution volumes of PSS standards were very similar in both UV and RI detectors. However, the increase in elution retardation for the PSS standards of low molecular weight appeared larger with HCl than with acetic acid treatments. This behavior may be due to protonation of some of the sulfonated functions of PSS, which caused an enhanced solute-stationary phase interaction for the low molecular-weight standards that presumably have a larger charge density than high molecular-weight standards. Despite treatment with the same amount of acids, the smaller standards were hence retarded while they diffused through the column gel pores during elution, whereas the larger PPS molecules did not significantly interact with the stationary phase and eluted at about the same elution volume. The retardation effect

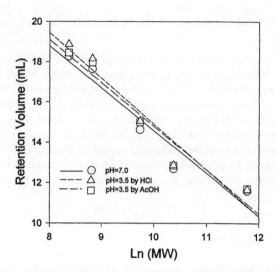

Figure 1 *HPSEC retention volumes by UV detector versus Ln of molecular weight of different polystyrenesulfonate (PSS) standards dissolved in control solution at pH 7 and in the same solution but with pH lowered to 3.5 by addition of either HCl or acetic acid*

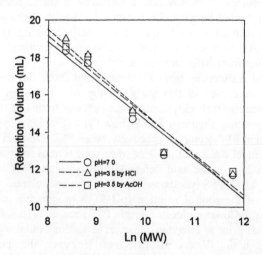

Figure 2 *HPSEC retention volumes by RI detector versus Ln of molecular weight of different polystyrenesulphonate (PSS) standards dissolved in control solution at pH 7 and in the same solution but with pH lowered to 3.5 by addition of either HCl or acetic acid*

observed for the low MW PSS standards seemed to be less pronounced in the case of acetic acid addition, probably because of lower acidity than with HCl. However, differences between HCl and AcOH treatments were well within the mentioned analytical error of the elution volume (see Materials and Methods). Hence the regression equations obtained with the different PSS standards and the UV detector were somewhat different when dissolved in the control solution at pH 7 (y = 35.45x - 2.09, r^2 = 0.93) or when this solution was modified either with HCl (y = 35.42x - 2.26, r^2 = 0.93) or with AcOH addition (y = 37.37x - 2.26, r^2 = 0.94). No significant differences in peak detection were observed between UV and RI detectors, as shown by the curves obtained for the three treatments by both detectors (Figures 1 and 2).

Elution volumes of PYR standards in the three different solutions of this study as well as the relative calibration obtained by RI detection are shown in Figure 3. No significant changes were observed in elution volumes of the MW standards of the nonionic uncharged PYR polymer when dissolved in the mobile phase at pH 7 or when this was modified with either HCl or AcOH additions. The relative regression equations were hence very similar: y = 88.9x - 5.99 (r^2 = 0.99), y = 88.9x - 6.00 (r^2 = 0.99), and y = 89.4x - 6.00 (r^2 = 0.99), for the mobile phase and for HCl and AcOH treatments, respectively. Unlike the PSS polyelectrolytes, the lack of ionizable functions in the PYR standards prevented any appreciable charge diversity in the polymer and thus no different solute interactions with the stationary phase were possible after acid treatments.

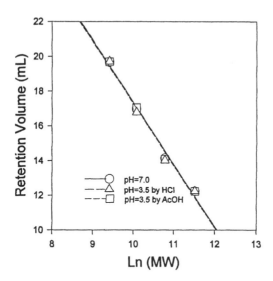

Figure 3 *HPSEC retention volumes by RI detector versus Ln of molecular weight of different of polysaccharide (PYR) standards dissolved in control solution at pH 7 and in the same solution but with pH lowered to 3.5 by addition of either HCl or acetic acid*

HPSEC chromatograms relative to the four different HSs as revealed by UV and RI detectors are shown in Figures 4 and 5, respectively. The molecular size distributions of the HAs in both UV and RI detecting modes are rather different from each other, as may have been expected from their different elemental (Table 1) and molecular composition

(Table 2). The substantial alteration in molecular-size distributions shown in both figures when passing from chromatograms of HAs dissolved in control solution at pH 7 to those dissolved in the HCl- or AcOH-treated solution are in line with the changes observed in previous HPSEC experiments.[21,22,27] The profiles of the UV-detected (280 nm) HPSEC chromatograms of HAs (Figure 4) are related to changes in the distribution of chromophores in humic size fractions following the modification of control solution with acids. It has been reported[1,21,22] that interactions of HSs with UV light and other macromolecules and colloids[32] do not follow the Beer-Lambert law. Large macromolecular polymers such as proteins and nucleic acids are subject to the phenomenon of chromism that is related to the interaction between one particular electronic excited state of a given chromophore and different electronic states of neighboring chromophores.[33,34] A variation in the reciprocal orientation between the transition dipole moments of an absorbing chromophore and the induced dipoles of neighboring chromophores may either increase (hyperchromism) or decrease (hypochromism) the molecular absorptivity of the UV spectrum.[33,34]

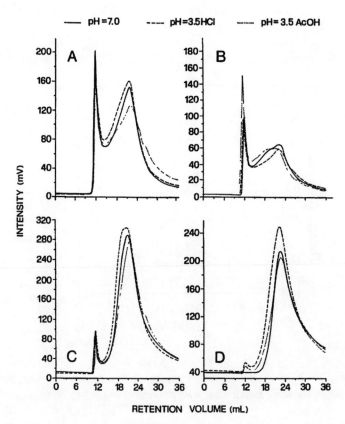

Figure 4 *HPSEC chromatograms by UV detector of four humic acids dissolved in control solution at pH 7 and in the same solution but with pH lowered to 3.5 by addition of either HCl or acetic acid*

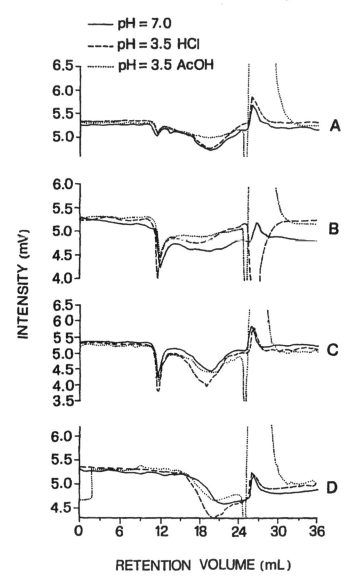

Figure 5 *HPSEC chromatograms with the RI detector (negative mode) of four humic acids dissolved in control solution at pH 7 and in the same solution but with pH lowered to 3.5 by addition of either HCl or acetic acid*

UV-detected chromatograms of HAs in solutions modified by acids additions were different from those in the control solution since both peak intensities and elution volumes were changed. Peak intensity of the lower molecular-size fraction increased with HCl addition in HA-A, HA-C and HA-D, whereas that of the high molecular-size fraction (at around V_0) also increased significantly for HA-B and HA-D with addition of acetic acid and of either acid, respectively. These changes are interpreted as hyperchromic and/or hypochromic effects that have taken place during treatments of humic solutions with acids. Following the alteration of conformational energies brought about by the formation of hydrogen bonds upon HCl or AcOH additions, chromophores interactions are different than in the control solution,[21,22] thereby increasing or decreasing the absorbance of the corresponding eluting fraction. Chromic effects in peak absorbances may be dependent on the chemical composition of HAs (Tables 1 and 2) and on the degree of conformational stability conferred by their content of hydrophobic components.[21,22] Changes in elution volumes of peaks were also noted in the HPSEC chromatograms (Figure 4) and were in accordance with the described changes of peak intensities.

The UV-detected chromatograms only provide a distribution of chromophores present in the humic samples and depend on the different molecular absorptivities of chromophores or from their mutual interactions. Conversely, chromatograms revealed by RI detector (Figure 5) are not related to chromophores but to the bulk of the eluting material and its real molecular size. These HPSEC chromatograms (recorded in negative mode) showed that humic samples were generally shifted to higher elution volumes when treated with acids, thereby confirming disruption of their molecular association as shown previously.[21,22]

Moreover, treatment with AcOH was more effective than HCl in disrupting the weakly bound humic conformations adopted by humic samples in the control solution, as shown by the appearance of a negative sharp hump eluted right before V_t in each chromatogram (Figure 5). Detection by RI failed to show an increase in bulk material at the V_0 with AcOH treatment for the HA-B sample and with either acid in the HA-D sample. This suggests that such increases of the V_0 peak intensity that were observed in the same samples by UV-detection (Figure 4) may be attributed to a hyperchromic interaction between chromophores following conformational rearrangements upon acid addition, rather than to a real molecular-size enhancement. A similar conclusion can be reached by comparing the UV- and RI-detected chromatograms for all humic samples (Figures 4 and 5).

Table 3 and 4 report the weight- (M_w), number-averaged (M_n) molecular weights and polydispersities (P) of humic samples calculated from HPSEC chromatograms in both UV- and RI-detecting modes by using calibration curves of either PSS or PYR standards. Close analysis of these data reveals that humic samples had very different HPSEC behavior from those of real polymeric standards. While changes in intercept values of calibration curves of PSS standards between control solution at pH 7 and those modified by acids addition to pH 3.5 were no larger than 7.5% for both UV and RI detectors (Figures 1 and 2), differences in M_w and M_n obtained by UV detection varied from -9.2% to 58.6% and from -9.9 to 48.5%, respectively (Table 3). Much larger (>1000) were the differences in M_w and M_n values for the HAs measured by RI detection upon solution modification before injection into the HPSEC system, except for HA-C that showed only 953 and 172% increase in M_w with HCl and AcOH additions, respectively (Table 3). While differences with changes in solution were small with PSS polymers and were only noticeable for the low MW standards, the HPSEC distribution of HAs was dramatically altered by solution composition and alteration was equally large in both high and low molecular size domains (Figure 4). Moreover, while identical behavior was noted for PSS standards in both UV and RI detectors (Figures 1 and 2), differences between the two detectors were of several orders of magnitude for the HAs (Table 3).

Table 3 *Weight- (Mw) and number-average (Mn) molecular weights and polydispersities (P) of control and acid-treated humic solutions calculated based on curves of polystyrenesulfonates (PSS) standards of known molecular weights revealed by either UV or RI detectors*

	UV detector					RI detector				
	Mw	Δ^d/%	Mn	Δ/%	P^e	Mw	Δ/%	Mn	Δ/%	P
HA-A										
Control[a]	164765		32078		5.1	19411		585		33
HCl[b]	169298	2.75	47630	48.5	3.6	237837	>1000	122735	>1000	1.9
AcOH[c]	171944	4.36	34526	7.63	5.0	259488	>1000	105404	>1000	2.5
HA-B										
Control	170104		38514		4.4	9961		321		31
HCl	154426	-9.21	45839	15.9	3.4	146914	>1000	83461	>1000	1.8
AcOH	169916	-0.11	34678	-9.9	4.9	200432	>1000	87906	>1000	2.3
HA-C										
Control	30409		16808		1.8	8497		167		51
HCl	48226	58.6	23203	38.0	2.1	89177	953	56640	>1000	1.6
AcOH	40459	33.0	18862	12.2	2.1	23156	172	13343	>1000	1.7
HA-D										
Control	77549		25980		2.9	17027		587		29
HCl	83856	8.13	33787	30.0	2.5	197417	1059	97684	>1000	2.0
AcOH	76327	-1.57	25762	-0.83	3.0	219442	1188	94982	>1000	2.3

[a] control: NaNO$_3$ solution at pH 7; [b] control with added HCl to pH 3.5; [c] control with added acetic acid (AcOH) to pH 3.5; [d] percent difference from control solution; [e] $P = M_w/M_n$

Even more striking were M_w and M_n results of HAs from both UV and RI detection when calculated from PYR standard calibration curves (Table 4). While the HPSEC behavior of PYR standards was identical when dissolved in either the control or in acid-added solutions (Figure 3), the relative M_w and M_n values for HAs revealed by UV detection differed from 1.7 to 20.8% and from -3.4 to 34.4%, respectively (Table 4). For the same samples, much larger differences were observed by RI detection: M_w values varied from 2.66 to 1013% whereas M_n values ranged from 0.73 to >1000% (Table 4).

The large variability of our results suggests that dissolved HSs cannot be regarded as covalently-bound polymers of well defined monomers such as the standards of known molecular weight used in this study. A plausible explanation of the diverse HPSEC behavior between humic matter and real polymers is that HSs molecules, though certainly heterogeneous, are not covalently bound in polymer-like macromolecules. They may be better described as supramolecular associations whose conformations vary widely in time and stability according to weak intermolecular forces such as multiple dispersive bondings (van der Waals, charge-transfer, etc.) or hydrogen bonding. In contrast to the random-coil polymeric model, regarding humic molecules in solution as supramolecular associations is in accordance with results of most recent experiments[20-23,27] and well explains previous findings and controversies.[19,35] Contrary to real polymers, HSs appear to have a macromolecular size that is subject to extensive variation depending on composition and pH of the mobile phase and the content and nature of humic components, whose weak intermolecular interactions (dispersive forces) only temporarily stabilize one of the possible conformational structures. The stability of a humic conformation is strictly a function of the chemical and physical parameters related to the solution in which humic matter is dissolved and to the method employed for its evaluation. In a neutral solution where the most acidic functions present on humic molecules are dissociated, intermolecular hydrophobic interactions are mainly responsible for conformational stability, whereas hydrogen bonding plays a more important role with progressive lowering of solution pH and contributes to stabilize a different conformation.[21] Shear strength arising during elution through a HPSEC column seemed also to affect the size distribution of humic samples.[27]

Table 4 *Weight- (Mw) and number-average (Mn) molecular weights and polydispersities (P) of control and acid-treated humic solutions calculated based on HPSEC calibration curves of polysaccharides (PYR) standards of known molecular weights revealed by either UV or RI detectors*

	UV detector					RI detector				
	Mw	Δ^d%	Mn	Δ%	P^e	Mw	Δ%	Mn	Δ%	P
HA-A										
Control[a]	141763		76011		1.9	176065		100742		1.7
HCl[b]	152831	7.80	85707	12.7	1.8	180752	2.66	103593	2.83	1.7
AcOH[c]	144194	1.71	73351	-3.4	2.0	194132	10.26	134544	33.5	1.4
HA-B										
Control	150107		82311		1.8	121214		7084		1.7
HCl	142264	5.22	82407	0.1	1.7	127526	5.20	77636	995	1.6
AcOH	168022	11.9	110682	34.5	1.5	161758	33.4	115373	>1000	1.4
HA-C										
Control	56692		45038		1.3	9625		54006		1.8
HCl	68497	20.8	49681	10.3	1.4	84308	775	57979	7.35	1.4
AcOH	61945	9.26	44931	-0.2	1.4	107170	1013	85977	59.1	1.2
HA-D										
Control	92393		61549		1.5	15625		90952		1.7
HCl	96932	4.91	65000	5.6	1.5	155438	894	91620	0.73	1.7
AcOH	106467	15.2	82764	34.4	1.3	168804	980	117907	29.6	1.4

[a] control: $NaNO_3$ solution at pH 7; [b] control with added HCl to pH 3.5; [c] control with added acetic acid (AcOH) to pH 3.5; [d] percent difference from control solution; [e] $P = M_w/M_n$

Despite the limitations inherent to column calibration, HPSEC provides deeper insight into the conformational nature of HSs, thereby proving to be a sensitive and very reproducible method for a rapid analytical evaluation of the conformational stability as well as the chemical reactivity of HSs from different sources. The concept of supramolecular association that arises from our results and that entails the phenomenon of self-assembly of small heterogeneous humic molecules points out the importance of intermolecular forces[36] in controlling the environmental and ecological behavior of humic substances.

ACKNOWLEDGEMENTS

This work was partially supported by the Italian Ministry of University and Scientific and Technological Research (MURST) through project no. 9807352092. The first author is grateful to the National Inter-University Consortium "Chemistry for the Environment" for a postdoctoral fellowship.

References

1. F. J. Stevenson, 'Humus Chemistry. Genesis, Composition, Reactions', 2nd Edn., Wiley, New York, 1994.
2. A. Piccolo, in 'Humic Substances in the Global Environment and Implications on Human Health', N. Senesi and T. M. Miano, (eds.), Elsevier, Amsterdam, 1994, p. 961.
3. T. M. Hayes, M. H. B. Hayes and L. V. Vaidyanathan, in 'Humic Substances, Peats and Sludges. Health and Environmental Aspects', M. H. B. Hayes and W. S. Wilson, (eds.), Royal Society of Chemistry, Cambridge, 1997, p. 208.
4. A. Piccolo, in 'Humic Substances in Terrestrial Ecosystems', A. Piccolo, (ed.), Elsevier, Amsterdam, 1996, p. 225.
5. I. D. White, D. N. Mottershead and S. J. Harrison, 'Environmental Systems, An Introductory Text', 2nd Edn., Chapman & Hall, London, 1992, p. 465.
6. C. Saiz-Jimenez, in 'Humic Substances in Terrestrial Ecosystems', A. Piccolo, (ed.), Elsevier, Amsterdam, 1996, p. 4.
7. M. H. B. Hayes, in 'Humic Substances, Peats and Sludges. Health and Environmental Aspects', M. H. B. Hayes and W. S. Wilson, (eds.), Royal Society of Chemistry, Cambridge, 1997, p. 3.
8. R. S. Cameron, R. S. Swift, B. K. Thornton and M. A. Posner, *J. Soil Sci.*, 1972, **23**, 342.
9. K. Ghosh and M. Schnitzer, *Soil Sci.*, 1980, **129**, 266.
10. R. L. Wershaw and G. R. Aiken, in 'Humic Substances in Soil, Sediment and Water', G. R. Aiken, D. M. McKnight, R. L. Wershaw and P. MacCarthy, (eds.), Wiley, New York, 1985, p. 477.
11. Y. Saito and S. Hayano, *J. Chromatogr.*, 1979, **177**, 390.
12. C. J. Miles and P. L. Brezonik, *J. Chromatogr.*, 1983, **259**, 499.
13. M. Berden and D. Berggren, *J. Soil Sci.*, 1990, **41**, 61.
14. Yu. -P. Chin and P. M. Gschwend, *Geochim. Cosmochim. Acta*, 1991, **55**, 1309.
15. Yu. -P. Chin, G. R. Aiken and E. O'Loughlin, *Environ. Sci. Technol.*, 1994, **28**, 1853.
16. R. Rausa, E. Mazzolari and V. Calemma, *J. Chromatogr.*, 1991, **541**, 419.
17. J. Peuravuori and K. Pihlaja, *Anal. Chim. Acta*, 1997, **337**, 133.

18. B. Ballarin, R. Seeber, D. Tonelli and S. Zappoli, *Annali Chimica (Rome)*, 1999, **89**, 211.
19. R. S. Swift, in 'Methods of Soil Analysis', part 3, Chemical Methods, SSSA Book Series no. 5, Madison, Wisconsin, 1996, p. 1011.
20. A. Piccolo, S. Nardi and G. Concheri, *Eur. J. Soil Sci.*, 1996, **47**, 319.
21. P. Conte and A. Piccolo, *Environ. Sci. Technol.*, 1999, **33**, 1682.
22. A. Piccolo, P. Conte and A. Cozzolino, *Eur. J. Soil Sci.*, 1999, **50**, 511.
23. A. Piccolo and P. Conte, *Adv. Environ. Res.*, 2000, **3**, 508.
24. I. P. Kenworthy and M. H. B. Hayes, in 'Humic Substances, Peats and Sludges. Health and Environmental Aspects', M. H. B. Hayes and W. S. Wilson, (eds.), Royal Society of Chemistry, Cambridge, 1997, p. 39.
25. W. W. Yau, J. J. Kirkland and D. D. Bly, 'Modern Size Exclusion Chromatography', Wiley, New York, 1979, p. 318.
26. J. Aho and O. Lehto, *Arch. Hydrobiol.*, 1984, **101**, 21.
27. P. Conte and A. Piccolo, *Chemosphere*, 1999, **38**, 517.
28. P. Conte, A. Piccolo, B. van Lagen, P. Buurman and P. A. de Jager, *Geoderma*, 1997, **80**, 327.
29. S. Rokushika, T. Ohkawas and H. Hatano, *J. Chromatogr.*, 1980, **190**, 297.
30. R. S. Swift and A. M. Posner, *J. Soil Sci.*, 1971, **22**, 237.
31. R. Beckett, Z . Jue and J. C. Giddings, *Environ. Sci. Technol.*, 1987, **21**, 289.
32. B. Lange and Z. J. Vejdelek, 'Photometrische Analyse', Verlag, Weinheim, 1979, p. 7.
33. C. R. Cantor and P. R. Schimmel, 'Biophysical Chemistry. Part II: Techniques for the Study of Biological Structure and Function', Freeman, New York, 1980, p. 399.
34. D. Freifelder, 'Physical Biochemistry', 2nd ed., Freeman, New York, 1982, p. 500.
35. R. L. Wershaw, *J. Contam. Hydrol.* 1986, **1**, 29.
36. J. -M. Lehn, 'Supramolecular Chemistry', VCH, Weinheim, 1995.

CHARACTERIZATION OF THE 'FLUORESCENT FRACTION' OF SOIL HUMIC ACIDS

M. Aoyama,[1] A. Watanabe[2] and S. Nagao[3]

[1] Faculty of Agriculture and Life Science, Hirosaki University, Hirosaki, 036-8561 Japan
[2] Graduate School of Bioagricultural Sciences, Nagoya University, Nagoya, 464-8601 Japan
[3] Department of Fuel Cycle Safety Research, Japan Atomic Energy Research Institute, Tokai-mura, Ibaraki, 319-1195 Japan

1 INTRODUCTION

The fluorescence of humic acids (HAs) is a general phenomenon. Thus far, it has been believed that the fluorescence of HAs is derived from the humic substances themselves, in particular from the condensed aromatic structures of the humic molecules.[1-3] However, Aoyama[4] provided evidence that the fluorescence was mainly due to the minor constituents of the HAs when high performance size exclusion chromatography (HPSEC) with UV and fluorescence detection was applied. Furthermore, Aoyama[5] succeeded in separating the fluorescent substances-rich fraction from the soil HAs using Sephadex G-25 gel chromatography. This method allowed collection of the fluorescent substances in large enough quantities for their characterization.

In the present study we separated the fluorescent substances-rich fraction from the HAs prepared from two types of soils by Sephadex G-25 gel chromatography and characterized the fraction using HPSEC, three-dimensional excitation emission matrix (3-D EEM) spectroscopy, UV-Vis absorption spectroscopy and solid-state cross-polarization magic-angle-spinning (CPMAS) ^{13}C NMR spectroscopy.

2 MATERIALS AND METHODS

2.1 HA Samples

The HA samples used were prepared from Entisol (Fujisaki soil) and Andisol (Takizawa soil). The HAs were extracted twice from the soil samples corresponding to 500 mg C with 150 mL of 0.1 M NaOH and separated by acidification to pH 1.0 with HCl. The HA precipitate was dissolved in 150 mL of 0.1 M NaOH and precipitated again by acidification to pH 1.0 with HCl. The dissolution-precipitation cycle was repeated twice. The resultant precipitate was dialyzed against distilled deionized water and freeze-dried. The total carbon and nitrogen contents of the HA samples were simultaneously determined using a Sumigraph-90A automatic analyzer. The total carbon and nitrogen contents and the C/N ratio are listed in Table 1.

Table 1 *HA samples used*

Sample	Origin	% T-C	%T-N	C/N ratio
Fujisaki	Brown lowland soil (*Udifluvent*)	43.3	5.09	8.5
Takizawa	Allophanic Andosol (*Melanudand*)	51.8	3.80	13.6

2.2 Sephadex Column Chromatography

To fractionate the HA samples by Sephadex column chromatography, 250 mg of each HA sample was dissolved in 25 mL of 0.1 M NaOH and adjusted to pH 8.0 with HCl, then distilled deionized water was added to bring the total volume to 50 mL. A 15 mL volume of the sample solution was applied onto a glass column (65 mm I. D. x 500 mm) packed with Sephadex G-25 (fine grade; Pharmacia) and eluted with distilled deionized water. The column effluent was collected with an Advantec SF-2120 fraction collector. All the procedures were carried out at 10°C. The Sephadex gel chromatography resulted in HA separation into two fractions. The slower-moving fraction (second fraction) was rich in fluorescent substances as described in a previous study.[5] Therefore, we referred to the first fraction as the 'humic fraction' and to the second fraction as the 'fluorescent fraction'. Both fractions were adjusted to pH 1.0 with HCl, and the resultant precipitates were washed with distilled deionized water and freeze-dried. The total carbon and nitrogen contents of the fractions were determined using the Sumigraph-90A automatic analyzer.

2.3 HPSEC

HPSEC was performed with the same equipment and operating conditions as used in our previous study[5] with the exception of the columns. In the present study, two Asahipack GS-220HQ columns (7.6 mm I. D. x 300 mm; Showa Denko) linked in series were used. The nominal molecular weight at void volume (V_o) of the columns was 3 kDa for pullulans. The sample solutions of the HAs and their fractions used for HPSEC were prepared as described in the previous study (pH 8.0, 50 mg C L^{-1}).[5] The chromatograms were monitored by a) absorbance at 280 nm, and b) by fluorescence at an excitation wavelength of 460 nm and an emission wavelength of 520 nm.

2.4 3-D EEM Spectroscopy

The 3-D EEM spectra were measured for the HA samples and their fractions dissolved in 0.01 M $NaClO_4$ solution (10 mg L^{-1}) using a Hitachi F-4500 scanning spectrofluorimeter.[6] The spectra were recorded over both the excitation and emission wavelength ranges of 200-600 nm. The relative fluorescence intensity (RFI) was expressed as the fluorescent intensity relative to that of quinine sulfate solution (0.01 mg L^{-1} in 0.05 M H_2SO_4).

2.5 UV-Vis Spectroscopy

The UV-Vis absorption spectra were measured for the HA samples and their fractions dissolved in 0.01 M NaOH solution (50 mg C L^{-1}) over the wavelength range 220-700 nm with a Hitachi U-2000 spectrophotometer equipped with a 1-cm quartz cell.

2.6 CPMAS ^{13}C NMR Spectroscopy

Solid state CPMAS ^{13}C NMR spectra of the HA samples and their fractions were recorded at 75.57 MHz on a Chemagnetics CMX-300 NMR spectrometer. The parameters used were: spinning rate 5,000 Hz; acquisition time 25.6 ms; contact time 1 ms; pulse delay time 700 ms; line broadening 50 Hz. The number of scans varied from 3,000 to 90,000 depending on the individual samples. The NMR spectra were divided into four chemical shift regions according to the chemical types of carbon as follows: 0-46 ppm (alkyl carbon), 46-110 ppm (*O*-alkyl carbon), 110-165 ppm (aromatic carbon) and 165-210 ppm (carbonyl carbon).[7] The percentage distribution of carbon among the chemical types was calculated by excluding contributions from spinning side bands.[8]

3 RESULTS AND DISCUSSION

3.1 Distribution of Carbon between the Humic and Fluorescent Fractions

For the Fujisaki HA, only 3% of the total carbon was recovered in the fluorescent fraction by the Sephadex G-25 column chromatography (Table 2). On the other hand, for the Takizawa HA, six times more carbon was recovered in the fluorescent fraction. Thus, the Andisol HA contained much more of the fluorescent fraction compared to the Entisol HA. The total carbon recoveries were around 80% irrespective of the origin of the HA samples.

Table 2 *Percentage distribution of carbon between the humic and fluorescent fractions*

Sample	Humic fraction	Fluorescent fraction	Recovery
Fujisaki	74.6	2.8	77.4
Takizawa	59.9	18.1	78.0

3.2 HPSEC

When HPSEC is applied to soil HAs, the substances responsible for the UV absorption can be regarded as humic substances, as described in the previous study.[5] For the whole sample of Fujisaki HA, the humic substances were eluted at the V_o and just after the V_o (Figure 1). In contrast, for the whole sample of Takizawa HA, a small peak appeared at the V_o, but most of the humic substances were eluted in an elution volume range larger than V_o. Thus, the humic substances in the Andisol HA were characterized by a smaller molecular size compared with those in the Entisol HA. When monitored by fluorescence, the fluorescent substances were eluted considerably later than the humic substances and resolved into many peaks for both of the whole HA samples. Each fluorescent peak appeared at the same elution volume irrespective of the HA samples used. However, the peak intensities of the fluorescent substances were significantly higher in the Takizawa HA than in the Fujisaki HA. These results confirm the previous study.[5]

The humic fraction resembled the whole sample in the elution profile of the humic substances (Figure 1). When monitored by fluorescence, peaks appeared in the same elution volume range as that for the humic substances, indicating that the humic substances in the soil HAs exhibited fluorescence. However, the peak maximum of fluorescent substances appeared significantly later than that of humic substances. Thus,

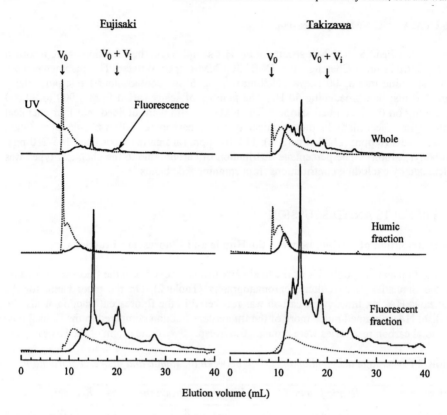

Figure 1 *Size exclusion chromatograms of whole HA samples and their fractions*

the fluorescence intensity increases with decreasing size of the humic molecules.

For the fluorescent fraction, the elution profiles were characterized by intense fluorescent peaks compared with whole samples and humic fractions (Figure 1). The peak intensities of the fluorescent substances were relatively stronger for the Takizawa HA than for the Fujisaki HA, indicating that the fluorescent intensity of the fluorescent fraction varied with the soils. In addition, the fluorescent fraction showed the presence of UV-absorbing substances. When monitored by UV absorption, a broad peak appeared earlier than the fluorescent peaks irrespective of the HA samples used. This fact suggests that the fluorescent fraction contained non-fluorescent substances along with fluorescent substances.

3.3 3-D EEM Spectra

For both of the whole HA samples, the fluorescence maximum was located at an excitation wavelength of 450 nm and at an emission wavelength of 535 nm (Figure 2, Table 3). The humic fraction had a fluorescence maximum at an excitation wavelength of 450-455 nm and at an emission wavelength of 535-545 nm. For the fluorescent fraction, the peak maximum was located at an excitation wavelength of 450 nm and at an emission wavelength of 530-535 nm irrespective of the HAs used. The fluorescent fraction of the Fujisaki HA had a minor fluorescent peak at an excitation wavelength of 360 nm and at an emission wavelength of 510 nm, whereas that of the Takizawa HA had no such fluorescent peak.

While the fluorescence maxima were located at nearly the same position, the RFI at the peak maximum differed widely between the humic and fluorescent fractions (Table 3). The RFI was 8-9 times stronger for the latter than for the former. In addition, the RFI was generally stronger for the Takizawa HA than for the Fujisaki HA. The fluorescent fraction of the Takizawa HA fluoresced more intensely (1.8 times at the fluorescent maximum) than that of the Fujisaki soil, agreeing with the HPSEC results.

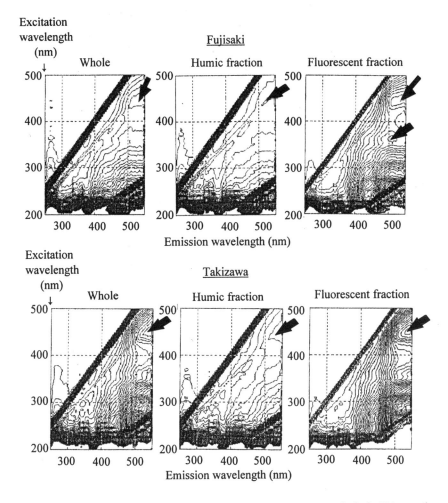

Figure 2 *Three-dimensional excitation emission matrix spectra of whole HA samples and their fractions. The arrows in the figure indicate the positions of the fluorescence maxima*

3.4 UV-Vis Absorption Spectra

The absorption spectra of the whole HA samples showed decreasing absorbance at the longer wavelengths and had a shoulder at around 280 nm (Figure 3). Similar absorption spectra were obtained for the humic and fluorescent fractions. However, the absorbance was generally higher (1.5-1.8 times higher at 280 nm) for the fluorescent than for the humic fractions. Swift et al.[9] observed that the optical absorptivity of HA in the UV and visible regions increased as the molecular weight decreased and attributed the increase to an increase in the degree of aromaticity. Therefore, it is suggested that higher absorbance in the UV region observed for the fluorescent fraction is due to a higher degree of aromaticity.

3.5 CPMAS ^{13}C NMR spectra

The CPMAS ^{13}C NMR spectra of the whole HA samples significantly differed between the two soils (Figure 4). The ^{13}C NMR spectrum of the Fujisaki HA was characterized by intense peaks of alkyl (0-46 ppm), O-alkyl (46-110 ppm) and carbonyl (165-210 ppm) carbon, whereas that of the Takizawa HA was predominated by a peak of aromatic carbon (110-165 ppm). The predominance of aromatic carbon in Andisol HAs has been recognized in the studies of Tate et al.[10] and Golchin et al.[7] using CPMAS ^{13}C NMR spectroscopy. The spectra of the whole sample and humic fraction were very similar to each other. In contrast, the fluorescent fraction showed a spectrum distinct from those of the whole sample and humic fraction for both HAs. However, the spectra were similar between the two HAs, and were characterized by a very intense peak of aromatic carbon at around 130 ppm and weak peaks due to alkyl and O-alkyl carbon.

When the percentages of the different types of carbon contained in NMR spectra were calculated (Table 4), the aromatic carbon accounted for nearly 75% of the total carbon in the fluorescent fractions. Thus, CPMAS ^{13}C NMR spectroscopy revealed the highly aromatic structure of the fluorescent fraction of the soil HAs as inferred from the intense UV absorption. This highly aromatic character would be responsible for the intense fluorescence of the fraction.

Although the ^{13}C NMR spectra of the fluorescent fractions were similar to each other, the fluorescent fraction of the Fujisaki HA had a slightly higher percentage of O-alkyl carbon and a slightly lower percentage of carbonyl carbon compared with the fluorescent fraction of the Takizawa HA. This fact suggests that the chemical structures in the fluorescent fraction varied slightly with soil type. The differences in the carbon chemistry appeared to be reflected by the differences in the HPSEC elution profile and 3-D EEM spectra.

Table 3 *Fluorescent properties of HA samples and their fractions*

Sample	Fraction	Solution pH	Fluorescence maximum		RFI
			Excitation (nm)	Emission (nm)	
Fujisaki	Whole	8.7	450	535	2.26
	Humic	8.3	450	535	1.01
	Fluorescent	8.2	450	530	8.41
Takizawa	Whole	8.5	450	535	5.63
	Humic	8.4	455	540	1.62
	Fluorescent	8.2	450	535	15.2

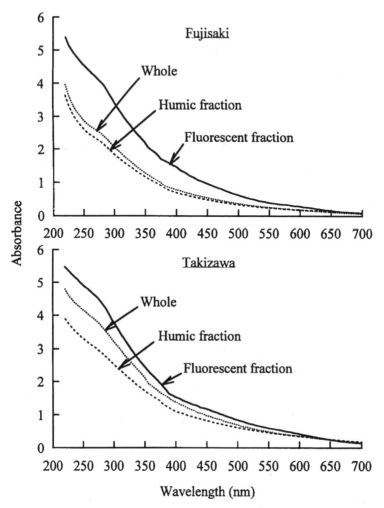

Figure 3 *UV-Vis absorption spectra of whole HA samples and their fractions*

Table 4 *Percentages of different types of carbon contained in CPMAS ^{13}C NMR spectra*

Sample	Fraction	Alkyl	O-alkyl	Aromatic	Carbonyl
Fujisaki	Whole	23.0	30.4	31.2	15.3
	Humic	23.5	31.7	27.7	17.2
	Fluorescent	10.4	12.2	74.1	3.4
Takizawa	Whole	11.9	15.0	65.3	7.7
	Humic	10.4	19.8	64.5	5.3
	Fluorescent	8.0	5.0	76.4	9.7

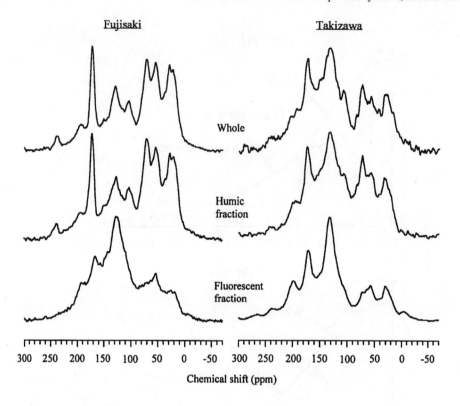

Figure 4 *CPMAS ^{13}C NMR spectra of whole HA samples and their fractions*

4 CONCLUSIONS

When Sephadex G-25 column chromatography was used for the separation of the humic and fluorescent fractions in the HA samples, the percentage of carbon recovered in the fluorescent fraction varied widely between the Entisol (3%) and Andisol (19%). The fluorescent substances were concentrated in the fluorescent fraction, and the RFI at the peak maximum was 8-9 times stronger for the fluorescent than for the humic fractions. Although the two HA samples significantly differed in their molecular size and carbon chemistry, the chemical characteristics of the fluorescent fraction were similar between the two HA samples. The fluorescent fraction of the soil HAs was characterized by the dominance of aromatic carbon. It was estimated from the CPMAS ^{13}C NMR spectra that approximately 75% of the carbon in the fluorescent fraction was aromatic carbon. The highly aromatic structure of the fluorescent fraction would be responsible for the intense fluorescence of the fluorescent fraction. Further research is needed to elucidate the origin of such an aromatic fraction.

ACKNOWLEDGEMENTS

We thank Mr. M. Ohnuma and Mr. M. Hoashi for technical assistance.

References

1. W. R. Seitz, *Trends Anal. Chem.*, 1981, **1**, 79.
2. T. I. Balkas, O. Basturk and A. F. Gaines, *Fuel*, 1983, **62**, 373.
3. N. Senesi, T. M. Miano, M. R. Provenzano and G. Brunetti, *Soil Science*, 1991, **152**, 259.
4. M. Aoyama, 'Proceedings of the 9th International Meeting of the International Humic Substances Society', Adelaide, in press.
5. M. Aoyama, in 'Understanding Humic Substances: Advanced Methods, Properties and Applications', E. A. Ghabbour and G. Davies, (eds.), The Royal Society of Chemistry, Cambridge, 1999, p. 179.
6. Y. Suzuki, S. Nagao, T. Matsunaga, H. Amano, A. V. Kovalev and Y. V. Tkatchenko, in 'The Role of Humic Substances in the Ecosystems and Environmental Protection', J. Drozd, S. S. Gonet, N. Senesi and J. Weber, (eds.), IHSS-Polish Society of Humic Substances, Wroclaw, 1997, p. 623.
7. A. Golchin, P. Clarke, J. A. Baldock, T. Higashi, J. O. Skjemstad and J. M. Oades, *Geoderma*, 1997, **76**, 155.
8. R. H. Newman, B. K. G. Theng and Z. Filip, *Sci. Total Environ.*, 1987, **65**, 69.
9. R. S. Swift, B. K. Thornton and A. M. Posner, *Soil Sci.*, 1970, **110**, 93.
10. K. R. Tate, K. Yamamoto, G. J. Churchman, R. Meinhold and R. H. Newman, *Soil Sci. Plant Nutr.*, 1990, **36**, 611.

INVESTIGATIONS OF HUMIC MATERIALS AGGREGATION WITH SCATTERING METHODS

James A. Rice,[1] Thomas F. Guetzloff[1,2] and Etelka Tombácz[3]

[1] South Dakota State University, Department of Chemistry & Biochemistry, Brookings, SD 57007-0896, USA
[2] Current Address: Chemistry Department, West Virginia State University, Institute, WV 25112, USA
[3] Department of Colloid Chemistry, Attila József University, Szeged, Hungary

1 INTRODUCTION

Natural organic matter consists of an ill-defined assemblage of organic molecules composed primarily of what are known as the humic materials. Humic materials are divided into three operationally defined fractions based on their solubility as a function of pH: fulvic acid is soluble in an aqueous solution at any pH value, humic acid is soluble in an aqueous solution whose pH > 7, and humin is insoluble in an aqueous solution at any pH value.

The components of a humic fraction can interact in different ways to form aggregates with physical and chemical properties different than those of the original components. Several mechanisms have been proposed to describe the aggregation of humic substances. A number of models treat humic materials as colloids and describe their aggregation in those terms.[1-3] For example, humic materials aggregate under low pH or high ionic strength conditions, or disaggregate at high pH values or in systems with low ionic strengths. Other models describe intermolecular aggregation processes resulting in the formation of micellar-like structures.[4-6] In all of these models the size, shape and aggregate morphology are very different from those of the unaggregated components as generalized in Figure 1.

Most chemical characterization methods used to explore the nature of humic materials (for example, NMR or mass spectrometry) implicitly assume that the material being studied is a pure substance. Because of its chemical heterogeneity, the best that usually can be obtained from a chemical characterization is an average description of the molecules that comprise the humic sample. While this approach has applications in the description of various chemical and geochemical processes, it is a distinct limitation when the objective is to describe the aggregation behavior of humic materials.

This paper discusses the application of light scattering techniques to examine aggregation induced by humic material concentration increases, presence and concentration of metal cations, pH and ionic strength effects. Analysis of scattering data with fractal geometry provides insight into the changes in morphology that occur as a result of aggregation. Within the particle-size range accessible by small angle X-ray scattering (SAXS) and laser light scattering (LLS), many of these aggregation processes do not produce a significant change in particle morphology.

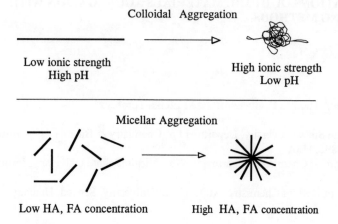

Figure 1 *Models of humic acid (HA) and fulvic acid (FA) aggregation*

2 MATERIALS AND METHODS

2.1 Samples

Humic acid (HA) was isolated from a peat soil using a traditional alkaline extraction procedure. The bulk peat material and extraction procedures are described in detail elsewhere.[7] Solutions of HA were made by dissolving enough material in aqueous NaOH to give a range of solution HA concentrations up to ~12.3 gm HA per liter.

2.2 Aggregation Studies

Carbon-14 labeled DDT (2,2-bis(4-chlorophenyl)-1,1,1-trichloroethane) was used as a probe to establish micelle formation in the samples. The solubilized probe was quantified via scintillation counting. Solution surface tension was measured with a du Nouy ring tensiometer. Other experimental details have been described elsewhere.[8]

2.3 Scattering Measurements

SAXS measurements were performed at the Center for Small-Angle Scattering at Oak Ridge National Laboratory (Oak Ridge, TN). The instrument has been described by Wignall et al.[9] An X-ray wavelength of $\lambda = 1.54$ Å was employed for all experiments. The methods employed have been described in detail elsewhere.[10,11] Dynamic laser light scattering (LLS) experiments were performed on an ALV-5000/E light scattering apparatus fitted with an argon-ion laser operated at 514.5 nm. For dynamic light scattering (DLS) data analysis, correlation functions were analyzed by cumulant and Contin analysis and the z-average particle radius was calculated using the method described by Martin and Leyvraz[12] and Martin.[13] The methods we employed have been described in detail elsewhere.[14]

2.4 Fractal Analysis of Scattering Data

The analysis of scattering data using fractal geometry is described in detail by Schmidt.[15] The application of fractal geometry to the study of humic materials has been reviewed.[16,17] The methods we have employed to extract the fractal dimension from scattering data obtained from humic acid solutions are described in detail elsewhere.[10,11] For this study only a single type of fractal, mass fractals, is relevant. Mass fractals are objects whose mass and surface follow the same scaling behavior.[18] A typical example of a mass fractal is a branching dendritic structure. In solution, humic acid has been shown to be a mass fractal.[10,11] Uncertainty associated with measurements of the fractal dimension is ± 0.1 units.

3 RESULTS AND DISCUSSION

Figure 2 shows the effect of changing pH and ionic strength on the size of the humic acid aggregates. As ionic strength increases at low pH, aggregate size increases as a result of inter-particle aggregation (as generalized in the legend of the graph). At this pH value the particles are compact, probably as a result of intra-molecular interactions such as hydrogen bonding. These types of interactions, which are in effect an intra-molecular aggregation, have been used to describe a variety of humic acid/contaminant interactions such as metal binding or hydrophobic organic contaminant solubilization.[19] At high pH values, humic acid aggregates initially are expanded, dendritic structures that are mass fractals but collapse as the ionic strength increases as depicted in the legend. As ionic strength continues to increase, inter-particle aggregation occurs and the aggregates increase in size.

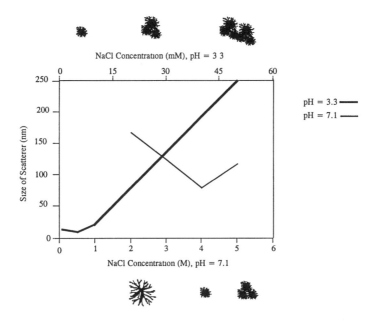

Figure 2 *Effect of pH and ionic strength on humic acid particle size. Graphics depict possible changes occurring during aggregation process*

Even when aggregated, the samples exhibit the scattering behavior of a mass fractal. All of these observations are in agreement with the behavior previously described by Ghosh and Schnitzer[1] (Figure 1).

Figure 3 shows the effect of increasing HA concentration on the solution's surface tension and the apparent solubility of DDT. Surface tension decreases until a point where it remains constant. The apparent water solubility of DDT remains essentially constant until the same point where a sudden increase is observed. The coincidence of these two points is definitive proof for the formation of a micelle[20] (Figure 1). The coincidence of these points represents the onset of micelle formation, and is referred to as the "critical micelle concentration" or CMC. For this sample, the CMC is ~8 gm HA/L, a value typical of humic acids.[8,21-23]

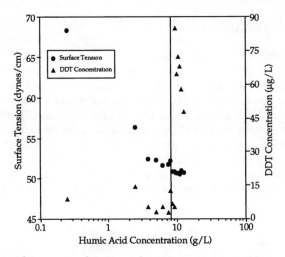

Figure 3 *Effect of increasing humic acid concentration on solution surface tension and apparent DDT solubility*

Both of these processes result in the formation of aggregates whose morphologies are quite different from those of the starting sample (Figure 1). For these reasons it should be possible to observe these transformations via changes in the fractal dimension.

Figure 4 shows the effect of pH on the fractal dimension (D) of the peat humic acid. Within the uncertainty of these measurements there is no difference in the fractal dimensions of the particles at any of the pH values. As shown in Figure 2, the extent of aggregation varies with pH. This is consistent with the colloidal behavior exhibited by humic acid. Yet there is no evidence of an expected transition from one particle morphology to another (Figure 1) in the SAXS measurements that might be expected.

Figure 5 is a plot of scattering data from humic acid solutions of varying concentrations. Within the uncertainty of these measurements there also is essentially no difference in the fractal dimensions of the particles at each concentration. While no difference would be expected between the first three solutions, the final concentration is well above the CMC (Figure 3) and should contain a significant number of aggregated particles whose fractal dimension should be markedly different than for the sample before the CMC (Figure 1). There is no evidence of an abrupt transition between D_C and D_A in Figure 5 that would be expected to accompany aggregation occurring at and beyond the CMC.

It can be shown with scattering techniques that pH, ionic strength and concentration increases will produce aggregation in humic acid solutions. But the changes in particle morphology that should accompany aggregation are not observed in the measured fractal dimensions. There are at least two possible explanations for this disparity. First, it could be that the characterization length-scale of SAXS is smaller than the humic acid aggregates. This would result in the SAXS "seeing" only the interior mass of the particle and not the entire particle. To assess this possibility it would be necessary to combine SAXS and static light scattering data for humic acid solutions of identical compositions in an attempt to determine the particle size limits in the characterization of humic acids by scattering techniques. A second possible explanation is that only a fraction of the components that comprise humic acid actually are involved in the aggregation phenomena. Thus, the aggregates might contribute little to the overall scattering interactions that occur in SAXS. In this case the aggregates' contribution to the overall scattering behavior could be lost in a much more intense background. This would be consistent with the heterogeneity associated with humic acid.[24,25] Both of these possibilities currently are under investigation.

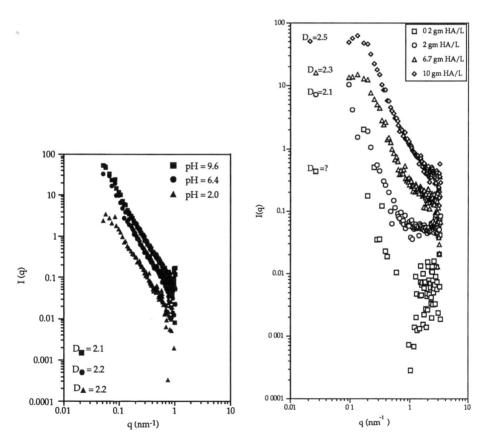

Figure 4 *Effect of pH on fractal dimention D values of a peat humic acid. 10 mM added NaCl*

Figure 5 *Effect of HA concentration on D for a peat humic acid. pH = 10.2*

4 SUMMARY

With dynamic light scattering and solubilization of a hydrophobic probe it can be shown that humic acids behave in the manner of both a colloid and a micelle-forming surfactant. Based on small-angle X-ray scattering data, these aggregation processes do not produce a change in particle morphology.

ACKNOWLEDGEMENTS

Funding for this work was provided by National Science Foundation under grants OSR-9452894 and the South Dakota Future Fund (ET and JAR) and the US Department of Agriculture via agreements 91-37102-6864 (TFG) and 98-35107-6515 (JAR). South Dakota EPSCoR and the South Dakota Future Fund provided funding for acquisition of the laser light scattering apparatus.

References

1. K. Ghosh and M. Schnitzer, *Soil Sci.*, 1980, **129**, 266.
2. E. Tombácz and I. Regdon, in 'Humic Substances in the Global Environment and Implications on Human Health', N. Senesi and T. M. Miano, (eds.), Elsevier, Amsterdam, 1994, p. 139.
3. E. Tombácz and E. Meleg, *Org. Geochem.*, 1990, **15**, 375.
4. R. L. Wershaw, *J. Contam. Hydrol.*, 1986, **1**, 29.
5. R. L. Wershaw, *Environ. Sci. Technol.*, 1993, **5**, 814.
6. T. F. Guetzloff and J. A. Rice, *Sci. Total Environ.*, 1994, **152**, 31.
7. J. A. Rice, PhD Thesis, Colorado School of Mines, 1987.
8. T. F. Gueztloff, PhD Thesis, South Dakota State University, 1996.
9. G. D. Wignall, J. S. Lin and S. Spooner, *J. Appl. Cryst.*, 1990, **23**, 241.
10. J. A. Rice and J. S. Lin, *Environ. Sci. Technol.*, 1993, **27**, 413.
11. J. A. Rice and J. S. Lin, in 'Humic Substances in the Global Environment and Implications on Human Health', N. Senesi and T. M. Miano, (eds.), Elsevier, Amsterdam, 1994, p. 115.
12. F. Leyvraz and J. E. Martin, *Phys. Rev. A*, 1986, **34**, 2346.
13. J. E. Martin, *Phys. Rev. A*, 1987, **36**, 3415.
14. E. Tombácz, J. Ren and J. A. Rice, *Phys. Rev. E*, 1996, **53**, 2980.
15. P. Schmidt, in 'The Fractal Approach to Heterogeneous Chemistry', D. Avnir, (ed.), Wiley Interscience, Chichester, 1989, p. 67.
16. N. Senesi, in 'Humic Substances in the Global Environment and Implications on Human Health', N. Senesi and T. M. Miano, (eds.), Elsevier, Amsterdam, 1994, p. 3.
17. N. Senesi and L. Boddy, in 'Interactions between Soil Particles and Microorganisms and their Impact on the Terrestrial Environment,' P. M. Huang, J.-M. Bollag and N. Senesi, (eds.), IUPAC, 2000, in press.
18. B. Mandelbrot, 'The Fractal Geometry of Nature', Freeman, New York, 1994.
19. L. M. Yates and R. von Wandruska, *Soil Sci. Soc. Am. J.*, 1999, **63**, 1645.
20. C. Tanford, 'The Hydrophobic Effect: Formation of Micelles and Biological Membranes', Wiley, New York, 1980.
21. M. Tschapek and C. Wasowski, *Geochim. Cosmochim. Acta*, 1973, **37**, 2459.

22. W. Rochus and S. Sipos, *Agrochim.*, 1978, **22**, 446.
23. K. Hayase and H. Tsubota, *Geochim. Cosmochim. Acta*, 1983, **47**, 947.
24. P. MacCarthy and J. A. Rice, in 'Humic Substances in Soil, Sediment, and Water: Geochemistry, Isolation, and Characterization', G. R. Aiken, D. M. McKnight, R. L. Wershaw and P. MacCarthy, (eds.), Wiley, New York, 1985, p. 527.
25. P. MacCarthy and J. A. Rice, in 'Scientists on Gaia', S. Schneider and P. J. Boston, (eds.), MIT Press, Cambridge, MA, 1991, p. 339.

APPLICATION OF MALDI-TOF-MS TO THE CHARACTERIZATION OF FULVIC ACIDS

G. Haberhauer,[1] W. Bednar,[1] M. H. Gerzabek[1] and E. Rosenberg[2]

[1] Department of Environmental Research, Austrian Research Centers, A-2444 Seibersdorf, Austria
[2] Institute of Analytical Chemistry, Vienna University of Technology, A-1060 Vienna, Austria

1 INTRODUCTION

Humic substances (HSs) are multifunctional aromatic components linked chemically and physically by a variety of aliphatic constituents.[1] A broad, polydisperse size distribution is characteristic of HSs.[2-4] The formation of HSs in a random transformation process results in a large variety of molecules from chemical and dimensional points of view. Agglomerations of such products by chemical and physical interactions form HSs clusters (Figure 1).[3] The size of such clusters seems to depend on several environmental factors.[5] HSs are divided by solubility into three fractions, humin, humic acids and fulvic acids (FAs).

Due to the complexity of HSs, molecular analysis is limited to average sum parameters. Structural properties of HSs are best described as distribution probabilities of certain functional groups. The deduction of structural models based on the available functional moiety information is restricted to hypotheses.[6-8]

The present knowledge of HSs structure is mainly derived from NMR,[9-11] vibrational spectroscopy[12-14] and pyrolysis mass spectrometry (py-MS) methods.[15] All these methods are powerful for determination of average amounts of certain functional groups or moieties and for studying HSs dynamic processes. Py-MS has been used extensively in the study of humics, primarily as a means of obtaining information about fragments after thermal degradation. Virtually all early mass spectrometric studies of humic substances were characterized by the extensive fragmentation produced by conventional electron impact ionization, leading to fragment ions at almost every nominal mass below mass-to-charge ratio m/z 200 and some few ions above m/z 200.

Several types of so called soft ionization mass spectrometry techniques have been applied to the characterization of HSs. While fast atomic bombardment,[16] laser desorption and laser ablation mass spectrometry did not yield useful results,[16] both electrospray ionization and MALDI-TOF MS seem to be more promising.[17,18]

Fievre at al.[19] reported mass spectra with ions at almost every nominal value, 200 < m/z < 2000 for FAs using electrospray ionization Fourier transform ion cyclotron resonance mass spectrometry (ESI FT-ICR MS). A broad and featureless m/z distribution was obtained by Brown and Rice using ESI FT-ICR MS.[18] Spray solution composition was found to have a dramatic effect on the ion distributions, with high-mass aggregates being formed in less polar spray solutions. Positive-ion spectra for each FA resulted in number-average molecular weights ranging from 1700 to 1900 Da. The reported ESI FT-ICR MS spectra were extremely complex, with ion distributions on the order of m/z ~ 500-3000. The presence of more than one ion at each nominal mass routinely was observed. Negative-ion ESI analysis of FA samples resulted in the observation of multiply charged ions. Coupling of size-exclusion chromatography to ESI mass spectrometry revealed similar results.[20]

A gas-phase hydrogen/deuterium exchange reaction to determine the number of active hydrogens in fulvic acid ions using ESI FT-ICR MS was described by Solouki et al.[21] The authors reported an average maximum number of active hydrogens for fulvic acid ions at m/z region 700-1000 to be 7-9. No significant differences were obtained between fulvic acids of different origin.

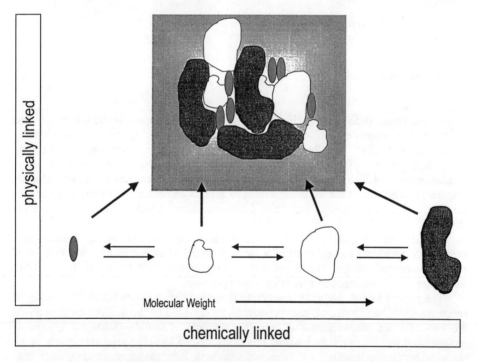

Figure 1 *Conceptual model for humic substances*

Matrix assisted laser desorption/ionization time of flight mass spectrometry[22] (MALDI-TOF MS) already has been applied to humic substances.[17,23,24] The principle of MALDI-TOF MS is shown in Figure 2. The energy of a laser pulse is absorbed by the co-crystallized sample/matrix mixture. This leads to desorption and ionization of both sample and matrix molecules. The ions are accelerated and separated according to their mass/charge ratio by a time-of-flight mass analyzer. Although the mechanism of ion formation is still not clear, it is thought to proceed via charge transfer processes after desorption. Another formation process for ions is the desorption of already preformed ions from the matrix.[25]

It is important to note that both molecular ions and cluster ions are detected. Cluster ions can consist of molecules adhering to each other by either specific interactions such as

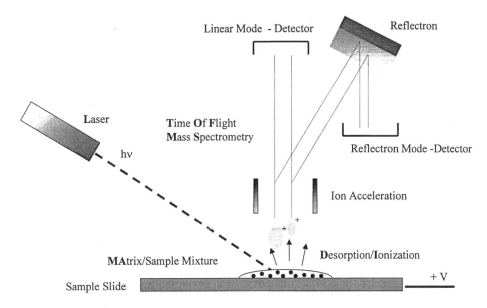

Figure 2 *Principle of MALDI-TOF MS*

hydrogen bond interactions or by non-specific interactions.[26,27]

The main advantages of MALDI-TOF MS (soft ionization, formation of singly charged ions and high mass range detection) are opposed by several limitations concerning the analysis of HSs. These are (1) impurities such as high inorganic salt concentration, which are known to reduce the ion yield;[28] (2) the low solubility of mainly uncharged HSs; and (3) the nature of HSs themselves. HSs consist of a vast number of macromolecules with a wide mass range distribution. This excludes the analysis of single molecules. Even if MALDI-TOF MS enables femtomol analysis[29] of pure substances, analysis of a mixture containing thousands of different molecules with each at low concentration certainly will produce a detection problem. Thus, the wide mass range distribution of HSs requires adoption and improvement of conventional sample preparation. The application of MALDI-TOF MS to HSs could give information on m/z distribution for use as a fingerprint identification.[17,24]

The aim of this work was to apply MALDI-TOF MS to a set of FAs of different origin and to study the effect of experimental parameters on the MALDI-TOF MS spectra.

2 MATERIAL AND METHODS

Three different types of fulvic acids were used. A soil fulvic acid and Suwannee River fulvic acid standard were purchased from the International Humic Substance Society <http://www.ihss.gatech.edu>. A third fulvic acid was extracted from the Ap-horizon of an Austrian agricultural soil (Chernozem) using sodium pyrophosphate procedures as described in ref. 30. Two different types of sample preparations were investigated. First the FA samples were dissolved in 75 % water, 25 % acetonitrile and 0.1% trifluoroacetic acid until saturation. Then the solution was applied onto the sample slides, either together with matrix solution (see below for composition) or without. Second, the FAs were mixed and suspended with an organic solvent like chloroform or acetonitrile. The suspensions were directly applied onto the sample slides. The samples were placed into the MALDI-TOF MS instrument after evaporation of the solvents.

All mass spectra were acquired on a Kratos Kompact MALDI III instrument operated at an acceleration voltage of 20 kV, using both linear and reflector mode. The samples were desorbed/ionized from the probe tip using a pulsed nitrogen laser (3 nsec pulse duration) with an output wavelength of 337 nm and maximum output of 6 mW. Fifty laser shots were used for each sample and the signals were averaged. All spectra were recorded using a positive ion mode. Calibration was performed with peptides and nucleosides with known molecular masses (angiotensin I, insulin, cytidine, ubiquitin and cytochrome C – all from Sigma) with an amount of 50 pmol. The matrices (sinapinic acid, gentisic acid all from Sigma) and standards were dissolved in 75% aqueous acetonitrile (Merck) with 0.1% trifluoroacetic acid (Fluka).

Aqueous size exclusion chromatography was conducted with a SigmaChromTM GFC-100 column filled with cross-linked polysaccharides (particle size 12-15 µm). The IHSS soil fulvic acid was dissolved in eluent (1 mg substance/mL solvent). The flow rate was 0.5 mL/min and the injection volume was 20 µL. A Beckman Instruments HPLC-system Gold equipped with a UV-Detector (280 nm) was used. A triethylamine – HCl (50 mM)/potassium chloride (100 mM) buffer system at pH 7.5 was used as eluent. Calibration of the column was carried out with the same protein standards as for MALDI-TOF MS.

Calculation methods, which are implemented in STATISTICA 5.1 of StatSoft Inc, (Tulsa, OK) were used for statistical analysis. Data from the m/z region 500 to 3000 were used for cluster analysis.

3 RESULTS AND DISCUSSION

MALDI-TOF MS allows the characterization and differentiation of humic acids (HAs) of different origin.[17,24] Repeatable m/z distributions of cluster and molecular ions were reported. However, due to the heterogeneity of the sample no specific single m/z ratio could be obtained.

As for HAs, MALDI-TOF MS investigations of FAs, the most soluble fraction of HSs, required a suitable sample preparation strategy. Interference of matrix compounds (e.g., sinapinic acid gives several broad signals on the m/z region 500 to 2000) can affect the FA spectra. This m/z range is of importance when analyzing molecular size distributions of fulvic acids. Simple subtraction of the matrix from the sample spectra may bias the results as was demonstrated for background subtraction in a fulvic acid laser desorption MS investigation.[18] Therefore, it was decided to work without any additional matrix compounds. Since FAs absorb energy at the wavelength of the laser, FAs can function both as matrix and sample.

Two different types of sample preparation were investigated. First, the sample was dissolved in a 75:25 water/acetonitrile mixture and then applied onto the sample plate. A very broad and featureless ion distribution from 500 up to 3000 m/z values was observed. However, high background noise and low ion intensities were obtained with this technique.

A second method was developed. The ground FA was mixed with an organic solvent. After application onto the sample plate, evaporation of the organic solvent yielded spots with a very rough surface. This sample preparation gave reproducible spectra of FA with higher intensities than with the above dissolution sample preparation approach. A broad

Application of MALDI-TOF MS to the Characterization of Fulvic Acids 147

and featureless ion distribution was obtained (Figure 3). The ion distribution ranged from 500 up to 3000 m/z for the IHSS standard soil FA. The observed ion distributions were located in a similar range as reported for ESI MS spectra of FAs.[18,20]

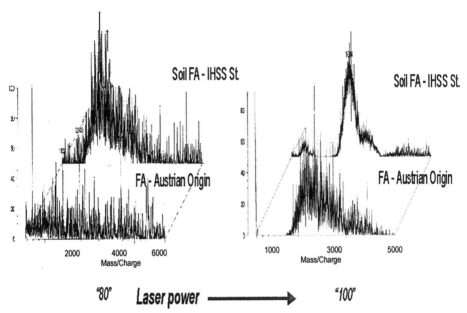

Figure 3 *Dependence of mass spectra of IHSS soil FA and Austrian FA on laser intensity*

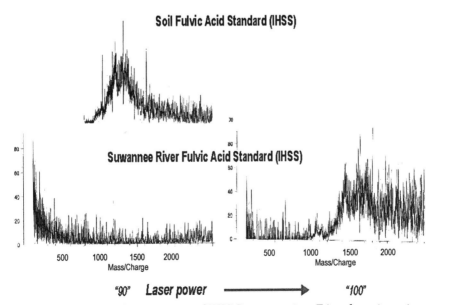

Figure 4 *Dependence of mass spectra of IHSS Suwannee river FA on laser intensity*

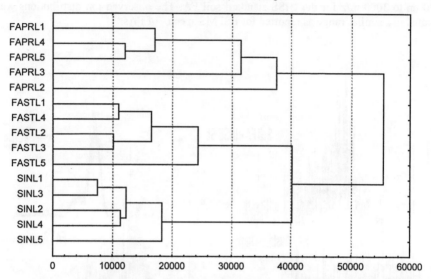

Figure 5 *Cluster analysis; FAPRL1-5 Austrian FA mass spectra (5 replicates); FASTL1-5 IHSS Soil FA mass spectra (5 replicates); SINL1-5 sinapinic acid mass spectra. All spectra were recorded with the same laser intensity*

It is important to note that the obtained ion distributions contain both molecular and cluster ions. Since no information on the mass distribution of the IHSS standard soil FA is provided, the contribution of cluster ions to the mass spectra can not be excluded.

However, using standardized conditions of data acquisition and sample preparation, repeatable ion distributions can be obtained. The intensities of single m/z values varied from sample acquisition to sample acquisition. This may be explained by the complexity of the sample and the contribution of more than one ion to each nominal mass. Changes in acquisition parameters like laser power showed a dramatic effect on the spectra. With increasing laser power cluster formation processes become more important and cluster ions with very high molecular masses up to 100 000 m/z (not shown) were detected.

This became even more evident when MALDI-TOF MS spectra of FAs of different origin were compared (Figures 3 and 4). Different spectra were obtained for IHSS standard soil FA and FA of Austrian origin when using the same laser power and sample preparation. Increase of the laser power yielded similar ion distributions for the FA of Austrian origin and the IHSS soil standard FA at lower laser intensity (Figure 3). The same behavior was observed when comparing the laser power dependence on the spectra of IHSS soil and river FAs. At lower laser power a broad ion distribution at around 1500 was observed for the IHSS soil FA. In contrast, the spectrum of the IHSS river FA showed ion signals at the region below only 500. Increasing the laser power produced similar spectra for the river FA and the soil FA (Figure 4).

The nature of FAs explains these results. Only the desorbable part of the sample will be analyzed by MALDI-TOF MS. Desorption and ionization of molecules and clusters depend on sample preparation and acquisition parameters. The different composition of FA of different origin requires different acquisition parameters to desorb and ionize similar compounds and aggregates with respect to molecular masses. These differences might also be due to the method of FA extraction from a soil matrix. A very mild extraction procedure was chosen for Austrian FA. This might have yielded a different FA fraction compared to IHSS soil FA.

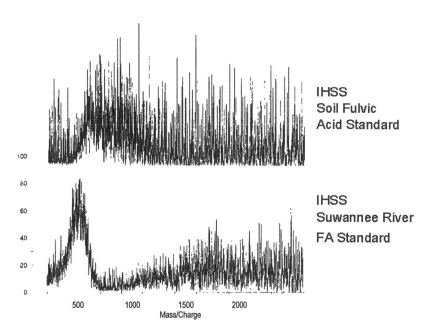

Figure 6 *Mass spectra of two FAs using positive ion reflector mode*

However, identification and fingerprint characterization of FAs are possible when the same sample preparation and sample acquisition methods are applied. The application of a cluster analysis to the spectra of FAs of different origin yielded well separated clusters (Figure 5).

Generally, MALDI-TOF MS spectra of FAs recorded with the reflector showed a shift to lower m/z values (Figure 6). Ion distributions around 600 m/z and 500 m/z were obtained for the IHSS soil fulvic acid and IHSS Suwannee river fulvic acid, respectively. Post source decay processes of molecular and cluster ions are known to contribute to reflector mode MALDI-TOF MS measurements and are probably responsible for that shift.[28]

Comparison of molecular mass to molecular size data is problematic due to several factors. Besides the size, other parameters like shape, specific interactions or composition of the eluent may influence the elution volume of certain substances in size exclusion chromatography (SEC).[5] Therefore, a set of suitable standards are required. Obviously, there is no set of FA standards for SEC. Protein standards were used for calibration of the column in our investigations. A broad molecular size distribution with a maximum corresponding to approx. 3 kDa (related to protein standards) was obtained for the IHSS standard soil FA.

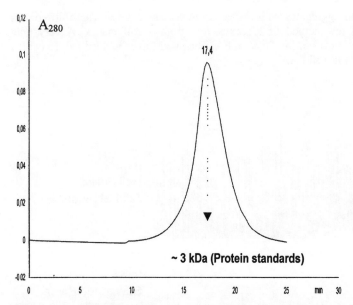

Figure 7 *Size exclusion chromatogram of IHSS soil FA (1 mg/mL; eluent: Tris-HCl (50 mM)/KCl (100 mM))*

MALDI-TOF MS of FAs of different origin yielded ion distributions in similar ranges. However, even if the formation of singly charged ions prevails it is known that multiply charged ions may also be formed.[28] The complexity from a chemical point of view of ill defined samples does not allow distinction between cluster ions and molecular ions and singly and multiply charged ions. Additionally, operational parameters influence desorption and yield of ions and result in different ion distributions. Thus, the m/z distribution of fulvic acids obtained from MALDI-TOF MS may not be considered as a direct molecular mass distribution.

4 CONCLUSIONS

Similar MALDI-TOF-MS spectra were obtained for FAs of different origin. These spectra exhibit a broad and featureless mass/charge distribution in the range up to 3000 m/z. Sample preparation and acquisition parameters influence the shape and m/z distribution of the spectra. The distribution seems to be more influenced by acquisition parameters than by the origin of the sample. However, by applying standardized sample preparation differentiation of MALDI –TOF MS spectra of FAs with different origin can be achieved. Thus, MALDI-TOF MS enables fingerprint characterization of FAs. Repeatable and distinctive mass/charge distributions can be obtained when standardized procedures are applied. Discrimination of cluster from single molecular ions is - due to the complexity of the analyzed material - not possible. Our results reflect the general problems that occur when analyzing complex mixtures with no prevailing substances. Additionally, the information gained is limited to the desorbable part of FAs. All these factors limit the use of the ion distributions obtained for FA molecular mass determinations.

However, MALDI-TOF MS offers the potential for desorption of certain fractions of

FAs. Further investigations will clarify if this information can be used to obtain additional insight into the composition and dynamics of FAs.

ACKNOWLEDGEMENTS

This research was partially supported by the Austrian Nationalbank (Nationalbank fonds - project, No.6761), which is gratefully acknowledged.

References

1. E. Tombácz and J. A. Rice, in 'Understanding Humic Substances: Advanced Methods, Properties and Applications', E. A. Ghabbour and G. Davies, (eds.), Royal Society of Chemistry, Cambridge, 1999, p. 69.
2. G. Davies and E. A. Ghabbour, (eds.), 'Humic Substances: Structures Properties and Applications', Royal Society of Chemistry, Cambridge, 1998.
3. R. L. Wershaw, *Environ. Sci. Technol.*, 1993, **27**, 814.
4. M. A. Schlautman and J. J. Morgan, *Environ. Sci. Technol.*, 1993, **27**, 961.
5. P. Conte and A. Piccolo, *Environ. Sci. Technol.*, 1999, **33**, 1682.
6. L. T. Sein, Jr., J. M. Varnum and S. A. Jansen, *Environ. Sci. Technol.*, 1999, **33**, 546.
7. H. -R. Schulten and M. Schnitzer, *Soil Sci.*, 1997, 162, 115.
8. H. -R. Schulten and P. Leinweber, *Biol. Fertil. Soils*, 2000, **30**, 399.
9. C. M. Preston, R. Hempfling, H. -R. Schulten, M. Schnitzer, J. A. Trofymow and D. E. Axelson, *Plant and Soil*, 1994, **158**, 69.
10. C. A. Fox, C. M. Preston and C. A. Fyfe, *Can. J. Soil Sci.*, 1993, **74**, 1.
11. Y. Inbar, Y. Chen and Y. Hadar, *Soil Sci. Soc. Am. J.*, 1989, **53**, 1695.
12. F. J. Stevenson and K. M. Goh, *Geochim. Cosmochim. Acta*, 1971, **35**, 471.
13. G. Haberhauer and M. H. Gerzabek, *Vibrational Spectroscopy*, 1999, **19/2**, 415.
14. J. Niemeyer, Y. Chen and J. -M. Bollag, *Soil Sci. Am. J.*, 1992, **56**, 135.
15. H. -R. Schulten, in 'Mass Spectrometry of Soils', T. W. Boutton and S. Yamasaki, (eds.), Dekker, New York., 1995, p. 373.
16. T. L. Brown, F. J. Novotny and J. A. Rice, in 'Humic Substances: Structures Properties and Applications', G. Davies and E. A. Ghabbour, (eds.), Royal Society of Chemistry, Cambridge, 1998, p. 91.
17. G. Haberhauer, W. Bednar, M. H. Gerzabek and E. Rosenberg, in 'Understanding Humic Substances: Advanced Methods, Properties and Applications', E. A. Ghabbour and G. Davies, (eds.), Royal Society of Chemistry, Cambridge, 1999, p. 121.
18. T. L. Brown and J. A. Rice, *Anal. Chem.*, 2000, **72**, 384.
19. A. Fievre, T. Solouki, A. G. Marschall and W. T. Cooper, *Energy & Fuels*, 1997, **11**, 554.
20. L. Persson, T. Alsberg, G. Kiss and G. Adham, *Rapid Commun. Mass. Spectrom.*, 2000, **14**, 286.
21. T. Solouki, M. A. Freitas and A. Alomary, *Anal. Chem.*, 1999, **71**, 4719.
22. F. Hillenkamp, M. Karas, R. C. Beavis and B. T. Chait, *Anal. Chem.*, 1991, **63**, 1193A.
23. M. Remmler, A. Georgi and F. -D. Kopinke, *Eur. Mass. Spectrom.*, 1995, **1**, 403.

24. L. Pokŏrna, D. Galdosŏva and J. Havel, in 'Understanding Humic Substances: Advanced Methods, Properties and Applications', E. A. Ghabbour and G. Davies, (eds.), Royal Society of Chemistry, Cambridge, 1999, p. 107.
25. V. Karbach and R. Knochenmuss, *Rapid Comm. Mass Spectrom.*, 1998, **12**, 968.
26. R. Knochenmuss, F. Dubois, M. J. Dale, and R. Zenobi, *Rap. Commun. Mass Spectr.*, 1996, **10**, 871.
27. E. Lehmann and R. Zenobi, *Angew. Chem.*, 1998, **110**, 3600.
28. J. A. Carroll and R. C. Beavis, in 'Laser Desorption and Ablation, Vol 30, Experimental Methods in the Physical Sciences', J. C. Miller and R. F. Haglund, (eds.), Academic Press, San Diego, 1997.
29. D. P. Little, T. J. Cornish, M. J. O'Donnell, A. Braun, R. J. Cotter and H. Köster, *Anal. Chem.,* 1997, **69**, 4540.
30. M. H. Gerzabek and S. M. Ullah, *Internat. Agrophys.*, 1989, **5**, 197.

SORPTION OF AQUEOUS HUMIC ACID TO A TEST AQUIFER MATERIAL AND IMPLICATIONS FOR SUBSURFACE REMEDIATION

D. R. Van Stempvoort,[1] J. W. Molson,[2] S. Lesage[1] and S. Brown[1]

[1] National Water Research Institute, Burlington, Ontario L7R 4A6, Canada
[2] Earth Sciences Department, University of Waterloo, Waterloo, Ontario N2L 3G1, Canada

1 INTRODUCTION

Aqueous humic substances (HSs) play key roles in controlling the aqueous concentrations, mobility, bioavailability and toxicity of hydrophobic contaminants in the environment. Because of their ability to bind hydrophobic compounds, commercial humic acids (HAs) dissolved in water may be useful agents for flushing organic contaminants from the subsurface.[1-3] Most laboratory tests have been conducted at the bench-scale, but recent studies suggest that use of high aqueous concentrations ($\geq$ 1 g/L) of commercial HSs would maximize the efficiency of the flushing process.[4,5] Little information is available on the sorption of commercially available bulk HAs by soils and aquifer materials and it is not clear whether sorption of HAs lessens the utility of HAs for subsurface remediation.

The batch and column tests reported in this paper were conducted to provide basic information on the sorption of concentrated Aldrich® humic acid (Aldrich Chemical Company) by a model aquifer material, including an evaluation of the kinetics of the overall sorption process. Aldrich humic acid (HA) was used to represent commercial HAs in this initial test because its properties are relatively well known.

1.1 The Role of Sorption

A consideration in the use of aqueous humic substances as flushing agents is the fact that they are not completely mobile in the subsurface environment. Previous work has shown that a significant proportion of aqueous HSs sorb on mineral surfaces, soils or aquifer materials. In some of these tests the net result was immobilization or retention of contaminants by the solid phase rather than mobilization.[6,7]

Studies of aqueous HSs sorption by soils and/or aquifer materials mostly are limited to low dissolved HSs concentrations (typically < 100 mg/L). The results indicate that the subsurface mobility of aqueous HSs may be strongly affected by sorption to mineral/solid surfaces,[2,7-15] particularly by sorption to positively charged surfaces of iron and aluminum oxides.[16-18] Only a few bench scale studies have been published on the dynamic transport of commercial HSs such as Aldrich HA[1,4] in soils or aquifer materials, or specifically on the sorption of these products by soil/aquifer materials.[2,19]

Sorption of aqueous natural HSs onto iron oxides decreases with increasing pH,[8,13] but is relatively unaffected by ionic strength in Na^+ solutions.[13,20] In contrast, the sorption of HSs by kaolinite is enhanced in Na^+ solutions of higher ionic strength.[20] HSs sorption by oxides can be enhanced by the presence of divalent cations such as Ca^{2+} and Mg^{2+}.[8] The sorption of relatively polar and hydrophilic HSs on hydrous oxides surfaces (including colloids) may be dominantly by ligand exchange.[9,13] This process is distinct from sorption of strongly hydrophobic organics in soils and sediments, which bind largely to organic matter. In fact, large amounts of solid organic matter in soils may impede the sorption of aqueous HSs.[16] Positively charged mineral surfaces (e.g., aluminum and iron oxides) strongly sorb HAs. Consumption of protons suggests that a complexation reaction has occurred.[21] In contrast, weaker sorption by negatively charged surfaces (e.g., silica and kaolinite) is not associated with proton consumption.[21]

Sorption of aqueous HSs on aquifer material in laboratory column studies generally is time-dependent, with tailing of breakthrough curves over long periods.[10,11] This partly is due to preferential sorption of relatively hydrophobic "subcomponents."[10] HSs sorption often appears to be largely irreversible and desorption is very slow or negligible during dilution.[10,13] Gu et al.[14] explained slow desorption in terms of time-dependent adsorption and displacement processes between different organic components. However, Avena and Koopal[19] found that HSs sorbed on iron oxide surfaces could be rapidly and reversibly desorbed by changing the pH.

Previous work shows that HSs sorption on mineral and sediment surfaces is nonlinear; the concentration of sorbed HSs reaches a plateau at relatively high aqueous levels.[9,11,13,18,20] For batch studies, this behavior generally has been modeled using the Langmuir model[9] or a modification of this model.[13] A different or more complex model[11,15,22] may be required to account for time-dependency (i.e., non-equilibrium), competitive behavior (e.g., different molecular weight fractions of HAs) and/or large differences in the rates of adsorption and desorption reactions.

Several investigators have studied HSs sorption on mineral surfaces, soils or aquifer materials, but these data cannot easily be extrapolated to subsurface remediation applications. As noted above, the aqueous HSs concentrations used were much lower than pertain to subsurface flushing applications. Most tests were conducted on the bench scale with columns a few cm in length, or batches on the order of one liter or less. The column tests sometimes were too short in duration for detection of slow sorption phases. In most batch tests the solids/solution ratio did not represent subsurface conditions. Some studies have shown that the solids/solution ratio is a major control of chemical species partitioning between aqueous and solid phases.[23,24] The mechanism of this "solids effect" is unclear. In conventional batch tests with relatively high solution/solids ratios, the partitioning coefficient may be overestimated, sometimes by several orders in magnitude.

1.2 Batch Tests to Complement the Pilot Scale Test

The batch experiments reported here complement pilot-scale tests with an aqueous humic product for enhanced flushing of hydrophobic contaminants from the subsurface.[3,25] In the pilot test, bulk Aldrich HA (sodium salt, technical product no. H16752, Lot No. 16206AN) in tap water (1 g/L, solution pH 8.5, with the product used as supplied, as in the field), was used to flush hydrocarbons from diesel fuel in a model aquifer (1.2 m x 5.5 m x 1.8 m deep). The model aquifer material was "Winter sand," a carbonate-rich sediment (77 ± 2 weight % as $CaCO_3$) purchased from a local aggregate supplier. This material was mainly medium to very coarse sand, granules and pebbles (> 99 % particles between 0.3 and 4.75 mm (Table 1)). Monitoring indicated pH values from 7.8 to 8.8 in this test system. Molson et al.[26] simulated the transport and sorption of the Aldrich HA in this pilot scale test using BIONAPL/3D, a numerical model recently developed at the University of Waterloo. This model provides a comprehensive, three-dimensional simulation of the complex processes involved in this experiment, including transport and sorption of HA, dissolution and transport of diesel contaminants and concurrent biodegradation.

The present batch test study used the same Aldrich HA and Winter sand. The analysis includes numerical modeling of the HA sorption data and a comparison of the results with simulations of the pilot scale test.

Table 1 *Particle size distribution of "Winter sand" as reported by the supplier*

Particle Size (mm)	Weight %
> 4.75	0
2.36 - 4.75	25.3
1.18 - 2.36	34.7
0.60 - 1.18	30.0
0.30 - 0.60	9.0
0.15 - 0.30	0.5
0.075 - 0.15	0.2
<0.075	0.3

2 METHODS

A "mini-well" batch technique was developed to have a test solids/solution ratio close to actual subsurface (aquifer) conditions (Figure 1). This helps to avoid the "solids effect."[23,24] For each batch, a cylindrical mini-well sampling tube was formed from a rectangular piece of 10 μm mesh stainless steel screen by crimping it lengthwise then crimping one end shut. Each finished tube was ~ 1 cm in diameter and 5 to 8 cm in length. Batches (400 to 700 g) of Winter sand were prepared with a sample splitting technique.

One to two liter volumes of 2 g/L and 3 g/L Aldrich HA (used as supplied) in Milli-Q water were prepared. Minor particulate HAs settled over the next 24-48 hr. The resulting nominal supernatant solutions (2 and 3 g/L) were collected by peristaltic pump and portions were diluted to prepare standard solutions at 1.5, 1, 0.5, 0.2 and 0.1 g/L. Subsequent analyses of 1 g/L and 3 g/L Aldrich HA were found to contain 301 and 969 mg total organic carbon/L, respectively. Based on its 39.03 % carbon content (HA product information provided by Aldrich), this indicates that approximately 20 % of the HA had precipitated during preparation of the batch solutions. Batch absorbance data for the pilot scale test suggested that a similar fraction of the Aldrich HA had precipitated.[25]

Duplicate batches and blanks were prepared with Milli-Q water for each HA concentration. Mason jars with lids were cleaned with a 2 % detergent solution (Contrad®, VWR CanLab in deionized water), rinsed with deionized water, dried and pre-weighed (with the mini-screen). The following were placed in each jar and weighed: solution estimated to saturate a sand sample, the sand sample with a mini-well (in the middle of the sand) and top-up solution to saturate the sand completely. The final ratios of sand to solution were 4 to 5 g/mL. The Mason jars were sealed and stored at 23 ± 2°C, and the batch solutions were sampled by glass mini-pipette from the mini-wells at 7, 14, 43, 94, 118 and 148 days. Their absorbances were measured at 500 nm with a Varian Model CARY3 uv/vis spectrophotometer. After measurement the samples were returned to the batches from which they were taken.

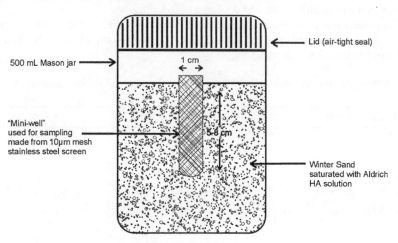

Figure 1 *Schematic of batch test apparatus*

The absorbance of HA standard solutions declined by 15 to 20 % over the course of the experiment, mainly over the first 50 days. This aging trend may be due to one or more of the following: precipitation of fine, particulate HAs (little was observed in this study), biodegradation (thought to be minimal because HAs are refractory)[27] or abiotic alteration of the Aldrich HA in solution. Chemical alteration processes might include HA structural and aggregates size changes over time. Because aging likely also has affected the HA in the batches, the same standards were retained for all batch sampling episodes to provide aging-corrected values.

3 RESULTS AND DISCUSSION

The trends over 148 days in the concentrations of aqueous HA in the batches indicated that Aldrich HA is sorbed by the sand (Figure 2). For the batches with relatively low initial Aldrich HA concentrations (0.1 and 0.2 g/L), 90 to 95 % of the aqueous HA was sorbed in 7 d. For batches with higher initial HA concentrations (0.5 to 3 g/L), the bulk sorption process was more sluggish, approaching apparent equilibrium by 148 days (Figure 2). The trends in concentrations indicated an initial fast sorption phase (typically > 50 % of total sorption), followed by a slow sorption phase. For nonlinear regression of these data we used an empirical Eq. (1) that incorporates both fast and slow sorption components. Here,

$$C_s = C_{max}[1-(F\exp(-k_1 t) + (1-F)\exp(-k_2 t))] \qquad (1)$$

C_s is the sorbed concentration (mg HA/g), C_{max} is the inferred maximum amount of sorbed HA (the final observed concentration, mg/g), F is the fraction of fast sorption, k_1 is the apparent fast sorption kinetic rate constant (time^{-1}) and k_2 is the apparent slow sorption rate constant (time^{-1}).

This equation assumes that the bulk HA behaves as a single substance. Eq. 1 is closely related to an equation applied to desorption test data.[28] Note, however, that the dissolved HA concentrations were not held constant in the batch sorption tests reported here. Application of Eq. (1) to the batch test results is an approximate, empirical fit that provides insight into the kinetics of the bulk sorption process as a cumulative result of adsorption and desorption reactions. Subsurface events also are affected by the rate of the aqueous transport process, which is mainly molecular diffusion. However, tests of the same data indicated that the parameters generated by Eq. (1) are nearly identical to those obtained with another equation (Eq. (10) in ref. 11) for the same application.

A curve-fitting program (GraphPad Prism® 3.00) was used with no weighting of data for nonlinear regressions based on Eq. (1). An example is shown in Figure 3. The calculated values of F, k_1 and k_2 for various batch tests are shown in Table 2. Based on these results, the fast sorption parameters F and k_1 decrease as the initial HA concentrations increase (Table 2). The fast sorption component likely is an initial adsorption-dominated phase.[19] Saturation of available mineral surface sorption sites was approached more rapidly at higher HA concentrations. Thus, a smaller fraction of dissolved HA was sorbed over the first few days and a larger fraction of the dissolved HA was available for subsequent slow sorption by some other mechanism. This resulted in lower F and k_1 values at higher dissolved HA concentrations.

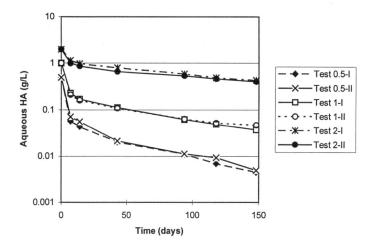

Figure 2 *Declines in aqueous Aldrich HA over time in batches (initially 0.5 to 2 g HA/L). Tests are designated 0.5-2 to indicate initial HA concentrations (g/L) and Roman numerals I and II indicate first and second duplicates. Data for tests with initial HA concentrations of 0.1, 0.2 and 3 gHA/L are not shown*

The calculated kinetic parameter for slow sorption, k_2, was found to be relatively constant at 0.02-0.03/day and appeared to decrease slightly with increasing initial HA concentration (Table 2). The slow sorption component largely may be a competitive exchange of dissolved HA molecules or aggregates for sorbed HA.[7,15,19] In essence, the slow sorption process may be a secondary net sorption process largely controlled by a reduced rate of desorption of HA over time. This may be attributed to replacement of relatively loosely-held HA molecules (e.g., low molecular weight) by more tightly-held, high molecular weight HA at the sorption sites. However, concurrent aging of the standard HA solutions (which was corrected for, see Methods) precludes a clear distinction between aging and slow sorption of HA in the batches.

Table 2 *Parameters obtained for nonlinear regressions (Equation 1) of selected batch test data*

Initial aqueous HA conc., g/L	F	k_1, day^{-1}	k_2, day^{-1}	r^2
0.5 (I)	0.885 ± 0.009	0.618 ± 0.081	0.028 ± 0.003	0.996
0.5 (II)	0.838 ± 0.017	0.733 ± 0.319	0.034 ± 0.005	0.994
1 (I)	0.813 ± 0.011	0.429 ± 0.035	0.022 ± 0.002	0.997
1 (II)	0.828 ± 0.010	0.483 ± 0.040	0.024 ± 0.002	0.998
2 (I)	0.556 ± 0.053	0.346 ± 0.139	0.018 ± 0.003	0.986
2 (II)	0.643 ± 0.036	0.355 ± 0.086	0.019 ± 0.003	0.991

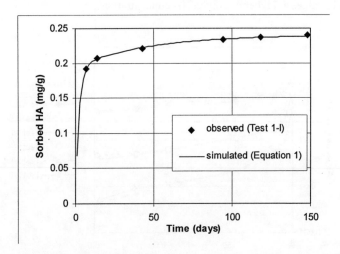

Figure 3 *Example of a nonlinear regression of batch test data using Eq. (1). In this case the initial aqueous HA was 1 g HA/L*

In line with previous studies, a Langmuir sorption isotherm for the final equilibrium systems (after 148 days) can be constructed with the data from all batch tests (Figure 4). This nonlinear isotherm follows Eq. (2), where C_w is the aqueous (dissolved) HA

$$C_s = Q_{max} K C_w / (1 + K C_w) \qquad (2)$$

concentrations (g HA/L), Q_{max} is the saturation sorbed HA concentration (mg HA/g) and K is the fitted sorption parameter (L/g). The parameters generated with GraphPad Prism from the fit shown in Figure 4 were Q_{max} = 0.423 ± 0.017 mg HA/g and K = 39.2 ± 7.3 L/g. The Langmuir equation (2) assumes a finite number of sorption sites that become saturated with increasing solute (HA) levels. Langmuir behavior indicates that the aquifer material has particularly high affinity for HA at lower concentrations.

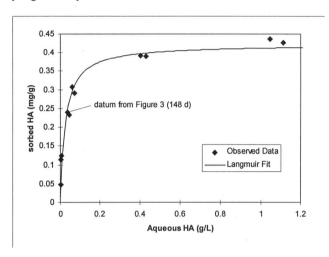

Figure 4 *Langmuir isotherm for sorption of Aldrich HA to Winter sand after 148 d of reaction. The Langmuir parameters for this fit are Q_{max} = 0.423 ± 0.017 mg HA/g and K = 39.2 ± 7.3 L/g*

Sorption of HS to soils and aquifer materials typically is reported per mass of organic C rather than per mass of total HA. The Langmuir parameters for the batch test results of this study can be restated in terms of organic carbon, given the TOC analysis reported for 1 g/L HA (see Methods). The resulting Langmuir parameters are Q_{max} = 0.127 ± 0.017 mg C /g sand and K = 11.8 ± 2.2 L/g C. The non-equilibrium 7 d data of this study give apparent Langmuir coefficients Q_{max} = 0.0744 ± 0.0043 mg C/g and K = 4.36 ± 1.00 L/g C. Comparison of the 7 d and 148 d parameters indicates that the sorption parameters strongly depend on the duration of the batch tests.

Although the above empirical modeling of the kinetics of the batch data using Eq. (1) provides a close fit of observed data, practical points to be considered include 1) the kinetic model is not based on a conceptual model of the sorption process, 2) data for each batch test must be fitted independently, and 3) the results cannot directly be related to the pilot scale test in which groundwater flow was important.

For these reasons further simulations of the batch tests using BIONAPL/3D were conducted. BIONAPL/3D is the same model that Molson et al.[26] employed for simulation of the pilot scale test data. This model assumes that Aldrich HA behaves as a single species that has representative bulk properties.

A Langmuir equation for the sorption for the batch test simulations was assumed with the above values for C_w and Q_{max} derived for the combined end of experiment data (Figure 4). Sorption was taken to include two components, one instantaneous and the other time-dependent, by fitting a first order mass transfer coefficient. Desorption and sorption rates were set equal (this assumes no hysteresis).

The numerical simulations provided a close fit to the observed time trends of both aqueous and sorbed (calculated) Aldrich HA in the batch tests (Figures 5 and 6). Thus, this model provides a useful way to simulate both the kinetics and the nonlinear (i.e., Langmuir) sorption behavior. In a parallel study, Molson et al.[26] have applied the parameters derived for the batch tests reported here to simulate the pilot scale test data using BIONAPL/3D. Only the kinetic parameters had to be adjusted (by about an order of magnitude (Figure 7)). The need for this adjustment is unknown. It perhaps results from the dynamic versus static conditions of these tests.

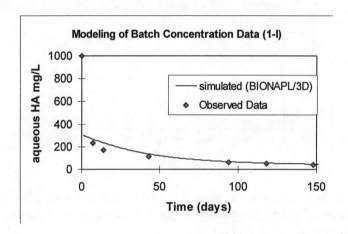

Figure 5 *Simulation by BIONAPL/3D of the trend of aqueous HA concentrations in batch test 1-I*

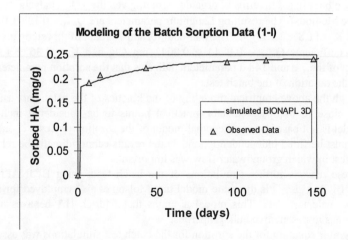

Figure 6 *Simulation by BIONAPL/3D of the sorbed concentration of Aldrich HA on Winter sand with the same batch test data as in Figure 5*

Overall close fit of the Langmuir parameters derived from the batch test to the pilot scale test data is a key finding of this study. This agreement suggests that the mini-well batch technique introduced in this paper is effective for investigating the sorption of concentrated HA to aquifer materials, enabling a direct application and prediction of Langmuir-type sorption on the pilot and field scales.

A word of caution is necessary at this point. The application of the BIONAPL/3D model in this study provided close simulation of two specific types of tests: 1) pulse input (time zero) of concentrated Aldrich HA in a static system (batch tests), and 2) input of concentrated Aldrich HA as a constant source input in a steady flow system (pilot scale test). For these cases it was not necessary to take hysteresis into account. Hysteresis would be important for field applications where the addition of a concentrated HA solution was eventually discontinued, since desorption of HA from aquifer solids generally is very slow or negligible.[10,13] The BIONAPL/3D code can be adapted to account for this effect as necessary.

For comparison, van de Weerd et al.[15] recently introduced a model that can simulate sorption of various components of a polydisperse mixture of HSs based on their molecular weights and with account of hysteresis. However, in its present form this model only can be applied to batch data and not to dynamic systems such as the above pilot test. By contrast and as demonstrated above, BIONAPL/3D can provide a "seamless" comparative analysis of batch and dynamic (column, pilot test) data.

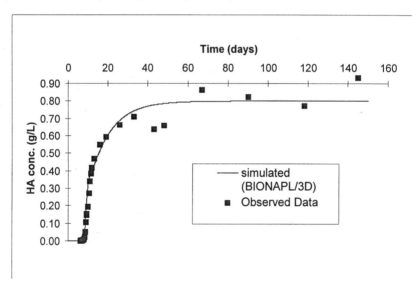

Figure 7 *Simulation of the breakthrough of Aldrich HA in the pilot scale test at monitoring point T7C4 in the model aquifer, 3 m downgradient of the HA input source (see ref. 26 for further details)*

3.1 Implications of Results for Subsurface Remediation

The results of this study indicate that if a commercial HA is added to water for use in subsurface remediation, a substantial fraction will become sorbed to soil/aquifer solids. If high concentrations of aqueous HA are used, the kinetics of this sorption will be relatively sluggish, requiring weeks or more to reach approximate equilibrium. The sorption kinetics will have to be taken into account in computer simulations of the subsurface transport of aqueous HAs. Given typical remediation schedules (weeks to years), the kinetics indicated by long-term testing (e.g., 148 d batch tests) are required for proper simulation. Worth noting is that delineation of the "early" phase of sorption (i.e., "instanteous" or "fast") as indicated by conventional 1-2 d batch and column tests may be irrelevant for modeling pilot and field-scale applications.

The steep slope of the Langmuir isotherm at low Aldrich HA concentrations (< 0.1 g HA/L) indicates that transport and breakthrough of dilute Aldrich HA solutions in field applications is severely retarded. In such circumstances the mobilization of contaminants might actually be retarded by the addition of HA to groundwater,[7] at least temporarily. Breakthrough of HA would become more rapid at higher levels of aqueous Aldrich HA (> 0.1 g/L) because only a finite number of surface sites are available for HA sorption.

If a concentrated solution of a commercial HA is used for subsurface remediation, the fraction that becomes sorbed ultimately will become a sink for hydrophobic contaminants. This could compromise the mobilization of contaminants in the aqueous HA. On the positive side, in the post-treatment phase when the "spent" HA carrier solutions have been extracted, the sorbed humics would increase the organic content (f_{oc}) of the soil/aquifer material and retard migration of the dissolved contaminant. Combined with natural or enhanced *in situ* bioremediation, this might substantially reduce the rate(s) at which residual NAPL and sorbed contaminants emerge from the contaminated zone.

This study reports tests with one commercial humic acid and one model aquifer material. There is a considerable amount of information on the chemical properties of Aldrich HA, but this product is relatively expensive. Practical field applications require more economical HA products. Further information also is required on how aquifer materials variations (e.g., grain size distribution, mineralogy, organic carbon content) and aqueous chemical parameters (ionic strength, pH, temperature) affect commercial HA sorption in the subsurface. These tests should be of sufficient duration (several weeks or longer) and should employ solids/solution ratios similar to subsurface conditions, as in the batch tests reported here.

ACKNOWLEDGEMENTS

Funding was provided by Environment Canada, the Panel on Energy Research and Development (PERD), the Centre for Research in Earth and Space Technology (CRESTech) and the Natural Sciences and Engineering Research Council of Canada (NSERC). Andrew Piggott reviewed an earlier version of this manuscript.

References

1. S. Abdul, T. L. Gibson and D. N. Rai, *Environ. Sci. Technol.*, 1990, **24**, 328.
2. C. Rav-Acha and M. Rebhun, *Water Research*, 1992, **26**, 1645.
3. S. Lesage, K. S. Novakowski, H. Xu, G. Bickerton, L. Durham and S. Brown, Proceedings, Solutions 1995, Int. Assoc. Hydrogeologists Congress, Edmonton, Canada, 1995, p. 6.
4. W. P. Johnson and W. W. John, *J. Contam. Hydrol.*, 1999, **35**, 343.

5. T. B. Boving and M. L. Brusseau, *J. Contam. Hydrol.*, 2000, **42**, 51.
6. E. J. M. Temminghoff, S. E. A. T. M. van der Zee and F. A. M. de Haan, *Environ. Sci. Technol.*, 1997, **31**, 1109.
7. K. U. Totsche, J. Danzer and I. Kögel-Knabner, *J. Environ. Qual.*, 1997, **26**, 1090.
8. E. Tipping, *Marine Chem.*, 1986, **18**, 161.
9. E. M. Murphy, J. M. Zachara and S. C. Smith, *Environ. Sci. Technol.*, 1990, **24**, 1507.
10. F. M. Dunnivant, P. M. Jardine, D. L. Taylor and J. F. McCarthy, *Soil Sci. Soc. Am. J.*, 1992, **56**, 437.
11. P. M. Jardine, F. M. Dunnivant, H. M. Selim and J. F. McCarthy, *Soil Sci. Soc. Am. J.*, 1992, **56**, 393.
12. Z. Filip and J. J. Alberts, *Sci. Total Environ.*, 1994, **153**, 141.
13. B. Gu, J. Schmitt, Z. Chen, L. Liang and J. F. McCarthy, *Environ. Sci. Technol.*, 1994, **28**, 38.
14. B. Gu, T. L. Mehlhorn, L. Liang and J. F. McCarthy, *Geochim. Cosmochim. Acta*, 1996, **60**, 2977.
15. H. van de Weerd, W. H. van Riemsdijk and A. Leijnse, *Environ. Sci. Technol.*, 1999, **33**, 1675.
16. P. M. Jardine, N. L. Weber and J. F. McCarthy, *Soil Sci. Soc. Am. J.*, 1989, **53**, 1378.
17. K. Kaiser and W. Zech, *Soil Sci. Soc. Am. J.*, 1998, **62**, 129.
18. Y. -H. Shen, *Chemosphere*, 1999, **38**, 2489.
19. M. J. Avena and L. K. Koopal, *Environ. Sci. Technol.*, 1998, **32**, 2572.
20. E. M. Murphy, J. M. Zachara, S. C. Smith, J. L. Phillips and T. W. Wietsma, *Environ. Sci. Technol*, 1994, **28**, 1291.
21. K. M. Spark, J. D. Wells and B. B. Johnson, *Austral. J. Soil Res.*, 1997, **35**, 103.
22. M. J. Avena and L. K. Koopal, *Environ. Sci. Technol.* 1999, **33**, 2739.
23. J. A. Celorie, S. L. Woods and J. D. Istok, *J. Environ. Qual.*, 1989, **18**, 307.
24. S. J. You, Y. J. Yin and H. E. Allen, *Science Total Environ.*, 1999, **227**, 155.
25. D. R. Van Stempvoort, S. Lesage and S. Brown, submitted to *J. Contam. Hydrol.*
26. J. W. Molson, E. O. Frind, D. R. Van Stempvoort and S. Lesage, submitted to *J. Contam. Hydrol.*
27. T. A. Jackson, *Soil Sci.*, 1975, **119**, 56.
28. G. Cornelissen, P. C. M. van Noort and H. A. J. Govers, *Environ. Sci. Technol*, 1998, **32**, 3124.

VALIDATION OF A ONE-PARAMETER CONCEPT TO ELUCIDATE THE SORPTION OF HYDROPHOBIC ORGANIC COMPOUNDS INTO HUMIC ORGANIC MATTER AND BIOCONCENTRATION PROCESSES

Juergen Poerschmann

UFZ-Centre for Environmental Research Leipzig-Halle, 04318 Leipzig, Germany

1 INTRODUCTION

1.1 Setting the Scene

Sorption and desorption processes in the environment generally are explained as adsorption or sorption phenomena. The latter refer to a partitioning process in which the hydrophobic organic sorbate is distributed between two immiscible or poorly miscible phases, e.g., an organic solvent, including humic organic matter (HOM), and water. Hydrophobic partitioning, comprising non-specific interaction forces and a substantial entropic term, is responsible for sorption into amorphous, polymeric HOM.[1] Partition coefficients (e.g., K_{HOM}) usually serve as a numerical measure of the partitioning process. Sorption is regarded as non-competitive and is associated with linear sorption/desorption isotherms. By contrast, adsorption is characterised by sorbate binding onto surfaces, interfaces or interior pores of a sorbent. Adsorption is competitive between sorbates due to constraints of available surfaces or specific sites.[2] It should be borne in mind that sorption generally is accompanied by adsorption because each bulk phase obviously has an interfacial area. However, in the case of prevailing partitioning processes the bulk property is much more relevant than the surface.

There has been a large body of contributions devoted to correlations between partition coefficients (K_{HOM}) and various molecular properties and descriptors, including molecular weight, ultraviolet absorbance at 280 nm, elemental composition, aromaticity, ^{13}C NMR signals of main structural fragments and molecular connectivity.[3-5] However, the overwhelming majority of these correlations between HOM binding of hydrophobic sorbates and HOM structure suffer from poor confidence intervals (e.g., R^2 about 0.7, as observed in ref. 4, sorption data measured by means of fluorescence quenching, see Section 2.3) and highly complex approaches. As has been known for some two decades, the sorbate's hydrophobicity is a good parameter to help predict and interpret partitioning. The sorbate's hydrophobicity is mostly related to its octanol-water coefficient (K_{OW}), Eq. (1),

$$\log K_{HOM} = a \cdot \log K_{OW} + b \tag{1}$$

where K_{HOM} is the partition coefficient related to the organic matter of the sorbent and a, b are arbitrary coefficients.

Basically, Eq. (1) is a linear free energy relationship describing the partitioning of a hydrophobic solute into the reference 1-octanol. There is strong empirical evidence in the literature that Eq. (1) is not valid across several "families" of hydrophobic sorbates (e.g., PAHs, PCBs, alkanes, chlorinated pesticides). In ref. 6 an attempt was made to correlate log (K_{HOM}/K_{OW}) versus log K_{OW} or "a molecular property." Unfortunately, the authors failed to mention the molecular property they had in mind. Although a good K_{HOM}-K_{OW} correlation exists across a defined sorbate "family," the coefficients a and b in the relationship (Eq. 1) have no fundamental physicochemical meaning. The partitioning process of organic solutes into octanol and dissolved HOM can be considered identical as far as the entropic contribution of solute dissolution is concerned because this expresses how the hydrophobic organic compounds are expelled from the water phase. However, the enthalpic contribution might be quite different for partitioning into octanol on the one hand and HOM on the other. What is needed is a concept that physicochemically underpins the arbitrary constants in Eq. (1).

The same is true in order to interpret and predict bioconcentration processes. The widely accepted hydrophobicity model considers bioconcentration as partitioning between the exposure water and the lipid phase of the aquatic organism, with no physiological barriers to impede the accumulation of the chemical.[7] Bioconcentration factors (BCF) have been estimated using the common Eq. (2), which is marred by the same shortcomings as outlined above for HOM-partitioning.

$$\log BCF = a \bullet \log K_{OW} + b \tag{2}$$

1.2 A One-Parameter Solubility Concept to Interpret and Predict Partitioning in both HOM and Lipid Tissues

Several years ago, a one-parameter concept based on the Flory-Huggins theory was introduced by Kopinke to explain the partitioning into HOM on the basis of Hildebrand solubility parameters.[8] The proposed concept involves both the enthalpic contribution (affinity of the target analyte for its amorphous, polymeric host) and the entropic contribution ("incompatibility" of the sorbate with water). Eq. (3) expresses the concept,

$$\ln \frac{K_{HOM,i}}{K_{OW,i}} = \frac{V_m}{RT}((\delta_i - \delta_{octanol})^2 - (\delta_i - \delta_{HOM})^2) - \ln\rho_{HOM} \tag{3}$$

where V_m is the molar volume of the analyte (mLmol^{-1}), $\delta_{i,\ HOM,\ octanol}$ is the solubility parameter of sorbate, HOM or octanol ((J cm^{-3})$^{0.5}$), ρ_{HOM} is the HOM density (to take into account the dimensions of K_{OW} (g mL^{-1}) and K_{OM} (g g^{-1})), R = 8.31 J K^{-1}mol^{-1}, T = 293 K and $\delta_{Ooctanol}$ = 21.0 (J cm^{-3})$^{0.5}$.

Solubility parameters are tabulated both for organic compounds and for amorphous polymers.[9-11] Solubility parameters refer to cohesive energy densities, which can be regarded as a measure of polarity. Cohesive energy densities can be calculated from the molar enthalpy of evaporation and the molar volume of the sorbate.[8] Therefore, only one parameter is unknown (hence the name "one-parameter approach"): the solubility parameter of the polymeric sorbent. The benefits of the proposed concept consist of the possibility to estimate partition coefficients (Eq. (3)) or to characterise the solubility parameter of amorphous polymers via Eq. (4).

$$\delta_{HOM} = \delta_i \pm \sqrt{(\delta_i - \delta_{Octanol})^2 + \frac{2.3RT}{V_m}(\log K_{OW} - \log K_{HOM} - \log \rho_{HOM})} \qquad (4)$$

In principle, the calculation of δ_{HOM} may be based on a single sorbate only because δ_{HOM} is a characteristic feature of the polymeric sorbent. However, due to the uncertainties encountered in calculating partition coefficients and uncertainties in tabulated solubility parameters (including the increment system to calculate δ values) and K_{OW} data, the calculation of δ_{HOM} should always be derived from studies of a multitude of sorbates. The solubility parameters of some particulates and dissolved HOM were shown to be within the narrow range of $\delta_{HOM} = 25.5 \pm 1.5$ (J cm^{-3})$^{0.5}$, which is between the value of polar cellulose (32.0 (J cm^{-3})$^{0.5}$) and non-polar polyethylene (16.3 (J cm^{-3})$^{0.5}$).[8] On the basis of which the δ_{HOM} was calculated, the solutes studied mainly included organic pollutants of medium hydrophobicity (1.5 < log K_{OW} < 5.0). It should be noted that the solubility parameters mentioned refer exclusively to reversible partitioning. When addressing sorbates that can exercise specific interactions within the HOM (e.g., hydrogen bonds occurring with nitrophenols), the δ_{HOM} value in Eq. (4) has to be replaced by a sum of solubility parameters reflecting both non-polar partitioning and specific interactions. However, this would lead to more complex relations in comparison with the one-parameter approach.

Assuming that partitioning phenomena prevail in bioconcentration processes, the solubility concept originally devoted to HOM may be extended to tackle this issue. Estimating BCFs for hydrophobic, non-ionic organic compounds is proposed here on the basis of analyte and matrix solubility parameter data via Eq. (5).

$$\ln \frac{BCF}{K_{OW,i}} = \frac{V_m}{RT}((\delta_i - \delta_{octanol})^2 - (\delta_i - \delta_{Lipid})^2) - \ln \rho_{Lipid} \qquad (5)$$

Therefore, this paper addresses the validation of the proposed concept to describe partitioning phenomena into HOM and bioconcentration. The HOM issue refers in particular to very hydrophobic sorbates using our own sorption data as well as to other hydrophobic pollutants of priority significance using sorption data published in the literature.[8] Very hydrophobic sorbates that include PAHs, PCBs and alkanes all possess log K_{OW} beyond 5. When applying the solubility concept, attention also should be paid to scrutinising methods to determine partition coefficients. As is well known, using different techniques to determine K_{HOM} can lead to biased results.

2 RESULTS AND DISCUSSION

2.1. Application of the Solubility

Table 1 gives K_{HOM} data of alkanes, PCBs and PAHs on a dissolved HOM purchased from Roth (Karlsruhe, Germany) along with calculated δ_{HOM} data. K_{HOM} was determined using Solid Phase Microextraction[12] (see Section 2.3). A strong K_{HOM}-K_{OW} correlation over the entire range of analyte hydrophobicity is observed across a given sorbate "family," as follows

PAHs: $\log K_{HOM} = 0.98 \log K_{OW} - 0.39$ \hspace{1em} ($R^2 = 0.988$) \hfill (6)

PCBs: $\log K_{HOM} = 0.93 \log K_{OW} - 0.54$ \hspace{1em} ($R^2 = 0.990$) \hfill (7)

Alkanes: $\log K_{HOM} = 0.87 \log K_{OW} - 0.64$ \hspace{1em} ($R^2 = 0.986$) \hfill (8)

Table 1 *Sorption coefficients for very hydrophobic organic compounds and calculated HOM solubility parameters*

Analyte	V_m (mL mol^{-1})	δ ($J^{0.5}$cm$^{-1.5}$)	Log K_{ow} [a]	Log K_{HOM}	δ_{HOM} ($J^{0.5}$cm$^{-1.5}$)
n-Heptane	148	15.3	4.51	3.36	23.8
n-Octane	164	15.5	5.05	3.62	24.2
n-Nonane	179	15.7	5.45 cal.	4.24	23.6
n-Decane	195	15.9	5.98 cal.	4.57	23.9
n-Undecane	211	15.9	6.51	5.20	23.5
n-Dodecane	227	16.0	7.04	5.48	23.8
n-Tridecane	243	16.1	7.57	6.00	23.7
PCB-1	172[b]	18.9	4.53	3.75	24.0
PCB-15	189[b]	19.0	5.58	4.52	24.6
PCB-28	204[b]	19.2	5.62	4.72	24.2
PCB-52	219[b]	19.5	6.09	5.13	24.4
PCB-118	234[b]	19.8	6.90	5.97	24.4
PCB-153	249[b]	19.9	7.16	6.06	24.8
Naphthalene	126	20.2	3.30	2.93	23.8
Fluorene	153	19.9	4.18	3.58	24.3
Phenanthrene	159	20.0	4.44	3.93	23.9
Pyrene	179	20.7	5.19	4.70	24.2
Chrysene	196	20.6	5.50	5.06	23.7

[a] Daylight Chemical Information Systems, Database Medchem, 1999, Irvine, CA; [b] From ref. 34 and increments given therein.

These regression lines provide strong evidence that a graduation of K_{HOM} data at a given hydrophobicity exists in the following order: PAH > PCB > alkane. As can be seen from Table 1, the calculated HOM solubility parameters are almost identical for all sorbates covering the three "families," thus stressing the usefulness of the one-parameter approach. The higher the compatibility between the hydrophobic sorbates and the HOM sorbent, as expressed by smaller differences in solubility parameters ($\delta_{Sorbate} - \delta_{HOM}$, see Eq. (3)), the higher the K_{HOM} values. Eq. (3) also indicates that the "compatibility" term between a defined sorbate and HOM reaches its maximum when both solubility parameters are identical. On the basis of 1) the similarity of solubility parameters across a defined "family" (ref. 11, see Table 1) and 2) the direct proportionality between the molar volume and the K_{OW},[13,14] two conclusions can be drawn: 1) Eq. (1) cannot be valid across several sorbate "families" due to different solubility parameters (as a rule of thumb: δ_{Alkane} = 15.5 $J^{0.5}cm^{-1.5}$, δ_{PAHs} = 20.0 $J^{0.5}cm^{-1}$), that is there are different compatibilities with the HOM and 2) the closer δ_{HOM} is to δ_i, the steeper is the regression line.

The average δ-value of 24.0 $J^{0.5}cm^{-1.5}$ is at the lower level of the range δ = 25.5 ± 1.5 $J^{0.5}cm^{-1.5}$ given in ref. 8 for HOM. Following Hildebrand's idea (see ref. 15) according to which the solubility parameter of a polymer is similar to major sub-units, this finding was expected. During further research (data not shown here), this coal-derived HOM was found to possess pronounced hydrophobic moieties in the polymeric backbone.

Our own findings are confirmed by literature data. When using pyrene partition coefficients measured on Suwannee River fulvic acid, Suwannee River humic acid and the commercial (coal-derived) Aldrich humic acid,[16] the solubility parameters of the three HOM sorbents studied can be calculated to be 27.3 $J^{0.5}cm^{-1.5}$, 26.5 $J^{0.5}cm^{-1.5}$ and 23.8 $J^{0.5}cm^{-1.5}$, respectively. The Aldrich HOM is assumed to be very similar to the HOM purchased from Roth according to both the manufacturer's information and our own data gathered from spectroscopic and pyrolysis work.

Once calculated, the solubility parameter of the sorbent may provide a basis to determine either solubility parameters of the analyte or partition coefficients. As an example, the solubility parameter of DDT (log K_{OW} = 6.00, log K_{HOM} = 5.19 measured using the Roth HOM, V = 228 mL mol^{-1}, δ_{HOM} = 24.0 $J^{0.5}cm^{-1.5}$) amounts to 19.5 $J^{0.5}cm^{-1.5}$, which is reasonable (see ref. 11). Furthermore, the log K_{HOM} of o-dichlorobenzene (log K_{OW} = 3.38, δ = 20.4 $J^{0.5}cm^{-1.5}$, V_m = 112 mL mol^{-1}, δ_{HOM} = 24.0 $J^{0.5}cm^{-1.5}$; for data sources see Table 1 and ref. 11) can be calculated to be 3.03, which is close to the experimental value of log K_{HOM} = 2.97.

There is some controversy between the results cited above and ours on the one hand, and those detailed in 17 on the other. Findings in ref. 17 result in quite different δ-data for the HOM being studied (Suwannee River humic acid) using perylene, pyrene and anthracene as solutes. We consider peculiarities of the fluorescence quenching method, detailed below, to be the main reason for these findings rather than steric effects as assumed by these authors.

To this point we have addressed sorption data into exclusively dissolved HOM. However, sorption of hydrophobic organic compounds on soils and sediments has been an matter of intense debate within the last few years. It is now widely acknowledged that more than one mechanism is involved when considering particulate HOM. Conceptual models to explain the deviations from sorption linearity include 1) at least two different HOM entities, that is, "hard" ("glassy") and "soft" ("rubbery") carbon, 2) the presence of charcoal-like carbonaceous material with high specific surface areas and 3) internal holes ("pore-filling" mechanisms).[18,19] Obviously, the solubility theory, which is based on the Flory-Huggins theory for amorphous polymers, cannot fully describe these effects in such heterogeneous sorbents.

Figure 1 shows calculated δ-values using literature-based partition coefficients from the last few years. The detailed source of the sorption data used on the basis of which the HOM solubility parameters were calculated is given in ref. 20. The aim of Figure 1 is to convey the idea that – notwithstanding the occurrence of phenomena other than partitioning – the large set of soils and sediments taken from different sources have solubility parameters in a very narrow range. This indicates that partition phenomena prevail. Moreover, the same narrow range also was observed with different dissolved HOM, implying the occurrence of basically similar partitioning processes with dissolved and particulate HOM.[12]

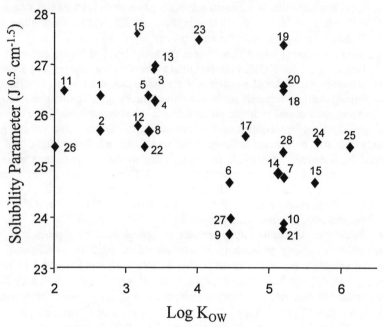

Figure 1 *Solubility parameters for soils and sediments (based on literature data) vs. analyte hydrophobicity*

2.2 Application of the Solubility Concept to Bioconcentration Processes

Table 2 gives δ-values of several biological matrices calculated on the basis of randomly selected BCFs from the literature. As found for HOM, quite similar solubility parameters result across all matrices studied. This provides strong evidence that (under equilibrium conditions) the application of the solubility concept also is very useful in estimating BCFs. In principle, the same route can be taken as with HOM: either calculating solubility parameters of the lipid tissues, the assumed locus of bioconcentration, or alternatively predicting BCFs after calculating this value. This approach simplifies the complex bioconcentration processes but may nevertheless provide a helpful estimate of bioconcentration.

With the solubility parameter of lipid tissues of aquatic organisms being close to that of the reference octanol ($\delta = 21.0 \ J^{0.5}cm^{-1.5}$, see Table 2), a similarity of lipid content-normalised BCFs at equilibrium and K_{OW} data is plausible for hydrophobic, non-ionic organic compounds (see Eq. (5)). In the literature, tricaprylin (or triolein) is suggested to be a better reference standard for simulating bioconcentration processes.[21-23] In our opinion, the SPME method to determine BCF data ought to be faster and less susceptible to error for volatile compounds than the cumbersome disk techniques. This approach assumes a strong correlation between SPME-fibre distribution coefficients and BCFs.[24]

Based on the conclusions drawn above, it makes sense to run biomimetic extractions with the polar polyacrylate (PA) fibre due to its compatibility in terms of solubility parameters with the lipid tissues. The solubility parameter of PA coatings was calculated to be about $\delta = 21.0 \ J^{0.5} cm^{-1.5}$ (ref. 25, which – despite some uncertainties in the calculation – is very close to that of octanol). Another approach would be to use home-made SPME coatings with dimyristoylphosphatidylcholine to simulate biomembranes.

Table 2 *Calculated solubility parameters of biomatrices based on bioconcentration factors (assumption: density 1 g mL^{-1} across all matrices)*

Matrix/Analyte	Analyte	δ_{Matrix} ($J^{0.5}cm^{-1.5}$)	Ref.
Lipid tissue trout/related to lipid content	Benzene, trichloroethane	20.9 / 21.0	35
Lipid tissue fathead minnow/related to lipid content	Hexachlorobenzene	21.1	36
Lipid tissues guppy (*poecilia reticulata*)/data related to lipid content	Chlorobenzenes, nitroaromatics	20.7 ± 0.9	37
Dimyristoylphosphatidylcholine[a]	Semi-volatiles	21.8 ± 0.5	38
Bovine serum albumin	Nitrobenzene/phenols	28.9 ± 1.0	39
	Ethylbenzene	28.3	40
	Propylbenzene	29.4	40

[a] The larger solubility parameter of choline (to simulate phospholipids) than of octanol is evident because solutes with $\delta_i > \delta_{octanol}$ show higher partition coefficients on the choline, whereas solutes with $\delta_i < \delta_{octanol}$ possess higher K_{OW} values.

2.3 Critical Evaluation of Methods to Determine Partition Coefficients - Fluorescence Quenching versus Solid Phase Microextraction

A good method for determining sorption coefficients ought to be applicable to both particulate and dissolved HOM. The latter is more difficult to handle because of the complicated phase separation. Established methods for measuring sorption on dissolved HOM include fluorescence quenching (FQT), the SPME method, the reverse phase method, the flocculation method, the solubility enhancement method and the headspace partitioning method.[26] The first two techniques are outlined briefly: FQT is the most widely used approach, while SPME is preferred by the author and is expected to become mainstream in the near future.

FQT, a genuine *in-situ*, non-invasive approach, is based on the premise that fluorophores completely lose their fluorescence activity when bound to HOM (see ref. 27). This assumption has been a matter of debate.[28] Due to the intensity of fluorescence, which is a concentration-related signal, a concentration-related K_{HOM} results. Problems arise from the HOM matrix, which might absorb both excitation and fluorescence light.

The SPME approach is based on the valid assumption that the fibre coating (e.g., 2.5 x 10^{-5} mL PDMS when considering a 7 µm fibre)[12] samples only the freely dissolved sorbate fraction rather than the fraction bound to dissolved HOM.[12] If the experimental conditions are properly adjusted,[24] the sampled sorbate fraction is negligible. Therefore, the sorption equilibrium between sorbates and HOM is not influenced. Because any analyte fraction sampled by the SPME fibre is related to the analyte's activity (e.g., the fibre uptake of 1 ppb pyrene is different from an aqueous solution and a water/methanol solution owing to their different activities), the partition coefficients are activity-related. Based on the fact that the pollutant activity is the more important characteristic than the concentration to assess the pollutant's environmental fate,[29] the application of the SPME approach ought to be beneficial. Another significant benefit of the SPME method is that multicomponent investigations are possible. Table 3 summarises the major characteristics of the two approaches under consideration.

Table 3 *FQT versus SPME to determine partition coefficients*

Criterion	SPME	FQT
Multicomponent analysis	Yes	No
Suitable sorbates	Hydrophobic sorbates in general	Sorbates with distinct fluorescence activity
Measured parameter	Activity, only "outer sphere" sorbates	Concentration, "outer" and "inner" sphere sorbates
Sources of errors	Sorbate losses in the system	Sorbate losses in the system, inner filter effects
Dissolved oxygen	No influence	Leads to errors
Sensitivity	High[a]	Very high
Time-resolved approach (sorption kinetics)	With restrictions	Possible
Disturbance of the sorption equilibrium	No (or negligible)[b]	No, non-invasive method

[a] Refers to SPME-GC/MS in SIM data acquisition (e.g. 0.5 ppb detection limit for phenanthrene using a 7 µm PDMS fibre); [b] Dynamic systems should be used for very hydrophobic sorbates.[24]

Recently, FQT and SPME were validated.[30] The investigations were triggered by indications that partition coefficients measured by FQT are higher by about one order of magnitude than SPME-based data (see refs. cited in ref. 30). The results of Doll's fundamental studies indicate that an "outer sphere" and an "inner sphere" of dissolved HOM need to be taken into account. The sorbate fraction, which is weakly bound to the "outer sphere" of the dissolved HOM and can readily be exchanged with the surrounding aqueous phase, is registered by FQT, but not by SPME. The "strong" binding of the sorbate to the "inner sphere" is measured by both methods. Obviously, the more significant the "outer sphere" bonding, the larger is the discrepancy of sorption data measured with both methods.

Our own results along with those of ref. 30 provide strong evidence that the findings published in refs. 31 and 32 are highly biased: partition coefficients obtained for the association of phenanthrene with dissolved HOM were found to be about one order of magnitude higher than the corresponding data for mineral-associated HOM. This was assumed to be explained by steric hindrance of the immobilised HOM (loss of "sorption activity").[32]

During our work to generate an *in-situ* permeable barrier (made up of HOM) in groundwater aquifers, sorption coefficients on dissolved HOM and on mineral-associated HOM were measured by means of SPME. Mineral-associated HOM (purchased from Roth, see Section 2.1) was coated on quartz sand using a procedure which will be detailed elsewhere.[33] Briefly, the simulated aquifer matrix was activated by means of a positively charged ferric hydroxide coating on the sand surface to immobilise dissolved HOM from solutions. Non-activated surfaces show poor sorption capabilities towards dissolved HOM because of the smooth surface and repulsion forces; at environmental pH values both the sand surface and the dissolved HOM possess negative surface charges.

Headspace-SPME was applied to avoid fibre fouling.[12] Sodium azide was added to prevent microbial activity and sodium chloride also was added to give a concentration of 0.02 mol L^{-1} to ensure constant ionic strength. The log K_{HOM} data for fluorene, phenanthrene and pyrene on the mineral-associated HOM were measured to be 3.79, 4.04 and 4.96. The results should be compared with partition coefficients measured on dissolved HOM (see Table 1: log K_{HOM} = 3.58, 3.93 and 4.70, respectively). The comparison provides strong evidence that there is no significant deviation in sorption data for a given solute depending on the state of HOM "aggregation." Therefore, the findings in ref. 31 appear to be associated with peculiarities of the fluorescence quenching technique.

3 CONCLUSIONS AND OUTLOOK

The one-parameter solubility concept is very useful for explaining and interpreting partitioning processes into HOM. Our own sorption data with very hydrophobic sorbates and literature data confirmed the value of the concept. HOMs of different genesis, maturity, source and so on have solubility parameters within a very narrow range δ_{HOM} = 25.5 ± 1.5 (J cm^{-3})$^{0.5}$. Hydrophobic humic acids possess solubility parameters at the lower level of this interval. Experimental partition coefficients into HOM agree very well with those predicted using this concept.

The concept also is very useful for interpreting bioconcentration as a partitioning process. Gas/solid and gas/liquid partitioning of hydrophobic organic compounds might also be explained by means of this concept. In this respect, the octanol-air coefficient should be the reference instead of K_{OW} data. Another field of application of solubility parameters is the selection of organic solvents to extract soils and sediments rich in HOM. According to the solubility theory, a solvent whose solubility parameter is close to that of the matrix (HOM) to be extracted is capable of "swelling" the HOM network. Diffusion is faster in such "swollen" polymers and the accessibility of the organic compounds is better as well. This will be detailed in a forthcoming contribution.

References

1. C. T. Chiou, D. E. Kile, D. W. Rutherford, G. Sheng and S. A. Boyd, *Environ. Sci. Technol.*, 2000, **34**, 1254.
2. C. T. Chiou, 'Soil Sorption of Organic Pollutants and Pesticides', Wiley Encyclopedia Series in Environmental Science, Vol. 8, R. A. Myers, (ed.), Wiley, New York, 1998, p. 4517.
3. S. K. Poole and C. F. Poole, *Anal. Commun.*, 1996, **33**, 417.
4. I. V. Perminova, N. Yu. Grechishcheva and V. S. Petrosyan, *Environ. Sci. Technol.*, 1999, **33**, 3781.
5. M. L. Brusseau, *Environ. Toxicol. Chem.*, 1993, **12**, 1835.
6. R. Seth, D. Mackay and J. Muncke, *Environ. Sci. Technol.*, 1999, **33**, 2390.
7. M. C. Barron, *Environ. Sci. Technol.*, 1990, **24**, 1612.

8. F. -D. Kopinke, J. Poerschmann and U. Stottmeister, *Environ. Sci. Technol.*, 1995, **29**, 941.
9. J. Brandrup and E. H. Immergut, 'Polymer Handbook', 3rd Edn., Ch. VII, Wiley, New York, 1991.
10. A. L. Horvath, in 'Studies in Physical and Theoretical Chemistry: 75', Elsevier, New York, 1992, p. 425.
11. A. F. M. Barton, 'CRC Handbook of Solubility Parameters and other Cohesive Parameters', CRC Press, Boca Raton, FL, 1985, p. 257.
12. J. Poerschmann, F. -D. Kopinke and J. Pawliszyn, *J. Chromatogr. A*, 1998, **816**, 159; *Environ. Sci. Technol.*, 1997, **31**, 3629.
13. M. J. Kamlet, R. M. Doherty, G. D. Veith, R. W. Taft and M. H. Abraham, *Environ. Sci. Technol.*, 1986, **20**, 690.
14. M. J. Kamlet, R. M. Doherty, P. M. Carr, D. Mackay, M. H. Abraham and R. W. Taft, *Environ. Sci. Technol.*, 1988, **22**, 503.
15. Y. -P. Chin and W. J. Weber, *Environ. Sci. Technol.*, 1989, **23**, 978.
16. Y. -P. Chin, G. R. Aiken and K. M. Danielsen, *Environ. Sci. Technol.*, 1997, **31**, 1630.
17. M. A. Schlautman and J. A. Morgan, *Environ. Sci. Technol.*, 1993, **27**, 961.
18. C. T. Chiou and D. E. Kile, *Environ. Sci. Technol.*, 1998, **32**, 338.
19. G. Xia and W. P. Ball, *Environ. Sci. Technol.*, 1999, **33**, 262.
20. J. Poerschmann and T. Gorecki, *Environ. Sci. Technol.*, submitted.
21. C. T. Chiou, *Environ. Sci. Technol.*, 1985, **19**, 57.
22. J. G. Burken and J. L. Schnoor, *Environ. Sci. Technol.*, 1998, **32**, 3379.
23. N. P. Bahadur, W. Y. Shiu, D. G. B. Boocock and D. J. Mackay, *J. Chem. Engn. Data*, 1999, **44**, 40.
24. J. Poerschmann, T. Gorecki and F. -D. Kopinke, *Environ. Sci. Technol.*, in press.
25. J. Poerschmann and T. Gorecki, *J. Microcol. Sep.*, submitted.
26. F. -D. Kopinke, J. Poerschmann and A. Georgi, in 'Applications of Solid Phase Microextraction', J. Pawliszyn, (ed.), Chromatography Monographs, Royal Society of Chemistry, Cambridge, 1999.
27. M. M. Puchalski, M. J. Morra and R. von Wandruszka, *Environ. Sci. Technol.*, 1992, **26**, 1787.
28. F. H. Frimmel and U. Kumke, in 'Humic Substances: Structures, Properties and Uses,' G. Davies and E. A. Ghabbour, (eds.), Royal Society of Chemistry, Cambridge, 1998, p. 113.
29. R. P. Schwarzenbach, P. M. Gschwend and D. M. Imboden, 'Environmental Organic Chemistry', Wiley, New York, 1993.
30. T. E. Doll, F. H. Frimmel, M. U. Kumke and G. Ohlenbusch, *Fres. J. Anal. Chem.*, 1999, **264**, 313.
31. Y. Laor, W. J. Farmer, Y. Aochi and P. F. Strom, *Wat. Res.*, 1998, **32**, 1923.
32. K. D. Jones and C. L. Tiller; *Environ. Sci. Technol.*, 1999, **33**, 580.
33. G. Balcke, S. Woszidlo and J. Poerschmann, in preparation.
34. F. C. Spurlock and J. W. Biggar, *Environ. Sci. Technol.*, 1994, **28**, 989.
35. S. H. Bertelsen, A. D. Hoffmann, C. A. Gallinat, C. M. Elonen and J. W. Nichols, *Environ. Toxicol. Chem.*, 1998, **17**, 1447.
36. W. M. Meylan, P. H. Howard, R. S. Boethling, D. Aronsson, H. Printup and S. Gouchie, *Environ. Toxicol. Chem.*, 1999, **18**, 664.
37. G. Schüürmann, *Umweltchem. Ökotox.*, 1997, **9**, 345.

38. W. H. J. Vaes, E. U. Ramos, C. Hamwijk, I. van Holsteijn, B. J. Blaauboer, W. Seinen, W. J. M. Verhaar and J. L. M. Hermens, *Chem. Res. Toxicol.*, 1997, **10**, 1067.
39. W. H. J. Vaes, E. U. Ramos, H. J. M. Verhaar, W. Seinen and J. L. M. Hermens, *Anal. Chem.*, 1996, **68**, 4463.
40. H. Yuan, R. Ranatunga, P. W. Carr and J. Pawliszyn, *Analyst,* 1999, **124**, 1443.

THE INTERACTION BETWEEN ESFENVALERATE AND HUMIC SUBSTANCES OF DIFFERENT ORIGIN

L. Carlsen,[1] M. Thomsen,[1] S. Dobel,[1] P. Lassen,[1] B. B. Mogensen[1] and P. E. Hansen[2]

[1] National Environmental Research Institute, Department of Environmental Chemistry DK-4000 Roskilde, Denmark
[2] Department of Life Science and Chemistry, Roskilde University, DK-4000 Roskilde, Denmark

1 INTRODUCTION

Pyrethroids have been widely used as agricultural insecticides for more than twenty years. In 1983, pyrethroids accounted for about a quarter of the world's foliage insecticide market.[1] Pyrethroids are highly lipophilic and have been shown to be extremely toxic toward aquatic organisms.[2] Due to their lipophilic nature, pyrethroids are capable of binding to organic material, which will reduce their bioavailability and toxicity.[2] Investigations of the sorption of pyrethroids by dissolved organic material (DOM) are therefore important for the understanding of the environmental fate of this class of compounds.

It is well known that hydrophobic organic compounds can sorb to DOM (e.g., humic substances, HSs) and thereby alter the fate of the hydrophobic compounds.[3-7] Humic substances are organic macromolecules originating from decomposition of plant and animal residues. HSs are complex, negatively charged polymers that have a large number of different functional groups attached to an aliphatic aromatic backbone structure. Thus, HSs contain both hydrophilic and hydrophobic sites. HSs are heterogeneous and have a range of molecular weights. Thus, fulvic acids (FAs) range from approx. 0.6 to 5 kDa and humic acids from approx. 1.5 to 500 kDa.[8] The size and chemical composition of HSs depend on the type and origin. HSs from marine environments are usually more aliphatic and less aromatic than those from freshwater, whereas HSs from soil are generally more aromatic and less aliphatic than those from freshwater.[8]

The ability of HSs to interact with hydrophobic organic compounds depends on different parameters such as the origin and composition of the humic substances[3,4,7,9,10] and the concentration.[5,11,12] Their binding ability also depends on the chemical structure of the hydrophobic organic components.[6,13-15] Most investigations have been made with relatively simple hydrophobic organic compounds like PAHs and other aromatic compounds. These investigations have shown a relationship between K_{DOM} and K_{ow} of the hydrophobic organic compound[6,14,16] as well as a relationship between K_{DOM} and the aromatic content of HSs.[4,7,10]

The purposes of the present study are 1) to determine the partitioning coefficient (K_{DOM}) for esfenvalerate as a model compound of the pyrethroids, between water and HSs from different sources; and 2) to correlate these data with different characterisation parameters of the HSs. Pyrethroids are complex molecules and it is therefore possible that K_{DOM} is affected by other parameters than simply the aromatic content of the humic materials. Two of the HSs were isolated in the present study. The HSs were characterised by liquid state ^{13}C-NMR spectroscopy, elemental analysis, ultraviolet-visible (UV-VIS) spectroscopy and size exclusion chromatography. Multivariate data analysis (partial least square regression, PLS-R) was used to determine the relationship, if any, between K_{DOM} values and the composition of the different HSs samples.

2 MATERIALS AND METHODS

2.1 Humic Substances

Three HAs, two FAs and two HSs containing both HAs and FAs were examined in the present study (Table 1).

2.1.1 Isolation of Humic Substances from the Water Pond and Gorleben Groundwater. Isolation of the two HSs was performed with diethylaminoethyl (DEAE)-cellulose according to Miles et al.[21] The DOC content of the water from the water pond was 11 mg/L and 100 mg/L of the water from Gorleben groundwater. The humic materials were eluted from the DEAE-cellulose resin with 0.5 M NaOH. The pH of the eluates were adjusted to approximately 9 with 1 M HCl. The HSs solutions were freeze dried to a smaller volume and dialysed (Spectrapor CE membrane MWCO 500). After dialysis, the solutions of HSs were freeze dried.

Table 1 *Humic substances used in the present study*

Humic substances	Names	Origin
Humic acids	Aldrich HA (Na$^+$)	Commercial cat. No.: H 1,675-2
	Purified Aldrich HA[17]	Commercial cat. No.: H 1,675-2
	Kranichsee HA[18]	Water from the raised bog Kleiner Kranichsee, Saxony, Germany
	Gohy-573-HA-(H$^+$)II[19]	From groundwater in Gorleben, Germany
Fulvic acids	DE72[20]	From Derwent Reservoir Derbyshire, England
	FA surface	From Soulaines surface water, France
Humic substances	Gohy-573-HS-(H$^+$)II	From groundwater in Gorleben, Germany
	Water pond HS	From an artificial water pond, Roskilde, Denmark

2.1.2 Stock Solutions of Humic Substances. Stock solutions of the different humic compounds with a concentration of 500 mg/L were adjusted to pH 8.3 with 0.1 M NaOH or 0.1 M HCl. The humic solutions were filtered through a 0.45 µm filter and diluted to concentrations 10, 20, 30, 40, 50, 75 and 100 mg humic substances L^{-1}.

2.2. Characterisation of Humic Substances

2.2.1 Elemental Analysis. Elemental analyses of HSs samples were performed on an EA 1110 CHNS analyzer.

2.2.2 UV-VIS Spectroscopy. Spectroscopic analysis were carried out on a Cary 50 UV-Visible spectrophotometer. Absorbance values at 250, 365, 465 and 665 nm were measured for calculating E_2/E_3 (the absorbance at 250 nm divided by the absorbance at 365 nm) and E_4/E_6 (the absorbance at 465 nm divided by the absorbance at 665 nm) ratios. The absorptivity at 272 nm also was determined.

2.2.3 ^{13}C-NMR Analysis. Spectra were recorded in 0.5 M NaOD (50 mg in 0.5 mL) on a 250 MHz Bruker instrument. ^{13}C NMR was not performed on the water pond HS due to insufficient amounts.

2.2.4 Size Exclusion Chromatography. The HSs size distribution was investigated on 300 mg/L and pH 8.3 solutions with a Sephadex® G-50 Medium gel (Code No. 17-0043-01, Pharmacia Biotech AB).

2.3 Determination of K_{DOM}

The partitioning coefficients, K_{DOM}, of esfenvalerate between HSs and water were determined using C-18 columns according to Landrum et al.[12] ^{14}C-esfenvalerate (Sumitomo Chemicals) with a total concentration of 2.6 µg/L and a nominal activity of 1.95 Bq/mL was added to the solutions of the HSs with concentrations 0, 10, 20, 30, 40, 50, 75 and 100 mg/L, respectively. The samples were allowed to equilibrate for 24 h. The single samples were added to a C-18 column (Isolute C-18, 500 mg/3 mL column reservoir). The freely dissolved esfenvalerate was retained on the column whereas the esfenvalerate bound to humic substances was eluted through the column. A small amount of HSs was retained on the columns, which subsequently were rinsed with water. Freely dissolved esfenvalerate was eluted from the columns with 1% ethyl acetate in hexane. ^{14}C-activity in the samples were determined by liquid scintillation counting on a Beckman LS 1801 counter using liquid scintillation counting cocktail Ultima Gold™ (Packard). The results were corrected for quenching, breakthrough of esfenvalerate and HSs retained on the C-18 columns.

3 RESULTS AND DISCUSSION

3.1 Characterisation of the Humic Substances

3.1.1 Spectroscopic and Elemental Analysis. Spectroscopic and elemental analysis have been widely used for characterisation of humic substances. These data are related in different ways to the HSs composition. Elemental analysis can give an indication of the composition of the humic material. In order to be able interpret the data correctly it is important that there are only few impurities in the sample. However, as any clean up procedure will alter the inherent characteristics of the humic material it may be difficult to get a precise picture of the elemental composition. H/C ratios are a simple method to get an indication of the overall structures of the humic material. Low H/C ratios indicate that the humic material predominantly consists of aromatic moieties and/or carbonyl and quinone groups, whereas high H/C ratios indicates aliphatic structures and primary amines.[17]

The E_4/E_6 ratio has been widely used by scientists for the characterisation of humic material. Chen et al.[22] showed that the E_4/E_6 ratio is mainly correlated to the average molecular weight and size and to a lesser degree to the oxygen content of the humic material. A high E_4/E_6 ratio indicates a low average molecular weight and size as well as high oxygen content. However, as the E_4/E_6 ratio contains information of more than one parameter the results may be difficult to interpret. The E_2/E_3 ratio has been shown to contain information concerning the degree of aromatic content, i.e., a low E_2/E_3 ratio indicates high degrees of aromatic content.[23] A high absorptivity at 272 nm (ε_{272}) should also indicate a high degree of aromatic content, as the $\pi \rightarrow \pi^*$ transition for aromatic compounds occurs at this wavelength.[24,25]

From the elemental analysis (Table 2) Gohy-573-HA and Purified Aldrich HA have the highest percentage of carbon whereas Water pond HS and Gohy-573-HS have the lowest percentage of C. The low percentage C of Aldrich HA (Na^+) compared to Purified Aldrich HA is due to high content of inorganic impurities of approximately 100 mg/g.[17] The low percentage C in Gohy-573-HS and Water pond HSs indicates a high percentage of oxygen. From the H/C ratios, Water pond HS and Gohy-573-HS appear to have the most aliphatic structures, whereas Kranichsee HA, Gohy-573-HA and Purified Aldrich HA appear to have the highest aromatic content or C=O structures. It should be emphasised that the elemental composition of Gohy-573-HA has been taken from the literature.[19]

The E_4/E_6 ratios (Table 2) indicate that Water pond HS has the lowest average molecular weight and/or highest oxygen content, the latter supported by the low carbon content in the elemental analysis. The E_4/E_6 ratio indicates that Aldrich HA (Na^+) has the highest average molecular weight. Purified Aldrich HA has a higher E_4/E_6 value compared to Gohy-573-HA and Aldrich HA (Na^+). This was not expected since HSs from soil are expected to have higher molecular weight compared to HAs from groundwater. However, due to impurities the spectroscopic data of Aldrich HA (Na^+) are difficult to interpret.

According to E_2/E_3 ratios the HAs with the largest fraction of aromatic carbon appear to be Purified Aldrich HA and Aldrich HA (Na^+). This is in agreement with the fact that commercial Aldrich HA has a high degree of aromaticity. The HSs with the lowest fractions of aromatic carbon determined by the E_2/E_3 ratios are Water pond HS and FA

Table 2 *Elemental and spectroscopic data*

Humic Substances	Elemental analysis				E_4/E_6	E_2/E_3	ε_{272}
	%C	%H	%N	H/C			$L\ (mg\ cm)^{-1}$
Aldrich HA (Na^+)	38.37	4.68	0.57	1.46	5.56	2.67	0.029
Gohy-573-HA-(H^+)II[19]	57.32	4.76	1.77	1.00	6.2	2.89	0.034
Purified Aldrich HA[17]	53.27	4.88	0.93	1.10	7.43	2.48	0.039
Gohy-573-HS-(H^+)II	34.03	5.05	1.14	1.78	8.27	3.30	0.016
Kranichsee HA[18]	49.34	4.07	1.60	0.99	8.82	3.10	0.024
FA surface	48.76	5.03	1.14	1.24	11.0	7.92	0.007
DE72[20]	49.06	4.18	0.60	1.02	11.8	3.00	0.022
Water pond HS	28.29	4.83	1.81	2.05	21.3	7.04	0.004

surface. HSs from fresh waters have a low input of organic material from higher plants compared to HSs derived from terrestrial sources.

From the ε_{272} it also is inferred that Purified Aldrich HA and Gohy-573-HS have the highest aromatic content. Water pond HS and FA surface have the lowest ε_{272}, which indicates the lowest degree of aromaticity, in agreement with the E_2/E_3 ratios and elemental analyses.

3.1.2 ^{13}C-NMR. The different fractions of carbons in the HSs determined by ^{13}C-NMR-spectroscopy are shown in Table 3. It can be seen that the HSs with the largest aromatic fraction determined by ^{13}C-NMR are Gohy-573-HA and Purified Aldrich HA. It was expected that Aldrich HA had a higher aromatic content compared to Gohy-573-HA. However, the difference is very small and could be due to experimental uncertainties.

Fulvic acids have the lowest content of aromatic carbon. FA surface is especially low in aromatic C. DE72 has a higher content of aromatic C compared to FA surface, which is in agreement with the E_2/E_3 ratios. The HSs with the largest fraction of aliphatic C is FA surface. Kranichsee HA and apparently Gohy-573-HS have the smallest fraction of aliphatic carbon. For Kranichsee HA, this is in agreement with the H/C ratio. However, the H/C ratio and aliphatic C obtained by ^{13}C-NMR are not in agreement for Gohy-573-HS, indicating the presence of impurities (*vide supra*). A further possible explanation could be a high content of hydroxyl groups, which contribute to the high H/C ratio. The relative amount of carboxylic and/or ester bonded carbon atoms is in agreement with the elemental analysis, the oxygen content decreasing in the order Gohy-573-HS, DE72 > Kranichsee HA, FA surface > Purified Aldrich HA, Gohy-573-HA.

3.1.3 Size Exclusion Chromatography (SEC). The results of the SEC are shown in Figure 1, which is divided into FAs (1a), HSs (1b) and HAs (1c). According to SEC, the MW range for the humic compounds are as expected: FAs < HSs < HAs. For the HSs and HAs, the chromatograms display two well separated peaks. The first peak represents a small fraction of high molecular weight molecules, which is eluted with the eluent front. The second peak represents a larger fraction of smaller molecules, which is separated according to size. The chromatograms of the fulvic acids are different, as FA surface only has one fraction of smaller molecules although DE 72 fulvic acid has a minor shoulder in

Table 3 *^{13}C-NMR-spectroscopy. The results are presented in %. It was not possible to record spectra of Aldrich HA (Na^+) and Water pond HS*

Humic Substances	Aliphatic C	Amino acids and carbohydrates C	Aromatic C (Phenol+ aromatic)	Carboxylic and ester C	Ketonic and aldehyde C
	0-65 ppm	65-95 ppm	95-165 ppm	165-195 ppm	195-215 ppm
Purified Aldrich HA	19	4	56	17	4
Kranichsee HA	17	20	35	25	4
Gohy-573-HA-(H^+)II	27	3	60	9	1
DE72	25	9	31	31	3
FA surface	51	14	7	22	6
Gohy-573-HS-(H^+)II	17	6	41	30	6

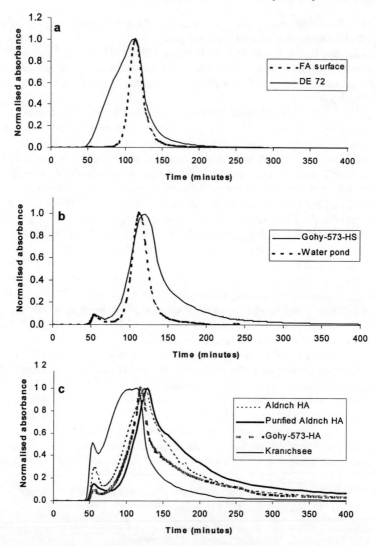

Figure 1 *Size exclusion chromatography of the humic substances*

the high MW area. In general the results from the size exclusion chromatography are in agreement with the E_4/E_6 ratios except for the Kranichsee HA and Water pond HS. From the E_4/E_6 measurement it was expected that Water pond HS would have the lowest MW range. However, according to the size exclusion chromatography this was not the case. As E_4/E_6 secondarily is correlated to the oxygen content of humic material,[22] it is possible that a high content of oxygen (see Table 2) in the Water pond HS has a stronger influence on E_4/E_6 than the average molecular weight compared to the other HSs. This could not be verified as it not was possible to record ^{13}C-NMR spectra of the Water pond HS. However, the elemental analysis showed a low C content and it is likely that the oxygen content is high. Kranichsee HA appeared to have the highest MW range according to SEC and the disagreement with the E_4/E_6 ratio possibly is due to the oxygen content.

The size exclusion chromatogram for Kranichsee HA significantly differs from the other humic acids, as the second peak has less tailing and a higher apparent average molecular weight. A possible explanation is that Kranichsee HA is less heterogeneous in size and composition due to is its origin from a raised bog, which has low diversity in flora and low decomposition.

3.2 The Partitioning Coefficient K_{DOM}

The partition coefficient was calculated as:

$$K_{DOM} = \frac{[esf]_{bound}}{[esf]_{free}[HS]_{free}} \qquad (1)$$

where $[esf]_{bound}$ and $[esf]_{free}$ are the concentration of esfenvalerate bound to HSs and the concentration of freely dissolved esfenvalerate, respectively. $[HS]_{free}$ is the humic substance concentration. It is assumed that the HSs are in excess compared to esfenvalerate and thus

$$[HS]_{total} \approx [HS]_{free} \qquad (2)$$

The partition coefficients K_{DOM} of esfenvalerate between HSs and water at different concentrations of humic substances are shown in Figure 2 and Table 4.

The effects of HSs concentration on K_{DOM} were examined at concentrations from 10-100 mg/L. K_{DOM} decreased with increasing concentrations of the humic substances (Figure 2 and Table 4). This is in agreement with previous observations for PAH's.[5,6,11,12,14] A possible explanation for the decrease in K_{DOM} with increasing HSs concentration is a change of HSs configuration with the concentration. The macromolecular configuration of humic substances depends on the concentration of the solution.[26] At high HSs concentration the molecules form rigid spherocolloids due to a decrease in the average molecular area.[26] This configuration change explains the decrease in K_{DOM}. When HSs are aggregated there will be less binding sites for esfenvalerate and a reduced K_{DOM}.

Table 4 shows that the HSs can be divided into three groups according to the K_{DOM} values at 10 mg/L. These groups are not in accordance to the division into humic acid, fulvic acids and humic substances. In the group with the highest K_{DOM} values are Aldrich HA (Na$^+$), Purified Aldrich and Kranischsee HA. The group with medium K_{DOM} values contains Water pond HS, FA surface and Gohy-573-HA and the group with the low K_{DOM} values is DE72 and Gohy-573-HS (Table 4).

In the group with the highest K_{DOM} values it is observed that the relative decrease in the K_{DOM} is highest for Kranichsee HA with a 93% decrease in K_{DOM} from 10 to 100 mg humic acid /L. This significant decrease in K_{DOM} indicates that Kranichsee HA has a better ability than Aldrich HA (Na$^+$) and Purified Aldrich HA to form rigid spherocolloids at higher HSs concentration. This can be explained by the low aliphatic content of Kranichsee HA, which results in less molecular flexibility when the concentration of HA increases. Gohy-573-HS apparently also has a better ability to form rigid spherocolloids compared to DE72 according to the highest relative decrease in K_{DOM}. The relative decreases in K_{DOM} for esfenvalerate ranged between 73 and 95% at concentrations of humic substances from 10 to 100 mg/L, and the largest decreases are observed between 10 and 30 to 40 mg/L.

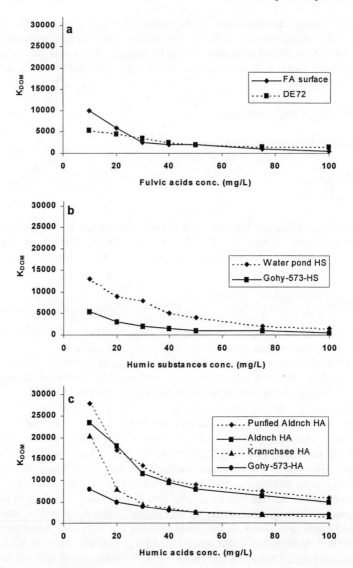

Figure 2 K_{DOM} *for esfenvalerate between water and humic substances as a function of the concentration of humic substance: (a) fulvic acids, (b) humic substances, (c) humic acids*

The partition coefficients K_{DOM} for esfenvalerate are not as large those for some other nonpolar organic compounds. Thus, K_{DOM} for pyrene between Aldrich humic acid and water has been determined as 75,000 at HA concentrations up to 80 mg/L[3] and K_{DOM} for DDT between Aldrich humic acid and water is 65,000 for concentrations at 20 mg/L HA.[12] This difference in the ability of esfenvalerate and compounds like pyrene and DDT to bind humic substances may well be due to the differences in configurations of the compounds.

PAHs such as pyrene are planar and aromatic, while esfenvalerate is more aliphatic and has polar ester and cyano groups attached (Figure 3). The planar structures of the PAH's are likely to fit better into the structure or binding sites of HSs compared to the more complex structure of esfenvalerate.

Table 4 *Average K_{DOM} values for esfenvalerate distribution onto humic substances at different concentrations*

Humic s	K_{DOM}							
Substance	Purified	Aldrich	Kranichsee	Water pond	FA	Gohy-573	DE72	Gohy-573
Mg/L	Aldrich	HA (Na$^+$)	HA	HS	surface	HA(H$^+$)II		HS(H$^+$)II
10	28,000	23,500	20,500	13,000	10,000	8,000	5,500	5,500
20	17,000	18,000	8,000	9,000	6,000	5,000	4,500	3,000
30	13,500	11,500	4,500	8,000	2,500	4,000	3,500	2,000
40	10,000	9,500	3,500	5,000	2,000	3,000	2,500	1,500
50	9,000	8,000	2,500	4,000	2,000	2,500	2,000	1,000
75	7,500	6,500	2,000	2,000	1,000	2,000	1,500	1,000
100	6,000	5,000	1,500	1,500	500	2,000	1,500	500

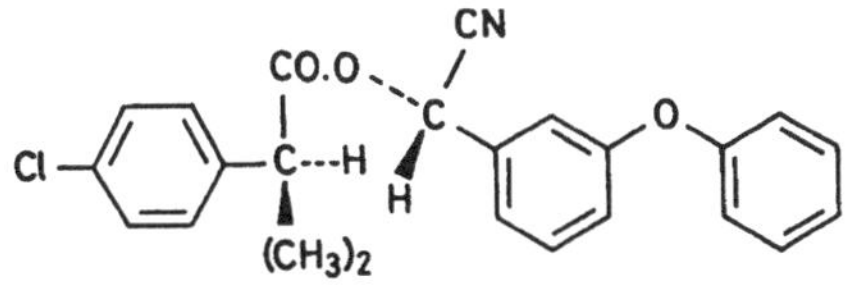

Figure 3 *Configuration of esfenvalerate*

Significant differences in the values of K_{DOM} are observed for esfenvalerate at 10 mg/L for the individual HSs. This difference in K_{DOM} for esfenvalerate between HSs from different locations and from water probably is due to differences in the number of hydrophobic sites. It is believed that HSs with higher K_{DOM} values for hydrophobic compounds have more hydrophobic sites. The lipophilic esfenvalerate will bind more strongly HSs with more hydrophobic character than to more hydrophilic HSs. Thus, HSs like DE72 and Gohy-573-HS have relatively low K_{DOM} values for esfenvalerate. This agrees with the observation that commercial HSs and HSs derived from soil bind more strongly to nonpolar organic contaminants compared to HSs derived from aquatic sources.[3,4,16,27] Soil and commercial HSs are known to have a larger fractions of aromatic carbon and less oxygen containing groups compared to aquatic humic substances.

A comparison of the different characterisation data and the K_{DOM} values showed that no simple correlation exists. Partial least squares regression was applied to include contributions from all characterisation data to the explanation of the patterns in K_{DOM}.

4 PARTIAL LEAST SQUARES REGRESSION

PLS-R allows us to investigate the simultaneous influence of all measured parameters on the variation in the partitioning of esfenvalerate to the different HSs. The sorption of esfenvalerate to dissolved HSs of different origin is a quantification of the latent properties consisting of the projected original characterisation parameters onto two dimensional space. What PLS does is to reduce the n-dimensional space, represented by the different characterisation parameters, into a two-dimensional space consisting of two principal components PC1 and PC2. The latter makes it possible to interpret the data graphically by picturing PC2 versus PC1 (Figure 4). Parameters having high positive PC1 values are proportionally related to K_{DOM}, whereas parameters having high negative PC1 values are inversely related to K_{DOM}.

K_{DOM} increases with increasing aromatic content, absorptivity and %C, as these three variables lie closest to K_{DOM}. Furthermore, K_{DOM} is inversely related to the polarity of the HSs expressed through the variables; (N+O)/C, %O, %N, amino and carbohydrates, ketones and aldehydes, carboxylic and ester groups and E_4/E_6, as these parameters have opposite sign. Figure 5 depicts the K_{DOM} values calculated by the model versus the measured K_{DOM} values. This PLS model for estimating K_{DOM} at 100 mg/L HSs has a correlation coefficient of 0.9. Investigation of models for estimating K_{DOM} at lower concentrations resulted in low correlation coefficients and low robustness of the models.

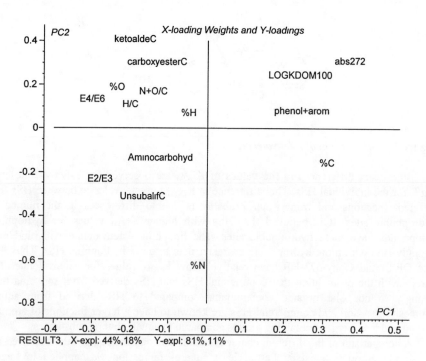

Figure 4 *PC1 explains 81 % of the variation, while PC2 explains 11% only of variation in the partitioning of esfenvalerate to HSs at a concentration of 100 mg/L*

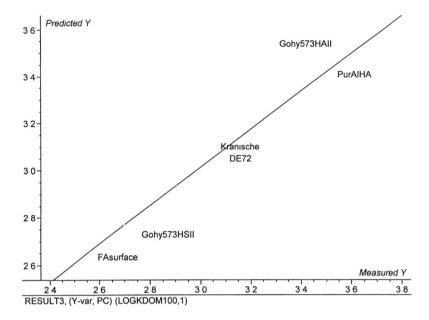

Figure 5 K_{DOM} values calculated from the model and compared to the measured values

The observed decrease in model performance was initiated by an increased significance of the polar characterisation parameters. It seems that high K_{DOM} values are favoured by hydrophobic interactions and may be limited by repulsive interactions between functional groups of HSs and esfenvalerate.

5 CONCLUSION

The partition coefficients K_{DOM} were measured for esfenvalerate between aqueous HSs of different origin and at different concentrations. This resulted in decreasing K_{DOM} with increasing HSs concentration due to a change in HSs configuration. The results showed different K_{DOM} values for esfenvalerate for the different HSs. K_{DOM} for nonpolar organic compounds are known to differ for HSs from different environments. The aromatic fractions of HSs are especially significant for the binding of the nonpolar organic compounds. For esfenvalerate, the same trend seems to prevail when modeling K_{DOM} at the highest HSs concentration (100 mg/L). For HSs with a variety of chemical compositions and size distributions, it has been possible to quantify K_{DOM} at [HSs] = 100 mg/L but not at lower concentration because of an increase in the significance of polar parameters. Esfenvalerate contains polar ester and nitrile groups. The strong dependence of K_{DOM} on polarity indicates several different binding mechanisms for esfenvalerate to more polar HSs at low humic concentrations, i.e. in the un-curled state. This aspect could not be verified because of the limited number of polar humic samples in the present study and will be subject to further investigations.

ACKNOWLEDGEMENTS

The HSs samples were kindly provided by: Dr. A. Maes, KUL, Belgium; Dr. K. H. Heise, FZR, Germany; Dr. J. Higgo, BGS, England; Dr. V. Molin, CEA, France; and Dr. G. Buckau, FZR, Germany.

References

1. J. J. Hervé, in 'The pyrethroid insecticides', J. P. Leahey, (ed.), Taylor & Francis, London, 1985, Chapter 6, p. 343.
2. I. R. Hill, in 'The pyrethroid insecticides', J. P. Leahey, (ed.), Taylor & Francis, London, 1985, Chapter 4, p. 151.
3. Y. Chin, G. R. Aiken and K. M. Danielsen, *Environ. Sci. Technol.*, 1997, **31**, 1630.
4. T. D. Gauthier, W. R. Seitz and C. L. Grant, *Environ. Sci. Technol.*, 1987, **21**, 243.
5. P. Lassen and L. Carlsen, *Chemosphere*, 1997, **34**, 817.
6. P. Lassen and L. Carlsen, *Chemosphere*, 1999, **38**, 2959.
7. S. Tanaka, K. Oba, M. Fukushima, K. Nakayasu and K. Hasebe, *Anal. Chim. Acta*, 1997, **337**, 351.
8. R. L. Malcolm, *Anal. Chim. Acta*, 1990, **232**, 19.
9. J. F. McCarthy, L. E. Roberson and L. W. Burrus, *Chemosphere*, 1989, **19**, 1911.
10. I. V. Perminova, N. Y. Grechishcheva and V. S. Petrosyan, *Environ. Sci. Technol.*, 1999, **33**, 3781.
11. C. W. Carter and I. H. Suffet, *Environ. Sci. Technol.*, 1982, **16**, 735.
12. P. F. Landrum, S. R. Nihart, B. J. Eadle and W. S. Gardner, *Environ. Sci. Technol.*, 1984, **18**, 187.
13. C. W. Carter and I. H. Suffet, in 'Fate of chemicals in the environment: Compartmental and multimedia models for predictions', R. L. Swann and A. Eschenroeder, (eds.), Symposium Series No. 225, American Chemical Society, 1983, p. 215.
14. J. F. McCarthy and B. D. Jimenez, *Environ. Sci. Technol.*, 1985, **19**, 1072.
15. N. R. Morehead, B. J. Eadie, B. Lake, P. F. Landrum and D. Berner, *Chemosphere*, 1986, **15**, 403.
16. C. T. Chiou, R. L. Malcolm, T. I. Brinton and D. E. Kile, *Environ. Sci. Technol.*, 1986, **20**, 502.
17. J. I. Kim, G. Buckau, G. H. Li, H. Duschner and N. Psarros, *Fresenius J. Anal. Chem.*, 1990, **338**, 245.
18. K. Schmeide, H. Zänker, K. H. Heise and H. Nitsche, in 'Effects of Humic Substances on the Migration of Radionuclides: Complexation and Transport of Actinides, First Technical Progress Report', G. Buckau, (ed.), FZKA 6124, 1998, p. 161.
19. J. I. Kim, G. Buckau, R. Klenze, D. S. Rhee and H. Wimmer, 'Characterization and complexation of humic acids', Commission of the European Communities, Nuclear Science and Technology, Luxembourg, EUR 13181 EN, 1991.
20. J. J. W. Higgo, J. R. Davies, B. Smith and C. Milne, in 'Effects of Humic Substances on the Migration of Radionuclides: Complexation and Transport of Actinides', First Technical Progress Report, G. Buckau, (ed.), FZKA 6124, 1998, p. 103.

21. C. J. Miles, J. R. Tuschall, Jr. and P. L. Brezonik, *Anal. Chem.*, 1983, **55**, 410.
22. Y. Chen, N. Senesi and M. Schnitzer, *Soil Sci. Soc. Am. J.*, 1977, **41**, 352.
23. J. Peuravuori and K. Pihlaja, *Anal. Chim. Acta*, 1997, **337**, 133.
24. J. M. Novak, G. L. Mills and P. M. Bertsch, *J. Environ. Qual.*, 1992, **21**, 144.
25. S. J. Traina, J. Novak and N. E. Smeck, *J. Environ. Qual.*, 1990, **19**, 151.
26. K. Ghosh and M. Schnitzer, *Soil Sci.*, 1980, **129**, 266.
27. C. T. Chiou, D. E. Kile, T. I. Brinton, R. L. Malcolm and J. A. Leenheer, *Environ. Sci. Technol.*, 1987, **21**, 1231.

ADSORPTION-DESORPTION INTERACTIONS OF ENVIRONMENTAL ENDOCRINE DISRUPTORS WITH HUMIC ACIDS FROM SOILS AND URBAN SLUDGES

E. Loffredo, M. Pezzuto and N. Senesi

Department of Agroforestal and Environmental Biology and Chemistry, University of Bari, 70126 Bari, Italy

1 INTRODUCTION

Several xenobiotic compounds largely diffused in the environment have the potential to interfere with the endocrine system of animals and humans by acting as hormone-like substances in the organism.[1] These compounds are known as "endocrine disruptors" (EDs) and are able to alter or disrupt the normal functioning of the endocrine system either directly, by blocking or imitating natural hormones, or indirectly, by interfering with the synthesis, storage, secretion, transport, catabolism and activity of various natural hormones.[2,3] Although the consequences of such interference on animal and human health are not yet completely understood, it is known that small disturbances in endocrine functions, especially during certain stages of life, such as development, pregnancy and lactation, can lead to serious and long-lasting effects.[4] Recent progress in both epidemiological and toxicological research indicates that exposure to potentially endocrine disrupting agents may interfere with human reproductive function and success, and that these agents may have a particularly important role in the etiology of a variety of endocrine mediated disorders.[5]

Compounds proven or suspected to act as EDs often may be introduced into the environment through common agricultural practices and by application, discharge and/or disposal of sewage and industrial effluents and sludges, and disposal of plastic and pharmaceutical residues. Different classes of herbicides, fungicides, insecticides and nematicides, several industrial chemicals, including polychlorobiphenyls (PCBs) phthalates, dioxins and some pharmaceutical products, like estrogenic compounds, have been identified as potential environmental EDs.[6,7] Despite the high inputs and potential toxicity of EDs and the increasing social concern for a reliable risk assessment of the actual exposure of animals and humans to EDs, relatively few data are available on the types and amounts of EDs introduced into soil and aquatic systems. Until now, attention on this issue mainly has focused on the contamination of water, and very little information is available on the role and contribution of sewage sludge application to the contamination of soil by EDs.

In recent years, soil amendment with sewage sludges has become a very common agricultural practice worldwide. The risk assessment of potential environmental hazards associated with the presence of EDs in sludge amendment and soil, as well as the definition of possible remediation measures, require an accurate evaluation and quantification of the soil response to these compounds. Adsorption onto soil solid phases unanimously is considered one of the most important processes that controls the fate, behavior and performance, including mobilization/immobilization, transport, bioavailability and toxicity, of organic xenobiotic compounds in soil.[8,9] Soil organic matter, and especially its water insoluble humified fractions, i.e., humic acids (HAs), are well known to play a key role in the adsorption of organic contaminants in soil.[10,11] The evaluation of the kinetics and extent of adsorption of EDs onto HAs, as well as the measurement of related desorption processes are key requirements for understanding the fate of EDs in soil.

The objective of this work is the study of adsorption and desorption processes of two ascertained ED compounds, bisphenol A (BPA) and 17-α-ethynilestradiol (EED), onto different HAs isolated from representative soils and sewage sludges used for soil amendment.

2 MATERIALS AND METHODS

2.1 Endocrine Disruptors

Two ED compounds have been examined in this work: (a) the xenoestrogen bisphenol A (BPA) [2,2-(4,4-dihydroxydiphenyl)propane]; and (b) the synthetic estrogen 17-α-ethynilestradiol (EED) [17α-ethynil-1,3,5(10)-estratriene-3,17β-diol]. Both compounds, of 99% purity, were obtained from Sigma-Aldrich. The molecular formulas of BPA and EED are shown in Figure 1.

BPA is an intermediate compound in the manufacture of epoxy resins and polycarbonates, and also is used in manufacturing adhesives, building materials, compact disks and electrical and electronic parts, and in agriculture as a fungicide. EED is a synthetic estrogen used for medical purposes, often in combination with progestogen as an oral contraceptive.

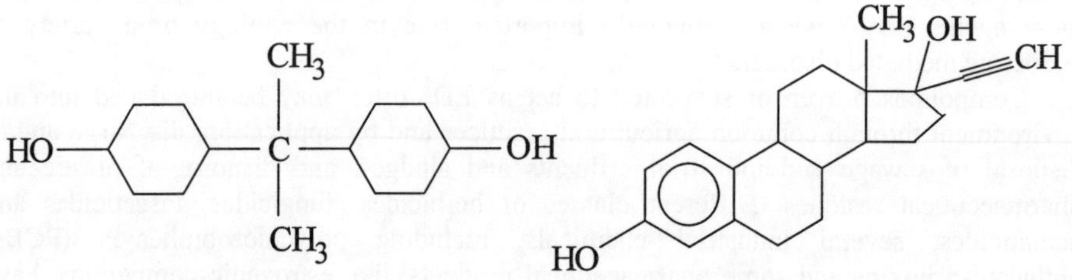

Figure 1 *Molecular formulas of bisphenol A (BPA, left) and ethynilestradiol (EED, right)*

2.2 Humic Acids

The HA samples were isolated from the surface (0-30 cm) and deep (30-90) horizons of two acidic sandy soils originating in Portugal (P) and Germany (G), and from two anaerobically-treated urban sewage sludges (PS and GS) used as amendments of these soils. The HA samples are consequently labeled as P30-HA and G30-HA (surface soil

HAs), P90-HA and G90-HA (deep soil HAs), and PS-HA and GS-HA (sludge HAs). Bulk soil and sludge samples were provided by the Universidade Nova de Lisboa, Lisbon, Portugal, and the Technical University of Dresden, Dresden, Germany, in the framework of the joint European Commission Research Project titled "Prendisensor."

Briefly, the isolation procedure of HAs consisted of the extraction of each substrate with 0.5 N NaOH + $Na_4P_2O_7$ solution, precipitation of the extracted material by HCl to pH ~ 2, mild purification by two successive NaOH-dissolution and HCl-precipitation steps, water washing of the precipitated HA, and final freeze-drying.[12]

The HA samples have been extensively characterized chemically and spectroscopically and their properties have been described in detail in a previous work.[13] Briefly, slightly different structural and functional properties are shown between soil HAs as a function of soil origin and depth, and between the two sludge HAs. In contrast, major differences are shown in the composition, structure and fuctionalities between soil HAs vs sludge HAs.[13] In particular, sludge HAs are richer than soil HAs in H, N, S and aliphatic and amide groups, and lower in O and carboxyl and carbonyl group contents, and in aromatic polycondensation and humification levels.[13]

Comparative spectroscopic analysis of laboratory-prepared model interaction products of both soil and sludge HAs with BPA and EED suggested that in any case relatively weak binding forces, such as hydrogen bonds, Van der Waals forces and hydrophobic bonding, possibly are involved in the adsorptive interaction.[13]

2.3 Adsorption Kinetics

A kinetic study was conducted to evaluate the adsorption rates and equilibration times of BPA and EED onto the various HAs examined. Aliquots of 10 mg of each HA were suspended either in 5 mL of a 10 mg L^{-1} aqueous solution of BPA or in 15 mL of a 10 mg L^{-1} 5% (v/v) ethanol-water solution of EED. The mixtures were mechanically shaken for nine time periods, 0.25, 0.5, 1, 2, 4, 8, 16, 24, and 48 h and successively centrifuged. The supernatant solutions were then analyzed by high performance liquid chromatography (HPLC) to determine BPA and EED concentrations using the same procedure adopted for obtaining adsorption isotherms as described in Section 2.4. All experiments were conducted in triplicate at a temperature of 20 ± 2 °C.

2.4 Adsorption Isotherms

Adsorption isotherms of BPA and EED onto each HA were obtained with a batch equilibrium method. Aliquots of 10 mg HA were added to 5 mL of aqueous solutions of BPA at concentrations of 1, 2, 4, 8, 12, 20 and 40 mg L^{-1}, and to 15 mL of 5% (v/v) ethanol-water solutions of EED at concentrations of 0.1, 0.2, 0.5, 1, 2 and 5 mg L^{-1} in glass flasks. All experiments were conducted in triplicate. Equilibration was achieved by mechanical shaking of mixtures for 24 h at 20 ± 2 °C in the dark. Suspensions were then centrifuged at 17,400 g for 15 min and the supernatant solutions were removed and stored in stoppered glass vials in the dark until further use.

The concentrations of free BPA in the supernatant solutions, i.e., the equilibrium concentrations, C_e, were measured by HPLC using a Thermo Separation Products Liquid Chromatograph equipped with a 15-cm Merck LiChrospher® 60 RP-Select B column, and ultraviolet (UV) detection at 280 nm. The mobile phase was an isocratic solution of 40% (v/v) acetonitrile in water.

In the case of EED, 2-mL aliquots of the supernatant solutions were subjected to a solid phase extraction (SPE) procedure using a Merck LiChrolut® Extraction Unit and Merck LiChrolut® EN cartridges (200 mg). The cartridges were previously conditioned with 2 mL of methanol and then with 2 mL of 5% (v/v) methanol in water. After the addition of sample solution, the cartridge was dried under vacuum, and the residue eluted twice with 1 mL acetone. Eluates were then analyzed by HPLC using a fluorescence detector operating at excitation and emission wavelengths of 280 and 306 nm, respectively. The amounts of BPA and EED adsorbed were calculated as the difference between the initial and the equilibrium concentration of BPA and EED in solution.

Experimental data for adsorption of BPA and EED by the HAs examined were tentatively fitted to the Freundlich equation (1) and the Langmuir equation (2),

$$x/m = K_f C_e^{1/n} \tag{1}$$

$$x/m = (KC_e b)/(1 + KC_e) \tag{2}$$

where x/m is the amount of BPA adsorbed in $\mu g\ g^{-1}$ and C_e is the equilibrium solution concentration of BPA and EED in $\mu g\ mL^{-1}$. The magnitude of adsorption, i.e., the adsorption capacity of the substrate, was estimated by the values of the distribution coefficient, K_d, which is defined as the mean value of the amount of BPA or EED adsorbed at each equilibrium concentration, and calculated according to

$$K_d = |(x/m)/C_e|_{mean} \tag{3}$$

2.5 Desorption Isotherms

Desorption isotherms of BPA and EED from each HA were obtained by measuring the sequential release of each compound immediately after its adsorption onto 10 mg of HA which was achieved using either 5 mL of a 20-mg L^{-1} aqueous solution of BPA or 15 mL of a 2-mg L^{-1} 5% (v/v) ethanol–water solution of EED. After 24-h equilibration, the mixtures were centrifuged and the equilibrium solution was carefully removed and replaced with the same volume of doubly distilled water (in the case of BPA) or 5% (v/v) ethanol-water solution (in the case of EED). The amount of dissolved compound present in the equilibrium solution that remained entrapped in the substrate was duly calculated and subtracted from the total amount of compound measured in the supernatant solution after each desorption step. The suspensions were shaken mechanically for 24 h in order to obtain a new equilibrium condition, and then centrifuged. The desorption procedure for BPA was repeated five times or until its concentration in the supernatant solution reached the detection limit of 0.1 $\mu g\ mL^{-1}$, and six times for EED. All experiments were conducted in triplicate. After each desorption step, the concentration of BPA and EED in the supernatant solutions was measured by HPLC as described above, and the amount of compound that remained adsorbed was calculated by difference.

For comparative purposes, desorption isotherm parameters, K_{fdes} and $1/n_{des}$, were calculated from the Freundlich Eq. (1) in all cases. The values of K_{fdes} provide an indication of the degree of reversibility/irreversibility of the adsorption process,[14] whereas the values of $1/n_{des}$ may be related to the rate of the desorption process.[15]

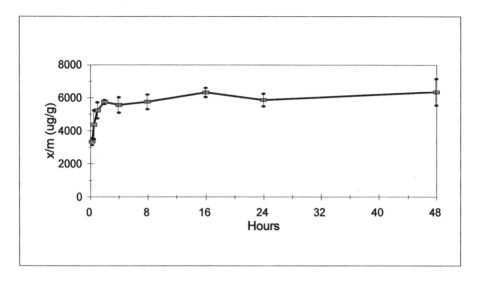

Figure 2 *Adsorption kinetics of bisphenol A onto humic acid sample G30-HA*

3 RESULTS AND DISCUSSION

3.1 Adsorption Rates

Results of kinetics measurements show that adsorption of BPA and EED is rapid onto all HAs examined and occurs almost completely (> 90%) in the first few hours. This is followed by a slow attainment of a steady-state equilibrium in less than 24 h. Figure 2 shows a representative adsorption kinetics curve for BPA on sample G30-HA.

The rapid adsorption would occur on the most accessible and/or most reactive sites of HA macromolecules, whereas the slower adsorption may reflect the interaction with less reactive and/or more sterically hindered sites.

On these bases, an equilibration time of 24 h was considered adequate for the adsorption isotherm experiments.

3.2 Adsorption Isotherms and Coefficients

On the basis of the calculated correlation coefficient (r) values over the whole concentration range tested, the best fit of experimental adsorption data of BPA was a Langmuir-type isotherm for soil HAs and a linear C-type isotherm for sludge HAs (Figure 3). In contrast, adsorption of EED onto all HAs examined was better described by nonlinear, generally L-shaped ($1/n < 1$) Freundlich isotherms (Figure 4). However, good correlation coefficient values, $r \geq 0.96$, were obtained when experimental adsorption data of BPA also were fitted in a nonlinear Freundlich isotherm. This result allowed the comparison of adsorption and desorption parameters calculated using the Freundlich equation for both BPA and EED with all HAs.

The values of the distribution coefficients, K_d, for BPA and EED adsorption onto the HAs examined are shown in Table 1. These values indicate that: (a) each HA exhibits a much higher (5 – 15 times) adsorption capacity for EED than for BPA; (b) the adsorption capacity of both surface soil HAs for BPA and EED is two or more times larger than that of the corresponding deep soil HAs; (c) the extent of adsorption of sludge HAs for BPA is generally higher than that of soil HAs, whereas their adsorption capacity for EED is a little lower than that of the corresponding surface soil HAs but much higher than deep soil HAs; and (d) the soil and sludge origin of HAs appears not to affect their adsorption capacity for BPA and EED. The Freundlich adsorption coefficients, K_{fads}, of HAs for BPA and EED given in Tables 2 and 3, respectively, are slightly different from the corresponding K_d values shown in Table 1, but both K_{fads} and K_d follow a similar trend. This result confirms the validity of both coefficients for evaluation of the adsorption capacity of HAs for BPA and EED and the appropriateness of use of K_{fads} coefficients in this adsorption-desorption study.

The different behavior exhibited by the various HAs examined in the adsorption of BPA and EED generally may be ascribed to their different structural and functional properties and chemical reactivity, and in particular to the prevalence of aliphatic or aromatic character and the different types and contents of reactive functional groups.

Table 1 *Distribution coefficients, K_d, for bisphenol A (BPA) and ethynilestradiol (EED) adsorption onto humic acids (HAs)*

Humic Acid	K_d - BPA (mL g^{-1})	K_d - EED (mL g^{-1})
P30-HA	273	1590
G30-HA	165	1780
P90-HA	59	828
G90-HA	87	889
PS-HA	286	1300
GS-HA	234	1710

The higher adsorption capacity for EED than for BPA shown by all HAs examined can be ascribed to the different physical properties of the two molecules, and in particular to the lower water solubility and polarity of EED compared to BPA. However, some effect of the different chemical structure of the two molecules on their adsorption extent onto HAs cannot be excluded.

3.3 Adsorption-Desorption Processes

Experimental data for desorption of BPA and EED from all HAs examined were well fitted in a nonlinear Freundlich equation, as shown by the calculated correlation coefficients, $r \geq 0.932$, which are given in Tables 2 and 3 together with those of the corresponding adsorption data, also fitted with a Freundlich model. Tables 2 and 3 also show, respectively for BPA and EED, the Freundlich adsorption and desorption coefficients, K_{fads} and $1/n_{ads}$ and K_{fdes} and $1/n_{des}$, onto the HAs examined, and the total amounts (averages of three replicates) of each compound that are desorbed after five (BPA) and six (EED) desorption steps, expressed as the percentage (%) of the initially adsorbed amount (100%). Figure 5 shows the % of initially adsorbed (100%) BPA (top) and EED (bottom) which remains adsorbed onto each HA after each desorption step.

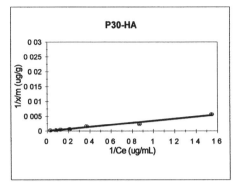
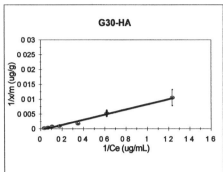
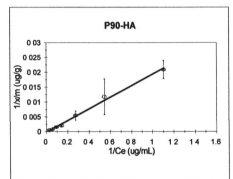
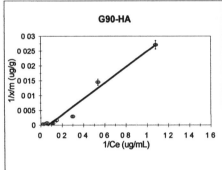
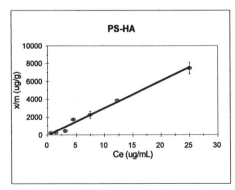
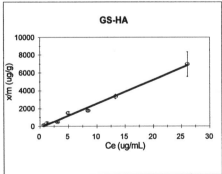

Figure 3 *Adsorption isotherms (Langmuir-type linearized for samples P30-HA, G30-HA, P90-HA, and G90-HA, and linear for samples PS-HA and GS-HA) of bisphenol A onto humic acids (HAs)*

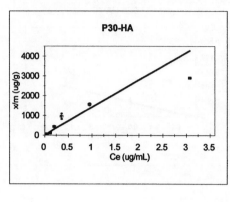
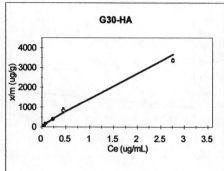
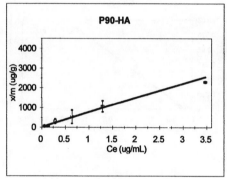
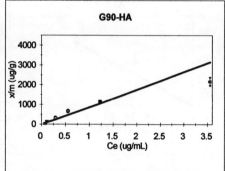
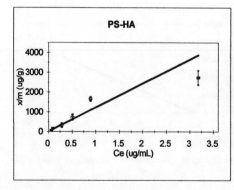
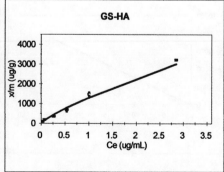

Figure 4 *Adsorption isotherms (nonlinear Freundlich-type) of ethynilestradiol onto humic acids (HAs)*

Table 2 *Correlation coefficients for Freundlich nonlinear isotherms, Freundlich adsorption and desorption coefficients, and total amount desorbed after a number of desorption steps indicated between parentheses, for bisphenol A onto humic acids (HAs)*

Humic Acid	Adsorption				Desorption			
	r	K_{fads} (mL g^{-1})	$1/n_{ads}$	r	K_{fdes} (mL g^{-1})	$1/n_{des}$	% desorbed	
P30-HA	0.973	326	0.85	0.932	44.3	1.55	100 (4)	
G30-HA	0.992	131	1.13	0.994	23.8	1.84	100 (3)	
P90-HA	0.996	51	1.08	1.000	2.8	2.14	100 (2)	
G90-HA	0.958	51	1.24	0.948	48.8	1.07	100 (3)	
PS-HA	0.967	258	1.04	0.946	1687	0.53	62.5 (5)	
GS-HA	0.983	204	1.07	0.973	72.7	1.64	100 (5)	

Table 3 *Correlation coefficients for Freundlich nonlinear isotherms, Freundlich adsorption and desorption coefficients, and total amount desorbed after a number of desorption steps indicated between parentheses, for ethynil estradiol onto humic acids (HAs)*

Humic Acid	Adsorption				Desorption			
	r	K_{fads} (mL g^{-1})	$1/n_{ads}$	r	K_{fdes} (mL g^{-1})	$1/n_{des}$	% desorbed	
P30-HA	0.958	1430	0.97	0.920	971	0.60	87.2 (6)	
G30-HA	0.994	1489	0.89	0.984	1102	0.78	88.4 (6)	
P90-HA	0.990	780	0.95	0.909	730	1.08	100 (4)	
G90-HA	0.961	852	1.03	0.963	721	0.82	93.3 (6)	
PS-HA	0.971	1227	0.99	0.972	1553	0.34	61.3 (6)	
GS-HA	0.993	1298	0.80	0.966	1100	0.55	81.8 (6)	

Data in Table 2 show that K_{fdes} values for BPA are generally lower or much lower than the corresponding K_{fads} values, with the exception of sample PS-HA, for which $K_{fdes} \gg K_{fads}$. These results suggest an easily reversible adsorption of BPA onto all HAs except PS-HA for which a high degree of irreversibility of adsorbed BPA appears to hold.[14] Data shown in Figure 5 (top) confirm a readily reversible adsorption of BPA onto all HAs except PS-HA. Desorption of BPA is completed (100%) after less than five desorption steps for soil HAs, whereas sample GS-HA needs five steps to attain complete reversibility, and sample PS-HA retains more than one third of initially adsorbed BPA at the end of the desorption experiment (Table 2 and Figure 5, top). For most HAs the values of $1/n_{des}$ are higher than the corresponding $1/n_{ads}$ values (Table 2), which indicates a fast rate of desorption of BPA from these substrates.[15] In contrast, $1/n_{des}$ of samples G90-HA and PS-HA is lower than $1/n_{ads}$, thus suggesting that the rate of desorption is lower than the rate of adsorption, i.e., an hysteresis effect occurring for these samples.[15] The degree of hysteresis (ω) quantified according to Eq. (4)[16]

$$\omega = \{[(1/n_{ads})/(1/n_{des})] -1\} \times 100 \qquad (4)$$

is much greater for PS-HA (ω = 96) than for G90-HA (ω = 16). This result indicates that the rate of BPA desorption is much lower for PS-HA than for G90-HA.

The K_{fdes} values of EED are generally slightly lower than the corresponding K_{fads} values, with the exception of sample PS-HA for which K_{fdes} is slightly higher than K_{fads} (Table 3). Thus, a reversible adsorption process is indicated also for EED on all HAs examined except PS-HA for which a certain degree of irreversibility would occur.[14] Data in Figure 5 (bottom) suggest, however, that adsorption of EED onto HAs is less readily reversible than that of BPA. With the exception of sample P90-HA, from which EED desorption is completed (100%) after four steps, desorption of EED is not completed after six steps from any other HA sample that still retain variable but generally low amounts (between 6.7 and 18.2%) of EED initially adsorbed at the end of the desorption experiment (Table 3 and Figure 5, bottom). Similar to the BPA experiment, sample PS-HA retains more than one third of EED initially adsorbed. Differently from BPA, for all samples except P90-HA the values of $1/n_{des}$ for EED are lower than the corresponding $1/n_{ads}$ values (Table 3), thus suggesting a lower desorption rate than the adsorption rate.[15] The degree of hysteresis, ω, calculated from Eq. (4) is, however, very variable among HA samples, ranging from 14 (G30-HA) to 145 (GS-HA) and 191 (PS-HA). These results indicate that the rate of EED desorption from sludge HAs is much lower than that from soil HAs.

The results discussed above indicate an adsorption-desorption behavior for BPA and EED that is different between sludge HAs, especially PS-HA, and soil HAs. Sludge HAs, especially PS-HA, appear to adsorb more tightly and partly irreversibly both compounds that are released with more difficulty and at a lower rate than from soil HAs. A marked difference also exists between BPA and EED in their adsorption-desorption interactions with the HAs examined. In particular, adsorption of BPA is looser and more readily reversible than EED, and its desorption is generally complete and faster than that of EED. The different behaviors discussed above may be attributed to the different chemical properties of the sludge HAs with respect to soil HAs, and the different physical properties, mainly solubility and polarity, of BPA and EED.

4 SUMMARY AND CONCLUSIONS

The adsorption of BPA and EED on all HAs examined is fast and occurs almost completely in the first few hours, reaching a steady-state equilibrium in less than 24 h. Although the highest correlation coefficients for BPA adsorption data onto soil HAs and sludge HAs are obtained using, respectively, the Langmuir equation and a linear model equation, a good correlation (r ≥ 0.96) also is obtained when adsorption data are fitted to a nonlinear Freundlich equation. For all HAs, EED adsorption data are best fitted in the latter equation. Apparently, the soil or sludge origins of HAs do not affect their adsorption capacity for BPA and EED, which is, however, markedly influenced by the soil horizon depth and the soil or sludge material from which HA derives. In particular, HAs from both surface soils adsorb BPA and EED to the highest extent with respect to the other HAs, whereas sludge HAs adsorb more BPA than all the corresponding soil HAs and a little less EED than the corresponding surface soil HAs.

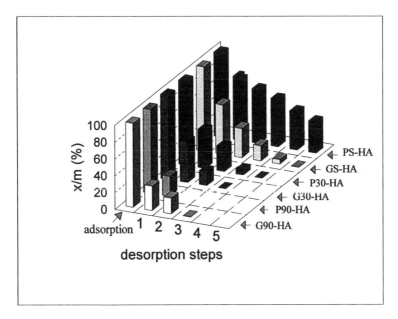

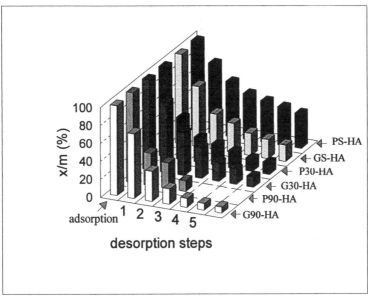

Figure 5 *Amounts (averages of three replicates) in percentage (%) of initially adsorbed (100%) BPA (top) or EED (bottom) which remains adsorbed onto the various humic acids (HAs) after each desorption step*

Adsorption-desorption data suggest a generally reversible adsorption process of BPA and EED onto all HAs examined except one sludge-derived HA, for which a certain degree of irreversibility holds. However, adsorption of BPA appears to be more readily reversible than that of EED, and its desorption, especially from soil HAs, is completed much easier and faster than for EED. This suggests that BPA is generally adsorbed onto HAs more loosely than EED. In general, release of BPA and EED from both sludge HAs is slower and occurs with more difficulty than from soil HAs, showing a marked hysteresis effect. These results suggest that sludge HAs adsorb these compounds more tightly than soil HAs.

The different structural and functional chemical properties, e.g., degree of aromatic polycondensation and humification, aliphatic-aromatic character and content of acidic and other reactive functional groups, primarily of sludge HAs with respect to soil HAs, and secondarily of surface soil HAs with respect to deep soil HAs, would explain the different behaviour and performances of HAs examined in the adsorption-desorption interaction with BPA and EED. On the other hand, the lower water solubility and polarity of EED with respect to BPA would support the higher extent and partial reversibility of adsorption, the slower and incomplete desorption, and the hysteresis effect of the former compound with respect to the latter.

In conclusion, adsorption-desorption interactions of endocrine disruptors with the HA fractions of soil organic matter appear to depend on the type of compound considered and to involve mostly HAs of the surface (0-30 cm) soil horizon. Finally, these processes are expected to be affected markedly by the addition of the HA fraction of sewage sludges used as soil amendments.

ACKNOWLEDGMENT

This research was supported by the EC Project "Prendisensor", Grant n° ENV4-CT97/0473.

References

1. Endocrine Disruptors Screening and Testing Advisory Committee, 1998, Final Report, August 1998. Available on line: www.epa.gov/opptintr/opptendo/whatsnew.htm.
2. R. J. Kavlock, G. P. Daston, C. Derosa, P. Fenner-Crisp, L. Earl Gray, S. Kaattari, G. Lucier, M. Lustre, J. M. Mac, C. Maczka, R. Miller, J. Moore, R. Rolland, G. Scott, M. Sheehan, T. Sinks and H. A. Tilson, *Environ. Health Perspect.*, 1996, **104**, 1.
3. L. H. Keith, in 'A Handbook of Properties', Wiley, New York, 1997.
4. T. Colborn and C. Clement, 'Chemically-induced Alterations in Sexual and Functional Development: The Wildlife/Human Connection', Princeton Sci. Publ., New Jersey, 1992.
5. C. Nolan (ed.), 'Ecosystem Research Reports Series No. 29: Endocrine-Disrupters Research in the EU', Report EUR 18345, Environ. Clim. Res. Progr., Office for Official Publication of the EC, Brussels-Luxembourg, 1998.
6. T. Colborn, F. S. vom Saal and A. M. Soto, *Environ. Health Perspect.*, 1993, **101**, 378.

7. L. H. Keith, in 'Annual Symposium on Waste Testing & Quality Assurance', July DATA, Alexandria, VA, 1997.
8. J. B. Weber, *Adv. Chem. Ser.*, 1972, **111**, 55.
9. F. J. Stevenson, 'Humus Chemistry: Genesis, Composition, Reactions', Wiley, New York, 1982.
10. N. Senesi and Y. Chen, in 'Toxic Organic Chemicals in Porous Media', Z. Gerstl, Y. Chen, V. Mingelgrin and B. Yaron (eds.), Ecological Studies, Vol. 73, Springer-Verlag, Berlin, 1989, p. 37.
11. N. Senesi and T. M. Miano, in 'Environmental Impact of Soil Component Interactions. Natural and Anthropogenic Organics', Vol. I, P. M. Huang, J. Berthelin, J. M. Bollag, W. B. McGill and A. L. Page, (eds.), CRC, Boca Raton, FL, 1995, p. 311.
12. M. Schnitzer, in 'Methods of Soil Analysis, Part 2, Chemical and Microbiological Properties', B. L. Page, R. H. Miller and R. D. Keeney, (eds.), 2nd Edn., Agronomy Monograph No. 9, Soil Science Society of America, Madison, 1982, p. 581.
13. E. Loffredo, M. Pezzuto, G. Brunetti and N. Senesi, *J. Environ. Qual.*, 2000 (submitted).
14. P. J. McCall, D. A. Laskowski, R. L. Swann and H. J. Dishburger, 'Test Protocols for Environmental Fate and Movement of Chemicals', Proc., 94th Annual Meeting, Washington, D.C. Assoc. Official Anal. Chem., Arlington, VA, 1981, p. 89.
15. J. J. Pignatello and L. Q. Huang, *J. Environ. Qual.*, 1991, **20**, 222.
16. L. Ma, L. M. Southwick, G. H. Willis and H. N. Selim, *Weed Sci.*, 1993, **41**, 627.

BINDING OF ORGANIC NITROGEN COMPOUNDS TO SOIL FULVIC ACID AS MEASURED BY MOLECULAR FLUORESCENCE SPECTROSCOPY

C. L. Coolidge and D. K. Ryan

Department of Chemistry, University of Massachusetts Lowell, Lowell, MA 01854, USA

1 INTRODUCTION

Humic substances are produced by microbial action on decaying plant and animal tissue, and are the largest fraction of organic carbon in soils and natural waters.[1] Fulvic acid, the humic fraction soluble at all values of pH, is highly functionalized and is known to contain many aromatic residues similar in structure to salicylic acid, benzoic acid, the phthalates, catechols and phenol.[2] Humic substances interact with environmentally important substances, notably metal ions and organic aromatic substances.[3-6] It is believed that many toxic environmental contaminants are rendered less harmful when associated with humic materials and that humics play a vital role in the fate and transport of pollutants in the environment.[7,8]

Aniline, the parent compound of the aromatic amines, is used as the starting material for the azo dyes and other industrial products[9] and may be released into the environment during their manufacture or degradation. Other aromatic amines form by the reduction of polynitroaromatics in munitions waste sites.

The aromatic amines have been shown to interact with humic materials by a variety of methods. Sorption isotherms of benzidine with soils and sediments showed increased sorption with decreasing pH, indicating that the cationic form is sorbed to a greater extent.[10] This result is consistent with an electrostatic attraction between the cationic amine and the carboxyl moieties in humic materials.

Recoveries of substituted anilines from several humic acids were found to diminish with time and Parris[11] interpreted these results as indicating the slow formation of covalent complexes between the amines and the humates studied. Hsu and Bartha[12] postulated that these covalent species form by the 1,4 addition of the amino functional group to quinone-like structures present in the humic substances. The immobilization of aromatic amines by humic materials was shown to follow second order kinetics[11,13] and to be inhibited by the presence of substituents ortho to the amine functional group.[11] Recently, Thorn et al.[14] used ^{15}N NMR spectroscopy to demonstrate covalent binding of aniline to humic substances. In another study[13] these authors observed a decrease in the rate of reaction of aniline with humic materials with decreasing pH as well as considerably diminished binding of aniline by humic substances pretreated with H_2S or $NaBH_4$. Both results are consistent with immobilization of aniline by nucleophilic attack by the amine on quinone like structures of the humics.

Fluorescence spectroscopy has been widely used to study the interactions of humic materials with metals and polyaromatic hydrocarbons (PAHs). Quenching of fluorescence is attributed to the formation of a nonfluorescent ground state complex between the fluorophore and the quenching agent. Studies with metals measure quenching of the innate fluorescence of humic substances by added metal, while those with PAHs generally monitor quenching of the more intense fluorescence signal of the multiple aromatic rings by added humic materials. These associations can be modeled by Stern-Volmer analysis or the nonlinear method introduced by Ryan and Weber.[15]

This paper introduces the use of fluorescence spectroscopy as a tool for studying the interaction of soil fulvic acid with aromatic amines. Aniline was found to cause considerable quenching of the 450 nm fulvic acid fluorescence peak over a concentration range of 2 - 200 mM. This quenching was studied at several pHs, and was greatest at pH 5.0. A Stern-Volmer model was employed to calculate a binding constant for the aniline-fulvic acid association. Quenching of the fluorescence of salicylic acid, used as a model fluorophore for fulvic acid, was also demonstrated.

Paraquat, an herbicide containing quaternary nitrogen atoms in aromatic rings, also quenched the fluorescence of the 450 nm soil fulvic acid emission peak over a concentration range of 50 – 4000 µM. Quenching was again pH dependent and greatest at pH 6.5, the highest pH value studied.

2 EXPERIMENTAL

2.1 General

Chemicals used in the experiments were reagent grade and water was purified using a Barnstead Nanopure water system. Aniline was obtained from Aldrich. Paraquat dichloride was obtained from Chem Service, Inc. Soil fulvic acid (SFA) was obtained from Dr. James Weber at the University of New Hampshire. The characteristics of this SFA have been described elsewhere.[16] A stock solution of SFA was made by dissolving 200 mg in 500 mL of 0.1 M $NaClO_4$ and filtering prior to use. Salicylic acid was prepared fresh as a 1000 µM stock solution in buffer prior to use and paraquat was prepared fresh as a 500 µM stock. All buffers were made with the correct ratio of acid to base as calculated using the Henderson Hasselbach equation, constraining the total phosphate or acetate concentration to 0.1 M. The pH of the resulting solutions was confirmed with a Model 701A pH meter from Orion Research. Glass and plasticware were cleaned with Alconox and rinsed with deionized water, then with 20% HNO_3 followed by a final rinse with deionized water. Figure 1 shows the structures of salicylic acid and paraquat dichloride.

2.2 Binding Studies

All incubations were carried out in 60 mL polypropylene bottles. A 10 mL constant volume was used for all studies. Buffer was added followed by an aliquot of SFA stock solution (final concentration 15 - 20 mg/L) or salicylic acid stock. Various volumes of the aniline quencher, prepared fresh as a 220 mM stock solution in buffer were added to give the desired final concentration, and vials were incubated for 3 - 48 hours before collecting the fluorescence spectrum. Ambient temperature was not controlled during these incubations.

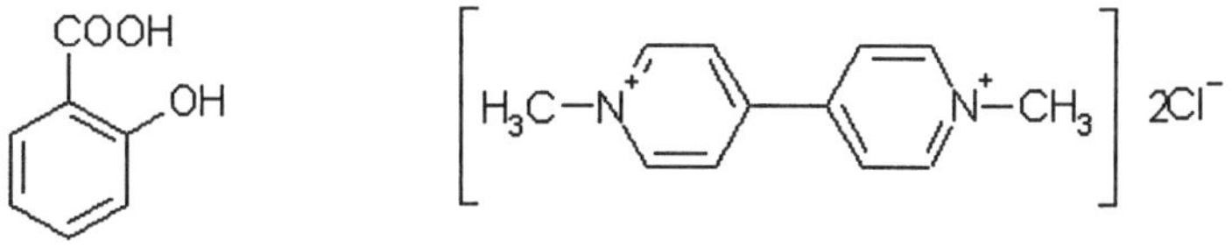

Figure 1 *Structure of salicylic acid and paraquat dichloride*

2.3 Luminescence Spectra

Spectra were collected with a Mark 1 spectrofluorometer from Farrand Optical Company, scanning at a rate of 100 nm/min. Emission, excitation, and synchronous offset scans were all employed. For the excitation monochromer, both entrance and exit slit widths were set to 10 nm and the emission monochromer slit widths were set to 5 nm. Data were collected with LabCalc Grams 386 (Galactic Industries) and stored on an IBM compatible PC. Files were translated to ASCII format and imported into SigmaPlot (Jandel Scientific) for analysis and graph production.

2.4 UV - Visible Spectra

A Hitachi U1100 spectrophotometer was used to determine absorbance at the wavelength of interest. All solutions were blanked against deionized water or buffer.

2.5 Data Treatment

Binding constants were calculated with the Stern-Volmer model, in which a plot of I_0/I vs. quencher concentration is expected to be linear with an intercept of 1.0 and a slope equal to K, the association constant. Aniline and paraquat do not absorb significantly at the excitation or emission wavelengths of either soil fulvic acid or salicylic acid, so no correction for inner filter effects was necessary.

3 RESULTS AND DISCUSSION

3.1 Aniline Binding to Soil Fulvic Acid

Figure 2 shows the quenching of the 450 nm excitation peak of 20 mg/L soil fulvic acid by increasing concentrations of aniline at pH 6.5. The excitation wavelength was 335 nm. The highest aniline concentrations show the tail of the 350 nm emission peak of aniline. The aniline fluorescence is negligible at 450 nm, the peak maximum of SFA. The location of the peak maximum appears unchanged throughout the titration.

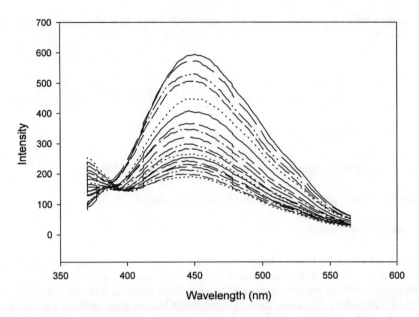

Figure 2 *Fluorescence emission spectra of 20 mg/L soil fulvic acid in 0.1 M phosphate buffer at pH 6.5 quenched with increasing aniline concentrations from 0 to 197.6 mM. The excitation wavelength is 335 nm*

This experiment was performed at five buffered pH values. To calculate relative fluorescence intensities, data are expressed as a ratio of quenched fluorescence intensity at 450 nm, I, in the presence of various aniline concentrations to the intensity of the 450 nm SFA peak in the absence of aniline, I_o. Figure 3 is a plot of relative intensity (I/I_o) vs. added aniline concentration at pH 8.0, 6.5 and 5.0. Data shown are the average relative intensities for two trials at pH 8.0, five trials at pH 6.5 and four trials for pH 5.0. The magnitude of quenching increases from pH 8.0 to 5.0, which indicates that the anilinium ion may be a more efficient quenching agent than neutral aniline. The pK_a of aniline is 4.70, and at pH 5.0 approximately 33% of the amino group of aniline possess positive charge with less than 0.1% carrying a formal positive charge at pH 8.0.

Figure 4 shows that the trend toward increasing quenching with decreasing pH is reversed at pH 4.0. Again, data shown are average relative intensities of four trails at pH 4.0 and two trials at pH 3.0. This shift likely is due to loss of charge on the carboxyl moieties of fulvic acid, whose pK_as are expected to fall in the pH 4.0 to 5.0 range. At pH values of 4.0 or below, the percentage of carboxyl groups with negative charge that can be electrostatically attracted to the increasing positive charge on the amino group of aniline is expected to fall off rapidly.

The Stern-Volmer model was used to calculate a binding constant for the association of aniline with SFA using the data from Figures 3 and 4. A plot of I_o/I vs. the concentration of the quenching agent is expected to be linear, with a slope equal to the binding constant and an intercept of 1.0. The weakness of this approach is that it assumes that the complex between the fluorophore and quenching agent has no residual fluorescence, that is, it is a "dark" complex. Fluorophores in the fulvic acid molecule that do not interact with the quencher may give rise to residual fluorescence that can cause curvature of the Stern-Volmer plot toward the x axis.[5]

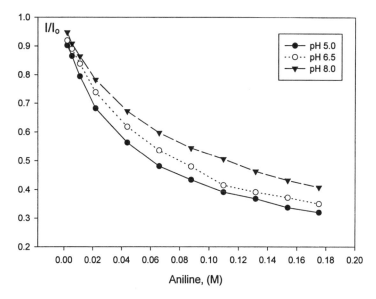

Figure 3 *Fluorescence quenching curves for 20 mg/L soil fulvic acid in 0.1 M phosphate buffer at pH 5.0, 6.5 and 8.0 quenched with increasing aniline concentrations. Fluorescence emission spectra were obtained at an excitation wavelength of 335 nm. Relative intensities I/I_o were calculated at an emission wavelength of 450 nm*

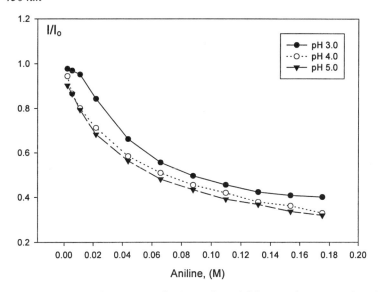

Figure 4 *Fluorescence quenching curves for 20 mg/L soil fulvic acid in 0.1 M phosphate buffer at pH 5.0, 4.0 and 3.0 quenched with increasing aniline concentrations. Fluorescence emission spectra were obtained at an excitation wavelength of 335 nm. Relative intensities I/I_o were calculated at an emission wavelength of 450 nm*

Figure 5 shows the Stern-Volmer plot of the quenching data obtained at pH 5.0. The plot is reasonably linear, but it does show some curvature toward the x axis at high aniline concentration. Relative intensities at low aniline concentration were the least reproducible, which may also contribute to less than perfect linearity. The shape of the Stern-Volmer plot shown in Figure 5 was similar to that obtained at all pH values. Table 1 summarizes the binding constants obtained at all pH values calculated with the Stern Volmer model. Highest binding was found at pH 5.0, with a value of K = 12.1 M^{-1}.

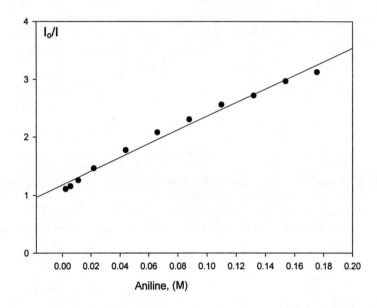

Figure 5 *Fluorescence quenching of 20 mg/L soil fulvic acid in 0.1 M phosphate buffer by aniline at pH 5.0 expressed in the form of a Stern Volmer plot of I_0/I vs. aniline concentration. I_0 is the fluorescence intensity in the absence of aniline and I is the fluorescence intensity in the presence of aniline. The excitation wavelength is 335 nm, and the intensities were measured at an emission wavelength of 450 nm*

Table 1 *Association constants for the aniline-soil fulvic acid complex calculated with the Stern-Volmer model. Data used are average intensities for two trials at pH 3.0, five trials at pH 4.0, four trials at pH 5.0, five trials at pH 6.5 and two trials at pH 8.0*

pH	K, M^{-1}	R^2	Intercept
3.0	9.5	0.968	1.04
4.0	11.3	0.988	1.12
5.0	12.1	0.985	1.15
6.5	10.8	0.988	1.10
8.0	8.2	0.993	1.07

In order to determine if electrostatic attraction between the anilinium ion and the buffer components could impact the observed results, the experiments were repeated in 0.1 M acetate buffer and with manual adjustment of pH with HCl or NaOH in 0.1 M NaClO$_4$. The results for acetate buffer shown in Figure 6 indicate nearly identical quenching results for acetate vs. phosphate buffer at pH 6.5 and 5.0. Quenching studies for the adjusted pH experiments (data not shown) again produced similar quenching profiles for those vials whose pH was manually adjusted vs. those buffered with 0.1 M phosphate buffer at pH values of 4.0, 5.0, 6.5 and 8.0.

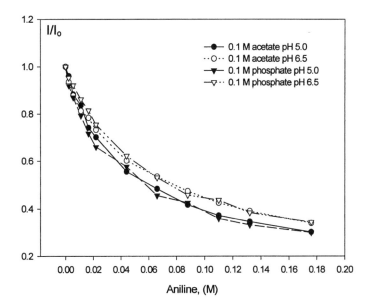

Figure 6 *Quenching of 20 mg/L soil fulvic acid by increasing concentrations of aniline from 2 – 178 mM studied in 0.1 M acetate at pH 5.0 and 6.5 and in 0.1 M phosphate at pH 5.0 and 6.5. The excitation wavelength was 335 nm and the emission wavelength was 450 nm*

3.2 Aniline Binding to Salicylic Acid

Supporting the results obtained for SFA, aniline also was shown to quench the fluorescence of 250 µM salicylic acid. Figure 7 shows the emission spectra of salicylic acid quenched by titrating with increasing aniline concentrations at pH 5.0. An excitation wavelength of 350 nm was used to eliminate self quenching of salicylic acid by inner filter effects and to eliminate inner filter effects produced by aniline absorbance. Two trials were performed at pH 4.0, 5.0 and 6.5. Relative intensities were recorded at an emission wavelength 415 nm. Binding constants for the association of aniline with salicylic acid were calculated with the Stern-Volmer model and the results are summarized in Table 2. pH effects were not noted for the interaction of aniline with salicylic acid. The pK$_a$ of salicylic acid is 2.97, which is strongly acidic for a carboxylic acid. Thus, greater than 90% of the salicylic acid is already anionic even at pH 4.0. To observe pH effects for salicylic acid, it is probable that pH values below 2.0 would be needed to observe loss of negative charge on the carboxyl group. The aromatic rings of salicylic acid and aniline may interact strongly enough that π - π interaction also is a significant effect.

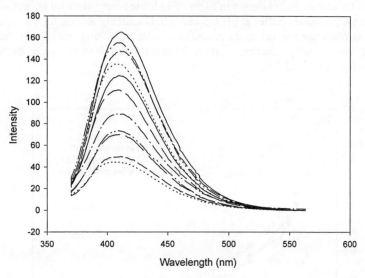

Figure 7 *Fluorescence emission spectra of 250 μM salicylic acid quenched with increasing aniline concentrations from 2 – 131.7 mM. The excitation wavelength was 350 nm*

Table 2 *Association constants for the aniline-salicylic acid complex calculated with the Stern-Volmer model*

pH	K, M^{-1}	R^2	Intercept
4.0	11.7	0.986	1.14
5.0	11.9	0.984	1.16
6.5	11.3	0.995	1.05

3.3 Paraquat Binding to Soil Fulvic Acid

20 mg/L samples of soil fulvic acid in 0.1 M phosphate buffer were exposed to a range of paraquat concentrations from 50 – 4000 μM paraquat for 24 hours, then fluorescence emission spectra were collected. The 450 nm emission peak of soil fulvic acid is quenched considerably by these concentrations of paraquat. Quenching was studied in duplicate at pH 3.0, 4.0, 5.0 and 6.5. The magnitude of quenching at each pH was determined from a Stern-Volmer plot of I_0/I vs. paraquat concentration. The results of these studies are summarized in Table 3. Because the nitrogen atoms in paraquat are quaternary, their charge is independent of pH. The magnitude of the association of paraquat and soil fulvic acid increases with increasing pH, which is consistent with an increase in negative charge on the carboxyl groups of soil fulvic acid with increasing pH. These data again support an electrostatic attraction between the quenching agent and soil fulvic acid.

Table 3 *Association constants for the paraquat-soil fulvic acid complex calculated with the Stern-Volmer model*

pH	K, M^{-1}	R^2	Intercept
3.0	114.3	0.986	1.02
4.0	159.6	0.992	1.02
5.0	205.0	0.986	1.06
6.5	270.0	0.996	1.04

4 CONCLUSIONS

This study shows that millimolar aniline concentrations quench the innate fluorescence of soil fulvic acid. Quenching is unaffected by buffer choice. The greatest degree of quenching of soil fulvic acid fluorescence by aniline occurred at pH 5.0 and the association constant at this pH was 12.1 M^{-1} as calculated with the Stern Volmer model. The fluorescence of salicylic acid is quenched by aniline over a similar concentration range, with a maximum binding constant of 11.9 M^{-1} at pH 5.0. Paraquat also quenched the fluorescence of soil fulvic acid in a manner consistent with electrostatic attraction. The highest degree of quenching for paraquat was observed at pH 6.5, where all carboxyl groups of soil fulvic acid are expected to carry a negative charge and thus attract the cationic nitrogen atoms of paraquat. The magnitude of the association between paraquat and soil fulvic acid is considerably larger than that observed between aniline and soil fulvic acid. If the observed quenching is largely attributable to electrostatic effects, the presence of two cationic nitrogen atoms in paraquat vs. one ionizable nitrogen in aniline should produce a stronger attraction with the carboxyl groups of fulvic acid. Paraquat also contains two aromatic rings vs. only one in aniline. These rings also may participate in the association of these quenching agents with soil fulvic acid via π - π interactions with the aromatic moieties of SFA.

The major limitations of fluorescence quenching for studying the interactions of organic nitrogen compounds with humic materials are that many of these test compounds exhibited background fluorescence over the same wavelength range as the emission peak of humic compounds or they absorbed significantly at the excitation or emission wavelength, causing considerable inner filter effects, or both. For compounds that are not affected by these problems, fluorescence spectroscopy appears to be a promising technique for monitoring the interactions of organic nitrogen compounds with humic substances.

References

1. F. M. M. Morel, 'Principles of Aquatic Chemistry', Wiley, New York, 1993, p. 237.
2. L. B. Sonnenberg, J. D. Johnson and J. D. Christman, in 'Aquatic Humic Substances', J. H. Suffet and P. MacCarthy, (eds.), Advances in Chemistry Series 219, American Chemical Society, Washington, D.C., 1989, p. 1.
3. D. K. Ryan, C. P. Thompson and J. H. Weber, *Can. J. Chem.*, 1983, **61**, 1505.

4. J. C. G. Esteves da Silva, A. A. S. C. Machado, C. J. S. Oliveira and M. S. S. D. S. Pinto, *Talanta*, 1998, **45**, 1155.
5. T. D. Gauthier, E. C. Shane, W. F. Guerin, W. F. Seltz and C. L. Grant, *Environ. Sci. Technol.*, 1986, **20**, 1162.
6. J. V. Goodpaster and V. L. McGuffin, *Applied Spectroscopy*, 1999, **53**, 1000.
7. E. U. Ramos, S. N. Meijer, W. H. J. Vaes, H. J. M. Verhaar and J. L. M. Hermens, in 'Aquatic Humic Substances', J. H. Suffet and P. MacCarthy, (eds.), Advances in Chemistry Series 219, American Chemical Society, Washington, D.C., 1989.
8. D. E. Kile, and C. T. Chiou, in 'Aquatic Humic Substances', J. H. Suffet and P. MacCarthy, (eds.), Advances in Chemistry Series 219, American Chemical Society, Washington, D.C., 1989, p. 131.
9. K. A. Thorn, W. S. Goldenberg, S. J. Younger and E. Weber, in 'Humic and Fulvic Acids. Isolation: Structure, and Environmental Role', J. S. Gaffney, N. A. Marley and S. B. Clark, (eds.), American Chemical Society, Washington. D.C., 1996, p. 299.
10. J. C. Means and R. D. Wijayaratne, in 'Aquatic Humic Substances', J. H. Suffet and P. MacCarthy, (eds.), Advances in Chemistry Series 219, American Chemical Society, Washington, D.C., 1989, p. 209.
11. G. E. Parris, *Environ. Sci. Technol.*, 1980, **14**, 1099.
12. T. S. Hsu and R. Bartha, *Soil Sci.*, 1974, **116**, 444.
13. E. Weber, D. L. Spidle and K. A. Thorn, *Environ. Sci. Technol.*, 1996, **30**, 2755.
14. K. A. Thorn, P. J. Pettigrew, W. S. Goldenberg and E. J. Weber, *Environ. Sci. Technol.*, 1996, **30**, 2764.
15. D. K. Ryan, and J. H. Weber, *Anal. Chem.*, 1982, **54**, 986.
16. J. H. Weber and S. A. Wilson, *Water Res.*, 1975, **9**, 1079.

FLOW FIELD-FLOW FRACTIONATION-INDUCTIVELY COUPLED PLASMA-MASS SPECTROMETRY (FLOW-FFF-ICP-MS): A VERSATILE APPROACH FOR CHARACTERIZATION OF TRACE METALS COMPLEXED TO SOIL-DERIVED HUMIC ACIDS

Dula Amarasiriwardena,[1] Atitaya Siripinyanond[2] and Ramon M. Barnes[2]

[1] School of Natural Science, Hampshire College, Amherst, MA 01002, USA
[2] Department of Chemistry, University of Massachusetts, Amherst, MA 01003-4510, USA

1 INTRODUCTION

Humic substances (HSs) are well known for their strong metal binding properties, partitioning of organic pollutants and adsorption onto clay materials. HSs polyelectrolytes have broad molecular size distributions and their structural conformations are affected by changes in pH, ionic strength and the identity of metal ions present in the surrounding environment. HSs macromolecules play vital roles in the regulation of natural and anthropogenic metal ions, pesticides and toxic organic molecules in terrestrial and aquatic environments. Among HSs, humic acids (HAs) and fulvic acids (FAs) are of particular interest owing to their prevalence and reactivity in nature. The presence of numerous metal-binding ligand sites and ability to form metal humate salts mean that humic acids can transport contaminants and nutritionally important metal ions in the environment.[1-3] The mobility of HAs in aqueous media partly depends on HAs' diffusion rates and to a large extent on their molecular weights.[4,5]

For the past few decades there has been intense interest in developing sensitive analytical procedures to enhance our understanding of trace elements complexed to macromolecules of varying molecular sizes like HAs in soil and aqueous environments. Despite its importance, trace elemental distribution information in HAs is still limited partly due to analytical challenges encountered during .separation of HAs based on molecular sizes. Classical separation techniques like dialysis, ultrafiltration, and gravitational and centrifugal settling methods can be used for the separation of humic acid molecular species, but they are not amenable to on-line coupling of sensitive trace element detectors for the determination of trace elemental distributions in HAs. The applicability of capillary electrophoresis (CE)[6,7] and polyacrylamide gel electrophoresis (PAGE)[8] for the separation of HA molecular fractions has been demonstrated. The latter method has been successfully used for off-line introduction of separated HA and FA gel channels by using a laser ablation (LA) sample introduction approach before the determination of trace elements in those HA and FA fractions by inductively coupled plasma-mass spectrometry (ICP-MS).

The more widely used techniques for metal-humic acid speciation studies include off-line graphite furnace atomic absorption spectrometric (GFAAS) detection after separation by high performance size exclusion chromatography (HPSEC)[9] and on-line detection of metal ions bound to aqueous HSs by ICP-MS after separation with an HPSEC column.[10-13] Application of HPSEC-ICP-MS for the determination of trace elements bound to soil and compost-derived humic substances and potential use of this approach in studies of metal complexing properties has been demonstrated.[14,15] Flow field-flow fractionation (FFFF), a relatively new technique, also has been successfully employed as a separation tool for humic materials, as reported by Beckett and coworkers[4,5] and Dycus et al.[16] In this study, FFFF was coupled to ICP-MS to investigate HA bound trace metals.

1.1 Field-Flow Fractionation

Giddings (1966) first proposed field-flow fractionation as an analytical separation method for macromolecules.[17] As an extremely versatile technique capable of separating and characterizing materials in the macromolecular (0.001 to 1 µm) and colloidal range (> 1 µm and beyond), fractionation is carried out in thin ribbon-like open channels. Separation driving forces can be generated by a number of fields and gradients leading to different FFF methods that are generally applicable to various materials. In general, the sub-techniques of FFF are flow-FFF (based on diffusion coefficient), sedimentation (based on both floating mass and diffusion characteristics of macromolecules), thermal (based on thermal diffusion) and electrical FFF (based on charge). Flow-FFF is the most universal method due to its wide application range (0.0003 to 2 µm).

The flow-FFF (FFFF) theory has been described at length in several papers.[4,16-19] A brief description of FFFF theory and related instrument operational principle is given here. Fractionation in the FFFF channel depends on diffusion coefficients and hence molecular weights. In a FFFF apparatus a thin plastic spacer defines a ribbon-like channel. An external flow field applied perpendicular to the separation axis induces separation, and field flow subsequently causes the components to migrate to the accumulation wall. A stream of carrier liquid, known as the channel flow, is introduced at one end of the channel, and a small-volume sample is injected. Just after the sample has entered the channel, the channel flow is stopped temporarily to allow the sample components to relax and reach equilibrium distributions. A second stream of liquid, known as a cross flow, is applied perpendicular to the channel to retard the sample macromolecules movements in the parabolic channel flow stream. This induced retardation provides the fractionation between different-sized macromolecules, owing to their diffusion coefficients (D, cm^2/s). Finally, the channel flow is reintroduced into the channel, and the separated macromolecules emerge from the channel at characteristic emergence times (t_r).[4,16,18,19] The principle of FFFF is illustrated in Figure 1. The diffusion coefficient can be related to the hydrodynamic (Stokes) diameter (d_s, cm) of the component by the Stokes-Einstein relationship in Eq. (1),[18,19]

$$D = kT/3\pi\eta d_s \tag{1}$$

where k is Boltzmann's constant (1.38 x 10^{-16} g cm^2/s^{2}K), T is the temperature (K) and η is the carrier liquid viscosity (g/cm·s). For random coil macromolecules, d_s is related to molecular weight M (Da) by Eq. (2),

$$d_s = A'M^b \tag{2}$$

where the constant A' depends upon the macromolecule-solvent system. The constant b depends on the molecular conformation in the solution. In normal mode FFFF the emergence or retention time (t_r, min) for well-retained components is approximated by Eq. (3), where w is the channel thickness (cm), V_c is the cross flow rate (cm^3/min), and V is

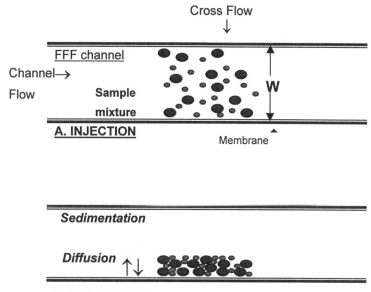
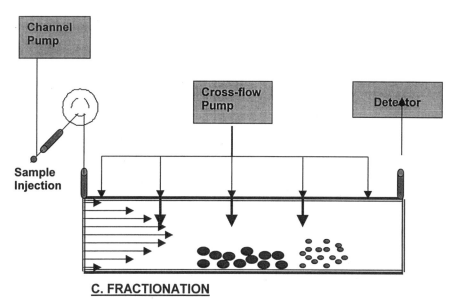

Figure 1 *Schematic showing the flow field-flow fractionation (Flow-FFF) separation mechanism: (a) injection (b) relaxation and (c) fractionation steps in the laminar parabolic flow stream in the channel. In Flow-FFF smaller HA macromolecules emerge before larger ones*

$$t_r = w^2 V_c/6DV \qquad (3)$$

the channel flow rate (cm^3/min). When D from Eq. (1) is substituted into Eq. (3), t_r becomes Eq. (4).

$$t_r = \pi\eta w^2 V_c d_s/2kTV \qquad (4)$$

Equations (3) and (4)[18-20] demonstrate that t_r readily can be controlled by experimentally altering the flow rates V_c and V. Thus, one can easily set up the experiment to suit the the sample characteristics as well as to achieve optimum resolution and analysis speed.[19]

There are several differences between FFF and chromatography based on the different types of driving forces used to facilitate retention. For instance, in chromatography highly localized forces at phase boundaries and surfaces are abundant. These very selective forces tend to cause irreversible adsorption and denaturation as well as structural disruptions, particularly with macromolecules. As a result, separation technique like HPSEC and ultrafiltration are prone to artifacts and interferences.[4,19] Conversely, the field-based FFF driving forces are relatively diffuse, locally mild and physically induced forces. The intensities of these forces seldom reach a level of strength capable of modifying molecular conformation, and thus FFF preserves the integrity of the molecular species under investigation.[21] Therefore, a separation principle operating without a solid support would be an alternative approach that diminishes the risk of denaturation of macromolecules.[21]

1.2 Flow Field-Flow Fractionation-Inductively Coupled Plasma-Mass Spectrometry

Field-flow fractionation-ICP-MS (FFF-ICP-MS) is a versatile hyphenated technique for size separation with subsequent elemental analysis. Promising preliminary results have been reported for coupling FFF to ICP-MS for soil and geological samples.[23,24] The concepts and initial experience in linking FFF separation techniques with ICP-MS were first described by Beckett in 1991.[18] Several interesting applications of FFF-ICP-MS have appeared since then. In these reports, trace metals bound to proteins,[20] natural suspended particulate matter,[25] soils,[23] clay minerals[24] and aqueous colloids[25,26] were analyzed and investigated.

Traditionally, total elemental quantification is used in humic acid characterization, but elemental distribution based on HA molecular sizes has become increasingly significant in understanding fate and transport properties of toxic and nutritionally important elements in soil and aqueous environments. For the simultaneous characterization of the molecular weight distribution and investigation of heavy metals bound to different size fractions of humic acids, an on-line coupling of flow field-flow fractionation (FFFF) and ICP-MS is presented in this work. In addition, the versatility of the FFF approach for determining physical characteristics of HA like hydrodynamic diameters and diffusion rates, and for investigation of aggregation properties of HA polyelectrolytes is demonstrated.

2 MATERIALS AND METHODS

2.1 Humic Acid Samples

Humic acids were extracted from an agricultural soil (HAMI-Chelsea, Michigan; the soil sample was kindly provided by Prof. B. Xing, University of Massachusetts, Amherst), and from Leonardite (HALN) obtained from the International Humic Substances Society (IHSS). HAs were extracted with the sodium pyrophosphate extraction method described by Chen and Pawluk.[27] The ash contents of lyophilized extracted HA samples were below 4.2% and HA samples were previously characterized by UV-visible and diffuse reflectance infrared spectroscopy (DRIFT) by Ruiz-Hass et al.[14] HA samples (2-3 mg) were dissolved in tris-(hydroxy methyl)aminomethane (TRIS)-nitric acid buffer (pH 7.3) solution (pH 7.6-8.0) before FFFF analysis.

2.2 Field-Flow Fractionation

A flow FFF system (Model F-1000-FO, FFFractionation LLC, Salt Lake City, UT, USA) equipped with a 3000 Da molecular weight cut-off (MWCO) ultrafilter membrane (FFFractionation LLC) was used. The FFF channel is 27.7 cm long, 2.0 cm wide, and 0.0254 cm thick. The instrument setup is shown in Figure 2. An HPLC pump (Model L-6010, Hitachi Instruments) controlled the channel flow rate. Another HPLC pump (Model 300, Scientific Systems) furnished the cross flow rate. A UV detector (Model L 4000, Hitachi) was set at 254 nm. Sensitivity was enhanced with the FFF frit outlet configuration and thereby minimized the fractionated HA zones mixing with the carrier fluid (eluent) at the outlet before UV detection. The FFF operating conditions are summarized in Table 1.

2.3 Flow FFF-ICP-MS

A commercial ICP-MS system (Sciex/Elan 5000a, Perkin-Elmer) was used as the FFFF detector. The coupling between FFFF and ICP-MS is physically simple and needs only a tube connecting the FFFF UV detector outlet with the ICP-MS nebulizer since their solution flow rates are compatible. Because of the similarity of the FFFF channel and ICP-MS sample flow rates typically used for analysis, the FFFF outlet was connected directly to the ICP-MS cross-flow nebulizer with poly-(tetrafluoroethylene) tubing (PTFE, 0.3 mm id). The experimental operating parameters are summarized in Table 1.

3 RESULTS AND DISCUSSION

3.1 Molecular Weight Distribution and Diffusion Coefficients

A 10 mM TRIS-nitric acid buffer was chosen as a carrier liquid for the fractionation of soil and Leonardite-derived humic acids. A sample volume as small as 20 µL was used for analysis. The buffer was selected because of its optimum fractionation abilities and because it does not contain alkaline elements, phosphorus or sulfur, which might cause interferences in the ICP-MS elemental determinations. Distilled deionized water also was found to be a suitable carrier liquid for analysis of trace elements bound to humic acids and colloidal organic matter. Details of analytical method development will be reported elsewhere.

A fractogram is a plot of analytical signal (in this case UV absorbance at 254 nm) against the retention time (t_r, here in minutes). As shown in Eq. (3), the retention time is inversely related to the diffusion coefficient (D) of the macromolecular species, but for a linear random coil polymer like humic acid the molecular weight is non-linearly related to its diffusion coefficient. Thus, smaller macromolecules tend to diffuse faster and are detected at earlier retention times than slower diffusing, large molecular species with longer retention times.

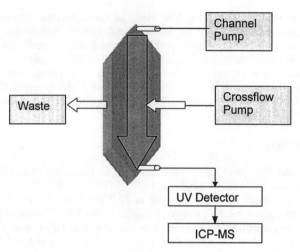

Figure 2 *Schematic of Flow-FFF-ICP-MS instrumental set-up*

Table 1 *Fractionation conditions and ICP-MS instrument operating parameters for Flow FFF-ICP-MS*

Flow FFF conditions : Normal Mode (FFFractionation)	
FFF channel dimensions/cm	27.7 × 2.0 × 0.0254
Carrier liquid	10 mM TRIS-HNO$_3$, pH 7.3 or water
Cross flow rate/mL min^{-1}	2.0
Channel flow rate/mL min^{-1}	1.0
Equilibration time/min	1.5
UV wavelength/nm	254
Membrane	3 kDa MW cut-off polyregenerated cellulose membrane
ICP-MS instrument settings and operating parameters	
RF generator frequency/MHz	37.5
RF forward power/W	1000
Torch	Sciex, short
Spray chamber	Ryton® Scott-type
Nebulizer	Cross –flow (Perkin Elmer)
Nebulizer, gas flow rate/ L min^{-1}	0.92
Auxiliary gas flow rate/ L min^{-1}	1.00
Outer gas flow rate/ L min^{-1}	12
Resolution	1 ± 0.1 at 10% peak maximum
Measurements per peak	1
Dwell time /ms	500
Isotope monitored (m/z)	^{63}Cu, ^{64}Zn and ^{208}Pb

Fractograms obtained for Leonardite (HALN) and Michigan (HAMI) HAs are shown in Figures 3a. The fractograms for both samples were obtained within 15 minutes. Reproducible, monomodal fractograms with different polydispersities were obtained and they are consistent with fractograms reported in the literature.[4,16] Hydrodynamic diameters calculated from Eq. (4) also were plotted, as shown in Figures 3b. Molecular weight (M_w) distributions of HAs were determined from the linear calibration function obtained from fractionation of a mixture (7, 41 and 80 kDa) of poly(styrenesulfonate) standards (PSS).

The distributions of M_w for HALN and HAMI are shown in Figures 3c. Nine replicates were performed for each sample. The apparent molecular weight of the Leonardite humic acid sample was 5080 ± 70 Da, and the soil-derived Michigan HA sample had a lower apparent molecular weight of 4670 ± 230 Da. The slightly larger molecular weight observed for the Leonardite-derived HA sample could be due to more polymerization and condensation or aggregation of smaller molecules during the formation of this HA material.[4]

Diffusion coefficients at peak maximum as well as the polydispersity of humic acid samples also were calculated, as shown in Table 2. The diffusion coefficients, $D_{at\,max}$, were calculated at the peak maxima from relationships (2) and (3). The polydispersity of each humic acid sample at peak width at half height is shown in Table 2. As expected, the larger HALN has a smaller diffusion coefficient (1.57×10^{-6} cm^2/s) than smaller Michigan soil HA (HAMI) (1.66×10^{-6} cm^2/s). The polydispersities of HA estimated by peak width at the monomodal fractogram peak half-height demonstrate that HALN has a wider molecular weight distribution (see Table 2) than the soil derived HAMI sample. Figure 3b illustrates the hydrodynamic diameter distributions for two HA samples. As indicated in Eq. (3), the hydrodynamic diameters are directly proportional to retention time (t_r) when other FFF parameters are kept constant experimentally. This is clearly demonstrated in fractograms 3a and 3b. However, molecular distributions of HALN and HAMI do not closely follow the particle diameter distribution pattern and time-axis fractogram (see Figures 3a and 3c). This is due to the non-linear theoretical relationship between M_w and d_s (and t_r) given by Eqs. (2) and (4).

Table 2 *Physical parameters obtained from flow-FFF for humic acids*

Samples	MW ± S.D., n=9	d (nm)	$D_{at\,max}$ (cm^2/sec x 10^6)	$W_{1/2}$ (min)
HAMI	4670 ± 230	2.46	1.66	1.88
HALN	5080 ± 70	2.64	1.57	2.05

3.2 Elemental Distribution in Humic Acids

Elemental distributions of ^{63}Cu, ^{64}Zn and ^{208}Pb in humic acids were detected on-line using FFFF-ICP-MS. Elemental ion fractograms of HALN and HAMI in 10 mM TRIS (nitric acid) are shown in Figure 3d. In elemental ion fractograms UV absorbance axis is replaced with ion intensities (counts per second) to represent the magnitude of the analyte cations detected (i.e., ^{63}Cu$^+$ where m/z = 63). Elemental fractograms demonstrate good conformity with UV-fractograms. All elements appear to be bound to the median of the HA molecular weight fractions or 2-4 nm hydrodynamic diameter HA macromolecules. Although results presented here are qualitative, ion fractograms show Zn and Cu readily bound to both HAs, consistent with results reported in the literature.[14,15] In the HAMI sample, Zn and Cu may be present in large molecular fractions (i.e.,10-20 kDa) although their signal magnitude is much smaller than at the peak maximum. In HALN perhaps a smaller fraction of Zn emerges at the void time (not shown in calculated molecular weight distribution elemental fractograms in Figure 3d) before the bulk of the HA material. This may indicate that some Zn is present in unbound form. (i.e., $Zn(OH)_4^{2-}$). These results demonstrate that FFFF-ICP-MS can be used effectively for speciation of HA bound trace metal fractions in soil and aqueous environments. With an appropriate calibration approach one can gather quantitative information about these metal humate molecular fractions.

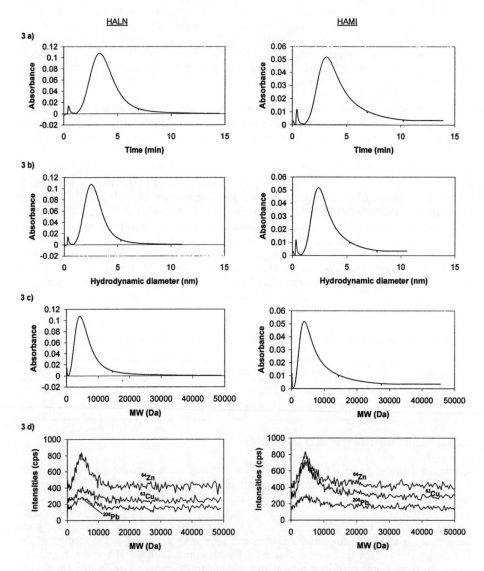

Figure 3 *Fractograms of Leonardite (HALN) and soil-derived (HAMI) humic acids with (3a) UV detection (3b) calculated hydrodynamic diameter distribution (3c) calculated molecular weight distribution and (3d) elemental fractograms with ICP-MS detection*

3.3 Studies of Humic Acid Aggregation Using Flow-Field Flow Fractionation

As humic acids are polyelectrolytes, slight alterations in pH and ionic strength can lead to changes in the structural conformation of humic molecules.[28] Aggregation of humic substances was studied with a series of increasing $CaCl_2$ solution concentrations from 0.002 to 0.8 M. A 1 mg/mL HAs (HAMI and HALN) aliquot was allowed to equilibrate with $CaCl_2$ solution at room temperature for 5 minutes. Shifting HA size distribution to higher molecular weights with increasing Ca^{2+} concentration is observed and fractograms obtained for HALN and HAMI are illustrated in Figure 4a. Humic acids tend to form larger aggregates as the $CaCl_2$ concentration increases. In Figure 4b the apparent molecular weight at peak maximum is plotted against the log calcium ion concentration. A linear increase of M_w was observed up to 0.2 M Ca^{2+}. Above 0.8 M Ca^{2+} the curve flattens. The lines in Figure 4b demonstrate a 912 Da increase in molecular weight per decade change in molar Ca concentration (M_W = 912 log [Ca] + 8220; r^2 =0.960). Similar information was obtained by FFF of the Leonardite HA sample (see Figure 4b, M_W = 827 log [Ca] + 6729; r^2 =0.995). The Ca^{2+} may act as a bridge between two carboxylate groups (-COOH) in humic acids leading to humic acid aggregates with larger molecular masses.[28]

Aggregation of humic acids depends not only on the ionic strength but also on the reaction contact time. Humic acids HALN and HAMI were allowed to react at room temperature at fixed ionic strength (0.02, 0.20 and 0.8M $CaCl_2$). Figure 4c shows that HA molecular aggregation is time dependent. Significant aggregation reactions are occurring within ~ 5 hours and tend to reduce the rate of reaction beyond this point. Conversely, as more time is allowed for equilibration of HAs with various molar concentrations of Ca^{2+} ions, the apparent M_W of the HAs is lower at higher [Ca^{2+}]. This switch in aggregation behavior with time is interesting and is worth further investigation. The formation of HA agregates via calcium-carboxylate bridges may be the dominant time dependent process at the early stage of reaction (i.e., ~ 2 hr for HALN). As the equilibration time increases the HA reaches its dispersed state. The slow process primarily controls the aggregation dynamics so that the reaction is more favorable at [Ca^{2+}] = 0.02 M than at [Ca^{2+}] = 0.8 M.

4 CONCLUSIONS

Reproducible, monomodal fractograms with different polydispersites were obtained for Leonardite-derived (HALN) and soil-derived humic acids (HAMI) by FFFF-ICP-MS. As the molecular mass increased the amount of corresponding humic acid molecular fractions decreased. These results are consistent with the literature.[4,16] The apparent molecular weights (M_w) of HALN and HAMI are 5080 ± 70 and 4670 ± 230 Da. These probably are better estimates than those obtained using chromatographic methods. This is because the separation mechanism in FFFF principally is based on mild physical forces and therefore the molecular integrity of HAs is maintained. The higher molecular weight observed for the Leonardite-derived sample might be related to the nature of its formation by polymer condensation and aggregation.

Diffusion coefficients of these two HAs were obtained and this information is useful to estimate the transport properties of humic substances bound to trace metals or organic contaminants in both soil and aqueous environments. By coupling a sensitive metal-specific detector like ICP-MS to FFFF, information about the distribution in HA molecular fractions of bound trace metals (Cu, Zn, and Pb) is achieved. Additionally, this study demonstrates how FFFF can be a valuable tool to investigate HA molecular aggregation changes resulting from ionic strength and kinetically controlled processes.

FFFF-ICP-MS should further be examined for its ability to provide metal-humate complex binding constants. Several divalent metal cations (Ba^{2+}, Sr^{2+}, Pb^{2+} and UO_2^{2+}) should be investigated to establish their ability to form bridged humic acid aggregates. This paper demonstrates how a FFFF approach could be used to acquire not only elemental speciation and molecular weight information under gentle separation conditions but also how important physical and chemical information about the nature, polydispersity and mobility characteristics of humic substances can be obtained.

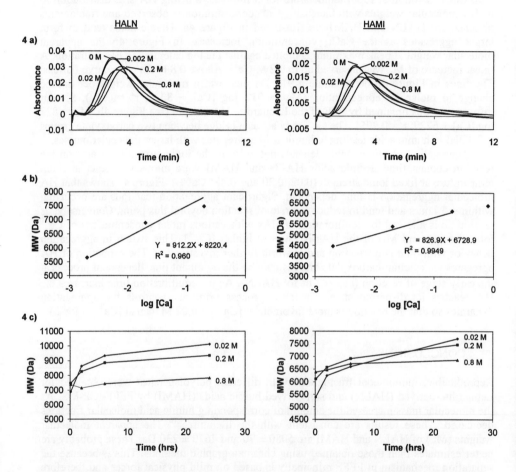

Figure 4 *Fractograms of Leonardite (HALN) and soil-derived (HAMI) humic acids (4a) equilibration with increasing concentration of calcium (0 M to 0.8 M) for 5 minutes (4b) a plot of apparent M_w versus log [Ca] (4c) plots of apparent M_w for HALN and HAMI after reaction with increasing concentration of calcium (0.02M, 0.2M, and 0.8 M) against the time (hrs)*

ACKNOWLEDGEMENTS

Funding was provided in part by the ICP Information Newsletter, Inc. (Hadley, Massachusetts). D. A. would like to acknowledge partial financial support from the Keck

Foundation. Technical support from FFFractionation, Inc. (Salt Lake City, UT) is greatly appreciated. A. S. thanks the Government of Thailand for his studentship awarded through the Ministry of University Affairs.

References

1. F. J. Stevenson, 'Humus Chemistry: Genesis, Composition, and Reactions', 2nd Edn., Wiley, New York, 1994.
2. G. Davies, A. Fataftah, A. Cherkasskiy, E. A. Ghabbour, A. Radwan, S. A. Jansen, S. Kolla, M.D. Paciolla, L.T. Sein, Jr., W. Buermann, M. Balasubramanian, J. Budnick and B. Xing, *J. Chem. Soc., Dalton Trans.,* 1997, 4047.
3. D. L. Macalady and J. F. Ranville, 'Perspectives in Environmental Chemistry', Oxford University Press, New York, 1998, Chapter 5, p. 94.
4. R. Beckett, J. Zhang and J. C. Giddings, *Environ. Sci. Technol.,* 1987, **21**, 289.
5. R. Beckett, J. C. Biglow, J. Zhang and J. C. Giddings, in 'The Influence of Aquatic Humic Substances on Fate and Treatment of Pollutants', P. McCarthy and I. H. Suffett, (eds.), ACS Advances in Chemistry Series No 219, American Chemical Society, Washington, DC, 1989, p. 65.
6. P. Schmitt, A. Kettrup, D. Freitag and A. W. Garrison, *Fresenius J. Anal.Chem.,* 1996, **354**, 915.
7. M. De Nobili, G. Bragato, and A. Mori, in 'Understanding Humic Substances: Advanced Methods, Properties and Applications', E. A. Ghabbour and G. Davies (eds.), Royal Society of Chemistry, Cambridge, 1999, p. 101.
8. R. D. Evans and J. Y. Villeneuve, *J. Anal. At. Spectrom.,* 2000, **15**, 157.
9. L. Zernichow and W. Lund, *Anal. Chim. Acta,* 1995, **300**, 167.
10. L. Rottmann, and K. G. Heumann, *Anal. Chem.,* 1994, **66**, 3709.
11. L. Rottmann, and K. G. Heumann, *Fresenius J. Anal. Chem.,* 1994, **350**, 221.
12. G. Rädlinger and K. G. Heumann, *Fresenius J. Anal. Chem.,* 1997, **359**, 430.
13. J. Vogl and K. G. Heumann, *Fresenius J. Anal. Chem.,* 1997, **359**, 438.
14. P. Ruiz-Haas, D. Amarasiriwardena and B. Xing, in 'Humic Substances: Structures, Properties and Uses', G. Davies and E. A. Ghabbour (eds.), Royal Society of Chemistry, Cambridge, 1998, p. 147.
15. S. A. Bhandari, D. Amarasiriwardena and B. Xing, in 'Understanding Humic Substances: Advanced Methods, Properties and Applications', E. A. Ghabbour and G. Davies (eds.), Royal Society of Chemistry, Cambridge, 1999, p. 203.
16. P. J. M. Dycus, K. D. Healy, G. K. Stearman and M. J. M. Wells, *Sep. Sci. Technol.,* 1995, **30**, 1435.
17. J. C. Giddings, *Sep. Sci.,* 1966, **1**, 123.
18. R. Beckett, *At. Spectroscopy,* 1991, **12**, 228.
19. A. Siripinyanond and R. M. Barnes, *J. Anal. At. Spectrom.* 1999, **14**, 1527.
20. J. C. Giddings, M. N. Benincasa, M. K. Liu and P. Li, *J. Liq Chromatog.,* 1992, **15**, 1729.
21. J. C. Giddings, *J. Chromatog.,* 1989, **470**, 370.
22. A. Exner, M. Theisen, U. Panne and R. Niessner, *Fresenius J. Anal. Chem.,* 2000, **366**, 254.
23. J. F. Ranville, D. J. Chittleborough, F. Shanks, R. J. S. Morrison, T. Harris, F. Doss and R. Beckett, *Anal. Chim. Acta,* 1999, **381**, 315.
24. D. M. Murphy, J. R. Garbarino, H. E. Taylor, B. T. Hart and R. Beckett, *J. Chromatog.,* 1993, **642**, 459.
25. M. Hassellöv, B. Lyvén, C. Haraldsson and W. Sirinawin, *Anal. Chem.,* 1999, **71**,

3497.
26. M. Hassellöv, B. Lyvén and R. Beckett, *Environ. Sci. Technol.,* 1999, **33**, 4528.
27. Y. Chen and S. Pawluk, *Geoderma*, 1995, **65**,173.
28. M. E. Schimpf and K. G. Wahlund, *J. Microcolumn Sep.,* 1997, **9**, 535.

EXAFS AND XANES STUDIES OF EFFECTS OF pH ON COMPLEXATION OF COPPER BY HUMIC SUBSTANCES

Anatoly I. Frenkel[1] and Gregory V. Korshin[2]

[1] Materials Research Laboratory, University of Illinois at Urbana-Champaign. Mailing address: Building 510 E, Brookhaven National Laboratory, Upton, NY 11973, USA
[2] Department of Civil and Environmental Engineering, University of Washington, Seattle, WA 98195-2700, USA

1 INTRODUCTION

The goal of this work was to further quantify the structure of the inner coordination shell of Cu^{2+} in its complexes with humic substances (HSs) and representative model compounds. Our previous XANES and EXAFS studies[1,2] have demonstrated that HSs molecules act as multidentate ligands when they bind Cu^{2+} and possibly replace both equatorial and axial water molecules from the inner coordination shell. The non-uniformity of the copper-binding sites in HSs manifests itself through a gradual change of XANES spectra associated with increased copper doses. This paper presents new XANES and EXAFS data relevant to the understanding of pH effects on the structure of the metal-binding sites in HSs.

2 MATERIALS AND METHODS

The HS sample used in the experiments was the fraction of Suwannee River NOM retained by XAD-4 resin and eluted by acetonitrile. It was isolated and purified by Dr. Jerry Leenheer of the U.S. Geological Survey. The preparation procedures, composition and properties of the sample are described elsewhere.[3] The copper concentration in the Cu^{2+}-HS system was 1.02×10^{-3} M and the Cu/C carbon molar ratio was 0.01. A model polymer (polystyrenesulfonic-co-maleic acid, PSM) also was employed. The copper concentration and molar Cu/monomer unit ratio in the PSM system were 1.41×10^{-3} M and 0.0081, respectively. The pH of the HS and PSM solutions was varied from 2.6 to 11.0. Measurements also were made for the $[Cu(H_2O)_6]^{2+}$ and $[Cu(OH)_4(H_2O)_2]^{2-}$ model systems, in which the concentration of copper was 0.1 M. The solution of the $[Cu(OH)_4(H_2O)_2]^{2-}$ complex was prepared by adding adequate amounts of sodium hydroxide to 0.1 M aqueous copper(II) perchlorate. The pH of the $[Cu(H_2O)_6]^{2+}$ and $[Cu(OH)_4(H_2O)_2]^{2-}$ solutions was adjusted to 2 and 14, respectively.

X-ray absorption measurements were carried out at the UIUC/Lucent Technologies beamline X16C of the National Synchrotron Light Source at Brookhaven National Laboratory. The X-ray energy was varied from 200 eV below to 1000 eV above the absorption K edge of Cu (E_K = 8979 eV) with a Si(111) double crystal monochromator. The fluorescence mode was used for data acquisition. Pre-edge and near-edge regions of the data (-30 eV < E-E_K < 40 eV) were acquired with a 0.5 eV energy increment. The EXAFS data in the range 40 eV < E-E_K < 1000 eV were acquired with a 2 eV increment. Several measurements were averaged for the same sample to improve the signal-to-noise ratio. To correct for a small angular drift of the monochromator, all data sets were aligned vs. their absolute energy and interpolated to the same energy grid before the averaging. Cu metal foil was measured in the transmission mode simultaneously with all other samples and was used as the reference for the alignment of energies.

3 RESULTS

3.1 EXAFS Data and Their Analysis

Following background removal and edge-step normalization, the k-weighted EXAFS signals $\chi(k)$ (where k is the photoelectron wave number) were obtained for the Cu-HS, Cu-PSM complexes and model compounds (Figure 1). By Fourier transforming the $k\chi(k)$ data to r-space, the effective radial pair-distribution function of nearest neighbors around central Cu ion (uncorrected for phase shift) can be visualized (Figure 2).

Analysis of the EXAFS spectra was performed with the routines previously employed.[2] Theoretical modeling was performed with FEFF6 computer code and the UWXAFS data analysis package.[4,5] In accordance with well-established literature data,[6] the inner oxidation shell was represented by a tetragonally distorted CuO_6 octahedron for both Cu-HS and Cu-PSM systems at all pHs. The structural parameters (including equatorial Cu-O_{eq} and axial Cu-O_{ax} bond lengths and their mean square disorders) were determined from the fit provided by the FEFF theory to the EXAFS data (Figure 3).

The equatorial Cu-O distances did not appear to undergo systematic changes with pH, although they were always somewhat shorter for the Cu^{2+}-HS system compared with those determined for PSM (Figure 4). The equatorial Cu-O distances were also not dramatically different for the model $[Cu(H_2O)_6]^{2+}$, pH 2 and $[Cu(OH)_4(H_2O)_2]^{2-}$, pH 14 compounds (1.98 ± 0.01 Å and 1.94 ± 0.01 Å, respectively).

In contrast to the equatorial distances, the axial Cu-O lengths exhibited noticeable variations with pH (Figure 4). For the model $[Cu(H_2O)_6]^{2+}$, pH 2 and $[Cu(OH)_4(H_2O)_2]^{2-}$, pH 14 systems, the difference in the Cu-O_{ax} lengths was large (2.29 ± 0.03 Å and 2.53 ± 0.11 Å, respectively). The changes of the Cu-O_{ax} bond length associated with the pH variations were not unidirectional (a minimum was observed at pH ca. 9), but they were conspicuously similar for the Cu-HS and Cu-PSM systems. Similar to the pattern seen for the equatorial bond lengths, the axial Cu-O distances were consistently shorter for the Cu^{2+}-HS system.

3.2 XANES Data Analysis

Edge-step normalized absorption coefficient data for the Cu-HS and Cu-PSM systems showed two important trends. First, a shoulder located between 8970 and 8976 eV was observed for all samples, but it became much more prominent and shifted to lower energies at high pHs. At the same time, the absorption peak maximum located in the range 8980 to 8984 eV gradually shifted to higher energies. The first derivatives of these

XANES spectra allowed us to discern these trends in more detail (Figures 5). The local minimum (1) in Figures 5 corresponds to the shoulder located in the range 8980 and 8984 eV in the XANES spectra. The zero in the XANES derivatives is denoted as (2) and corresponds to the maximum of the absorption coefficient. It also shifted towards higher energies with increasing pH. These two characteristic points indicate significant structural changes occurring in both systems. Of these, the shift of the absorption coefficient maximum (2) to higher energies (by ca. 2.5 eV) when the pH was increased from 2 to 14 is most notable. A detailed examination of the first XANES derivatives for the 8980 – 8984 eV region indicates that the effects of pH in the two systems are not identical. The pH-dependence of the energies corresponding to the locations of features (1) and (2) in Figures 5 is shown in Figures 6.

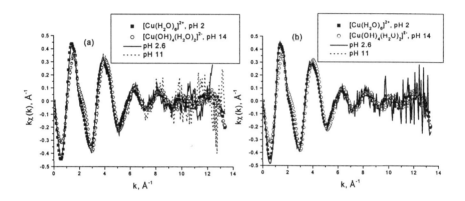

Figure 1 *k-weighted EXAFS data for $[Cu(OH)_4(H_2O)_2]^{2-}$, $[Cu(H_2O)_6]^{2+}$, and Cu^{2+} complexes with (a) HS and (b) PSM. Only the data for the lowest and highest pH are shown for clarity*

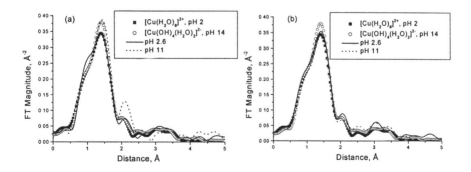

Figure 2 *Fourier transform magnitudes of the k-weighted EXAFS data shown in Figure 1 for (a) $[Cu(OH)_4(H_2O)_2]^{2-}$, $[Cu(H_2O)_6]^{2+}$ and Cu^{2+}-HS and (b)PSM complexes*

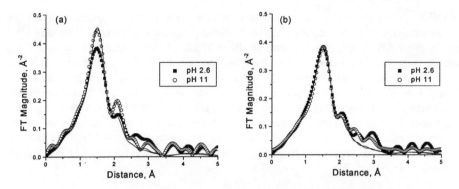

Figure 3 *Fourier transform magnitudes of the k-weighted EXAFS data (symbols) and theory (solid) for the Cu^{2+} complexes with (a) HS and (b) PSM at pH 2.6 and 11*

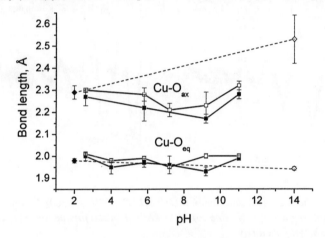

Figure 4 *$Cu-O_{eq}$ and $Cu-O_{ax}$ bond lengths obtained from the EXAFS analysis of the Cu^{2+}-HS and PSM systems at varying pHs. The data for Cu-HS and Cu-PSM complexes are shown by black and white squares, respectively. The data for the $[Cu(H_2O)_6]^{2+}$ and $[Cu(OH)_4(H_2O)_2]^{2-}$ complexes are shown by black and white rhombs, respectively. The straight dashed lines connecting the data for the reference complexes serve as guides to the eye*

4 DISCUSSION

The profound changes of the XANES spectra for the Cu^{2+}-HS and PSM systems associated with pH variations are an important finding of this study. Somewhat similar observations were made in independent studies of pH effects on K-edge XANES spectra of copper[6] and other targets (e.g., sulfur and vanadium),[7,8] although the direction of shifts in the XANES spectra of sulfur and vanadium was opposite to that found for copper.

As discussed previously,[2,6,9] the location of features (1) and (2) in the XANES spectra is likely to be affected by the tetragonal distortion of the inner shell of the Cu^{2+} complexes.

Although advanced XANES analysis was shown to be very useful in quantifying the distortion parameters in the CuO_6 octahedra in relatively simple cases,[2,6] the interpretation of the data for the Cu^{2+}-HS and PSM systems requires more sophisticated structural modeling.

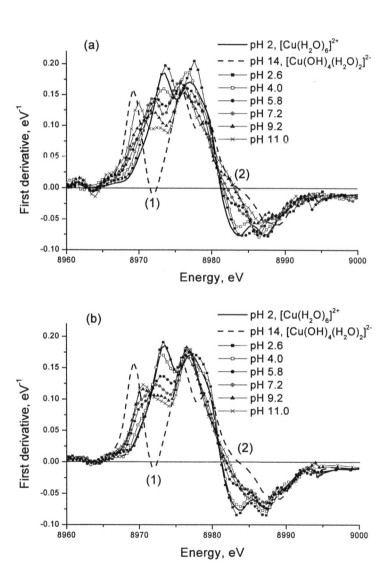

Figure 5 *First derivatives of the XANES data for (a) $[Cu(OH)_4(H_2O)_2]^{2-}$, $[Cu(H_2O)_6]^{2+}$, and (b) Cu^{2+}-HS and PSM complexes*

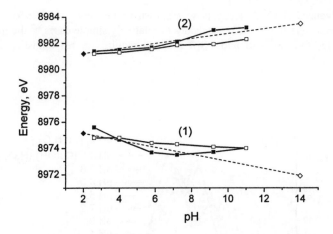

Figure 6 *Shifts of the characteristic energies (1) and (2) in the XANES derivatives (Figure 5) as a function of pH. The data for Cu-HS and Cu-PSM complexes are shown by black and white squares, respectively. The data for $[Cu(H_2O)_6]^{2+}$ and $[Cu(OH)_4(H_2O)_2]^{2-}$ complexes are denoted by black and white rhombs. The dashed lines connecting the data for the reference systems serve as guides to the eye*

In parallel with the presumption of the tetragonal distortion changes induced by the increase of pH, an alternative qualitative hypothesis to explain the observed shifts of the binding energies to the higher values (Figure 6) can be proposed. The energy corresponding to the maximum of the absorption coefficient is directly related to the binding energy of the $1s$ electrons excited to the $4p$ state by photoabsorption. The positive shift of the $1s \rightarrow 4p$ transition energy measured by XANES may be viewed as caused by a decrease in the electron density of non-resonant electrons and the concomitant increase of the binding energy of the photoelectron. That is, the shift of the $1s \rightarrow 4p$ transition energy to higher values indicates less ability of the non-resonant electrons to screen the positive charge of the nucleus. Our results suggest that the electron density on the Cu^{2+} ions gradually decreases as the pH increases. However, the mechanism of the proposed charge transfer currently is unclear. We believe that the combination of EXAFS/XANES studies and *ab initio* modeling of these systems is a powerful tool to elucidate the mechanisms of the observed changes of the local structure and electronic properties of Cu-HS complexes.

ACKNOWLEDGEMENTS

We are grateful to Dr. Patrick Frank (Stanford University) for useful discussions. A. I. Frenkel acknowledges support by DOE grant DEFG02-96ER45439 through the Materials Research Laboratory at the University of Illinois at Urbana–Champaign. G. V. Korshin acknowledges support from the U.S. Environmental Protection Agency (Grant #R826645).

References

1. G. V. Korshin, A. I. Frenkel and E. A. Stern, *Environ. Sci. Technol.*, 1998, **32**, 2699.
2. A. I. Frenkel, G. V. Korshin and A. L. Ankudinov, *Environ. Sci. Technol.*, 2000, **34**, 2138.
3. J. -P. Croué, G. V. Korshin and M. M. Benjamin, 'Characterization of Natural Organic Matter in Drinking Water', AWWA Research Foundation and American Water Works Association, Denver, CO, 2000.
4. S. I. Zabinsky, J. J. Rehr, A. Ankudinov, R. C. Albers and M. Eller, *Phys. Rev. B*, 1995, **52**, 2995.
5. E. A. Stern, M. Newville, B. Ravel, Y. Yacoby and D. Haskel, *Physica B*, 1995, **208/209**, 117.
6. L. Palladino, S. Della Longa, A. Reale, M. Belli, A. Scafato, G. Onori and A. Santucci, *J. Chem. Phys.*, 1993, **98**, 2720.
7. P. Frank, B. Hedman, R. M. L. Carlson and K. O. Hodgson, *Inorg. Chem.*, 1994, **33**, 3794.
8. P. Frank, B. Hedman and K. O. Hodgson, *Inorg. Chem.*, 1999, **38**, 260.
9. J. Garcia, M. Benfatto, C. R. Natoli, A. Bianconi, A. Fontaine and H. Tolentino, *Chem. Phys.*, 1989, **132**, 295.

MAIN CONCLUSIONS OF THE EC-HUMICS PROJECT: "EFFECTS OF HUMIC SUBSTANCES ON THE MIGRATION OF RADIONUCLIDES: COMPLEXATION AND TRANSPORT OF ACTINIDES"

G. Buckau,[1] P. Hooker,[2] V. Moulin,[3] K. Schmeide,[4] A. Maes,[5] P. Warwick,[6] C. Moulin,[7] J. Pieri,[8] N. Bryan,[9] L. Carlsen,[10] D. Klotz[11] and N. Trautmann[12]

[1] Forschungszentrum Karlsruhe, Institut für Nukleare Entsorgungstechnik, D-76021 Karlsruhe, Germany
[2] British Geological Survey, Keyworth, Nottingham NG12 5GG, UK
[3] CEA/Centre d'Etudes des Saclay, Fuel Cycle Division, DESD/SESD/LMGS, F-91191 Gif-sur-Yvette, France
[4] Forschungszentrum Rossendorf, Institut für Radiochemie, D 01314 Dresden, Germany
[5] Katholieke University Leuven, Landbouwinst.-Laboratorium voor Colloïdchemie, B-3001 Heverlee, Belgium
[6] Loughborough University, Department of Chemistry, Loughborough, Leicestershire LE11 3TU, UK
[7] CEA, Fuel Cycle Division, DPE/SPCP/Laboratory of Organic Analysis, F-91191 Gif-sur-Yvette Cedex, France
[8] Université de Nantes, Laboratoire Bio- et Radiochimie, F-44322 Nantes, France
[9] University of Manchester, Department of Chemistry, Manchester M13 9PL, UK
[10] National Environmental Research Institute, Department of Environmental Chemistry, DK-4000 Roskilde, Denmark
[11] GSF-National Research Center for Environment and Health, Institute of Hydrology, D-85764 Neuherberg, Germany
[12] University of Mainz, Institute of Nuclear Chemistry, D-55128 Mainz, Germany

1 INTRODUCTION

Colloid mediated actinide migration has received much deserved attention with regard to potential host formations and repository types for radioactive waste disposal.[1-5] Laboratory experiments and natural chemical analogue studies show the marked influence of humic colloids on the chemical behavior of actinide ions.[6-8] In fractured media the concentration of dissolved organic carbon is very low and thus emphasis is focused on inorganic colloids.[9,10] Humic colloids are of main concern in clay formations.[11] Clay as a backfill material in crystalline rock also can lead to the release of humic colloids.[12]

At an OECD-NEA workshop on the influence of natural organic substances in performance assessment of radioactive waste disposal, key questions were 1) the mobility of humic substances, 2) their long-term stability and 3) the distribution of radionuclides between mobile and stationary humic substances.[13] These and other issues that determine the influence of humic substances on actinide migration are the subject of the project described.

The overall approach is a combination of studies of different types of systems (Figure 1). The designed systems use purified humic substances and controlled experimental conditions (pH, ionic strength, temperature, etc.). The experimental conditions are varied and basic data with process understanding are obtained.

Batch and column experiments are conducted with near-natural systems, i.e., natural sediments and groundwater conditioned over long times under an inert gas atmosphere. Actinide transport properties are investigated in these systems. The objective is to deviate as little as possible from natural conditions, and thus variation of experimental conditions is not possible. Comparison of data from designed and near-natural systems is used for data verification and process understanding.

Designed Systems	Near-Natural Systems	Real System
Purified Humic Acid "Humic Gel"	Real groundwater and Sediments	Analysis of Data
Defined pH, I, ...	Batch, Column, and Diffusion experiments	
Parameter Variation		
Basic Process Understanding	Complex Systems	Data not Available from Lab Investigations
Basic Data	Verification of Data and Processes	Verif. of Processes + Data
Rationalization of Processes by **Models** Implementation of Models and Data through **Codes** **Demonstration** by Migration Case Studies		

⇩

Trustworthy Safety Assessment

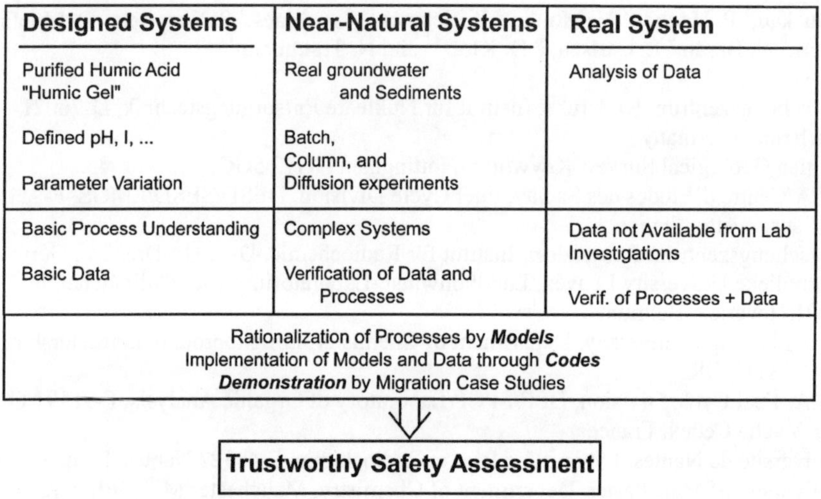

Figure 1 *Overall approach of the project*

Real systems analysis is used to verify or reject the findings from the other two systems. Another objective is to obtain data that cannot be determined from laboratory approaches. Such information includes, for example, the origin and mobility of humic colloids in natural aquifer systems.

Process understanding from investigations of all three types of systems is rationalized by models. Models and data are implemented in transport codes and the outcome is demonstrated by migration case studies. By this overall approach, the project contributes to the scientific acceptance of methodologies and data for safety assessment of radioactive waste disposal by developing the basis for trustworthy actinide transport predictions. Scientific acceptance is the prerequisite for political and public acceptance of radionuclide protocols and practices.

To the extent possible, the data generated rest on intercomparison between different groups and experimental methods. Models are established that rationalize the developed process understanding. Models are kept as simple as possible and black-box fitting without process understanding is avoided. Humic colloids in natural aquifer systems are shown not to sorb or decompose but to be stable and mobile over time-frames relevant to radioactive waste disposal. For the first time, introduction of a kinetic concept allows batch and column experiments to be described consistently and up-scaling of column experiments is successful. Humic colloid mediated actinide transport is shown to be much more significant than concluded from previous estimates based on equilibrium concepts.

Technical project reports have been published for each year of the project.[14,15] These reports give detailed results of individual contributions from the different partners. Details of project achievements are given in a final report.[17] This paper summarizes the project achievements.

2 MATERIALS AND METHODS

Different isolated, purified and characterized natural humic and fulvic acids were used for the study of basic metal ion humic colloid complexation properties. These HSs are Aldrich HA, Gohy-573 HA and FA, Derwent FA, and Kranichsee HA and FA.

Purified Aldrich humic acid serves as a reference substance within the project and also as a reference to results from previous EC-supported activities.[18,19] The Gorleben humic and fulvic acids also have been introduced and used throughout previous EC-supported activities. They originate from the groundwater Gohy-573 from 136 m depth in the Gorleben aquifer system. The Derwent fulvic acid originates from the Derwent Reservoir (Derbyshire, UK). The Kranichsee humic and fulvic acids are isolated from the bog "Kleiner Kranichsee", Saxony, Germany.

Characterization of these HSs was conducted with a number of different methods, including elemental analysis, analysis of inorganic impurities, IR spectroscopy, capillary electrophoresis, ultracentrifugation, field-flow-fractionation, atomic force microscopy, photon correlation spectroscopy and gel permeation chromatography. The characterization results verify that the substances used are not contaminated or altered by their treatment.

In order to mimic the behavior of surface bound HA in natural sediments, humic acid coated silica gel was prepared. For a clearer understanding of the behavior of surfaces with different functional groups, silica gel also was coated with organic molecules with specific functional groups.

Designed laboratory systems were investigated to obtain basic data for the behavior of the ternary system consisting of actinide ions, HSs and minerals. Binary systems also were investigated as a prerequisite for interpretation of results from the ternary system. This includes the interactions between minerals and actinide ions or HAs. Minerals investigated were hematite, goethite, FeO, kaolinite and Al_2O_3. Furthermore, the rock phyllite and its main constituents (quartz, muscovite, chlorite and albite) were studied, as well as ferrihydrite, which is formed as a secondary mineral due to weathering of phyllite.

For the purpose of actinide transport investigations by batch and column experiments, natural groundwaters and sediments were conditioned and the hydraulic properties of columns were characterized. Sediments varied from coarse irregular pebble to fine sand. The column lengths varied between 25 cm and 10 m. The most thoroughly investigated systems are Gorleben groundwater and sediment. However, a large number of other systems also have been investigated. Adequate handling (inert gas atmosphere (1 % CO_2, 99% Ar)) and long and carefully monitored conditioning is essential in order to obtain stable systems. It was shown that column systems require a minimum of three months of conditioning. However, longer conditioning time is prefered. Experiments on unstable systems give results that are disturbed by chemical and hydraulic transformations and thus are untrustworthy.

3 RESULTS AND DISCUSSION

Humic colloid-mediated actinide and technetium transport depends on two key factors: 1) the radionuclide-humate interaction and 2) the stability and mobility of humic colloids. Various aspects of these issues are discussed in sections 3.1-3.6. Model and code development and application is discussed in sections 3.7 and 3.8.

3.1 Actinide Humate Complexation

The complexation of Eu(III), Th(IV), Np(IV), Tc(IV), Np(V) and U(VI) by different humic and fulvic acids has been investigated by different experimental methods under varying experimental conditions. The Charge Neutralization Model has been applied for evaluation of complexation data.[20] Comparison also has been made with other evaluation approaches. In cases where the loading capacity is not available, operational mass based constants also are used. An overview of actinide humate complexation constants is given in Table 1.

Table 1 *Actinide-humate complexation constants for different actinide ions. The results are based on application of the Charge Neutralization Model*[20]

Ion	$\log \beta$, L/mol
An^{3+}	6.2
An^{4+}	8.5
AnO_2^+	3.6
AnO_2^{2+}	6.1

3.1.1 Critical Assessment of Experimental Methods. The quality of data is directly related to the experimental methods applied where different artifacts may occur. Such artifacts often are not recognized or not adequately regarded. Two different experimental methods were critically examined to measure uranium(VI) humate complexation at pH 4.0 and 5.0 and at ionic strengths of 0.004 and 0.1 M ($NaClO_4$). It was shown that electrophoretic ion focusing leads to incorrect results due to experimental difficulties. Anion exchange chromatography also was investigated for its applicability. It was shown that the anion exchanger also contains a small number of cation exchanging sites. The consequence is that if not corrected for, an increase in the stability constant is found with decreasing metal ion concentration. If this effect is corrected for, a single uranium(VI) humate metal ion concentration independent complexation constant is found ($\log \beta = 6.08 \pm 0.15$ L/mol).

The Np(V) humate system was studied by UV/Vis spectroscopy at pH 7 - 9 in 0.1 mol/L $NaClO_4$. Comparison with published results shows that results based on direct spectroscopic speciation with three different HAs gave indistinguishable experimental data. On the other hand, application of indirect speciation by anion exchange, electrophoretic ion focusing or dialysis led to inconsistencies.

These studies show that care must be taken in assessing experimental results. Results from different experimental methods should be consistent. Experimental artifacts become increasingly troublesome with decreasing metal ion concentrations.

3.1.2 Intercomparison of Results. Trustworthy results preferably rest on intercomparison from different laboratories applying different experimental methods. The humate and fulvate complexation of U(VI) has been studied at different pH and ionic strengths (Table 2). The studies with Aldrich humic acid and Kranichsee humic and fulvic acids were conducted at uranyl ion concentrations up to saturation of the humate and fulvate ligands. In contrast to this, a study on Derwent fulvic acid was done at negligible loading of the fulvate ligand. Despite differences in humic materials and in physico-chemical and experimental conditions (especially the degree of loading), application of the Charge Neutralization Model leads to consistent results. These results also are in excellent agreement with published results that apply three different experimental methods at pH 4 and ionic strength 0.1 M, with log ß = 6.16 ± 0.13.[21] The overall conclusion is that reliable data are available for U(VI)-humate interactions under relevant physicochemical conditions.

Table 2 *U(VI) humate and fulvate complexation constants by time-resolved laser-induced fluorescence spectroscopy, solvent extraction and anion exchange*

HSs	Method	pH	Ionic Strength, M	LC mmol/g	log β, (L/mol)
Aldrich HA	Anion Exchange	4	0.1	0.20	6.08
		4	0.004	0.17	5.97
		5	0.1	0.34	6.05
		5	0.004	0.31	6.16
Derwent FA[a]	Solvent Extraction	3.9	0.1	0.20	6.13
		4.1	0.03	0.20	5.92
		5.9	0.1	0.48	6.04
		7.2	0.1	0.65	6.61
Kranichsee HA	TRLFS[b]	4	0.1	0.14	6.35 ± 0.22
Kranichsee FA		4	0.1	0.13	6.21 ± 0.20

[a] From the charge Neutralization Model, calculated with loading capacity (LC) estimated from the anion exchange results and from (ref 21) results; proton exchange capacity (PEC) set at 6 meq/g FA; [b] Time-resolved laser-fluorescence spectroscopy.

3.1.3 Mixed Species. Ternary complex formation by Eu(III) and U(VI) was investigated by time-resolved laser-induced fluorescence spectroscopy. The binary hydroxo and carbonate complexes were investigated as a prerequisite for interpretation of the spectroscopic data on the mixed species.[22] For U(VI), only the hydroxo complexes are characterized as the carbonate complexes do not fluoresce. Including mixed species, the species distribution can be calculated for different conditions. An example is given in Figure 2. In the neutral pH range the uranyl ion predominantly is found as a mixed humate hydroxo complex. Investigations on the ternary system europium(III)/carbonate/humic acid resulted in a spectrum with a different band shape and a longer lifetime (170 ms) than those found for binary complexes. This supports the existence of a ternary Eu(III) complex.[23] These results underline the necessity to improve the database for mixed complexes, especially hydroxo and carbonate species under near neutral pH conditions as found in natural groundwater.

3.1.4 Characterization of Humate Complexes. Characterization of actinide humate complexes contributes to our understanding of the interaction process. EXAFS allows direct determination of the local binding environment of complexed ions. An example is the EXAFS investigation of the uranyl humate interaction at pH 2. Kranichsee humic and fulvic acids as well as Aldrich HA and the synthetic humic acid M1 were investigated. Experimental results are shown in Table 3. Comparison of the EXAFS results with published mean values of bond-distances for defined crystalline uranyl carboxylate complexes shows that the uranyl ion is monodentate coordinated to carboxylic groups of

HAs. Further investigations in this direction will be required to obtain the required detailed process understanding. This also includes assessment of the reasons for the different kinetic availabilities (see below).

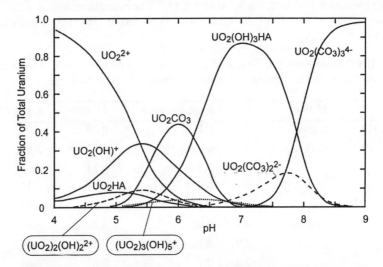

Figure 2 *U(VI) species distribution as a function of pH (ambient atmosphere, [U]=1 mg/L, [HA]=1 mg/L, I=0.1 M $NaClO_4$). Uranyl-humate complexation constants applied are log β UO_2HA = 5.4 and log β $UO_2(OH)_3HA$ = 6.7*

Table 3 *EXAFS data for U(VI) humate complexation at pH 2*

Humic/Fulvic Acid	N	U-O(axial) R, Å	σ^2, Å^2	N	U-O(equatorial) R, Å	σ^2, Å^2
Kranichsee HA	2	1.78	0.001	5.2	2.39	0.012
Kranichsee FA	2	1.78	0.002	5.3	2.39	0.012
Aldrich HA	2	1.78	0.001	5.3	2.40	0.012
M1	2	1.78	0.002	5.2	2.38	0.014

3.1.5 Thermodynamic Data; Enthalpy and Entropy. Thermodynamic parameters for Eu(III)- and U(VI)-humate and fulvate complexation have been measured. Investigations were conducted between 20 and 60°C and the results were evaluated with the Charge Neutralization Model and the Scatchard and Schubert approaches. One or two complexing sites are assumed depending upon the experimental results. Results from different evaluation approaches are in acceptable agreement with each other. The Eu(III)-humate complexation is found to be endothermic and entropy driven whereas U(VI)-humate complexation is exothermic. These data were used to improve metal ion humate interaction process understanding.

3.1.6 Humate Complexation Model Development. When the metal ion concentration is increased, the concentration of the metal-humate complex increases until saturation of the ligand occurs. The saturation value can be expressed as the fraction of relevant functional groups that can be loaded with the investigated metal ion under given experimental conditions. This is the loading capacity (LC). Because metal binding is competitive, the LC decreases with increasing counterion concentration, e.g. Na$^+$. Accordingly, it also decreases with increasing proton ion concentration. This shows the co-inclusion of counterions during the complex formation. In addition to this, the LC depends on the nature of the complexing ion. Figure 3 shows that the LC is a function of pH for different metal ions. The LC not only decrease with increasing proton ion concentration, but also the dependence of LC on pH varies systematically with the nature of the complexing ion identity. The values for Np(IV) should be regarded with caution because the experiments are conducted at pH 1 and 1.5 where the FA is highly protonated.

Figure 4 shows complexation constants from the same investigations. The complexation constants in this figure were all evaluated by the Charge Neutralization Model and represent data from loading of around 1 % of the humate ligand up to near saturation. The complexation constants of calcium and the actinide ions follow a systematic trend with increasing effective ionic charge. The value for copper is larger than expected solely from its ionic charge. This indicates a stronger local binding contribution with the humate ligand for this transition metal ion than for the other ions.

One approach to a mechanistic model has been developed within the project. This model is based on humic colloids as penetrable dispersed polyelectrolyte microgels where the metal ion complexation is governed by the exchange of metal ions with polyelectrolyte counterions. The main driving forces for the exchange process are identified to be 1) dehydration of the complexing metal ion and 2) release of water molecules due to double layer relaxation.

An important question in assessing the appropriateness of a polyelectrolyte model approach is the mass and size of humic entities required for developing polyelectrolyte properties. The calculated influence of different energy contributions to the binding of Ni(II) as a function of molecular mass of dispersed polyelectrolyte microgels is shown in

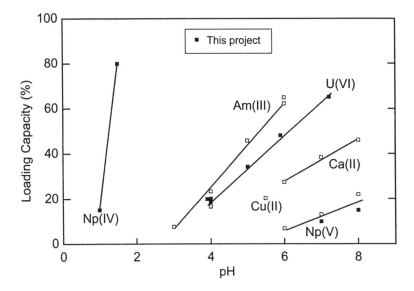

Figure 3 *Loading capacity of humic and fulvic acids (Gohy-573 HA and FA, Derwent FA and Dachau FA) as a function of pH and nature of the metal ion (in 0.1 M NaClO$_4$ or KClO$_4$ medium). Data from the present project and from refs. 20, 21 and 24-26. Unpublished data for Ca(II): Czerwinski (kczer@mit.edu)*

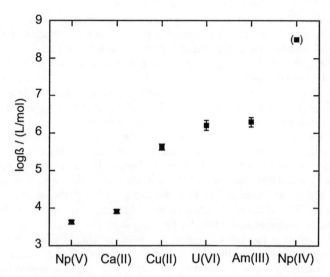

Figure 4 *Complexation constants evaluated by the Charge Neutralization Model for systems described in Figure 3. The complexation constant for Np(IV) is given without error estimates and in parentheses. It originates from experiments at pH 1 and 1.5 where fulvic acid is strongly protonated*

Figure 5. The enthalpy change on binding increases with increasing ligand molecular mass.

The entropy changes are cooperative so that negative ΔG values result over the entire molecular mass range. It is important to note that 1) with increasing molecular mass the ΔG values become progressively more negative and 2) the polyelectrolyte contribution is almost fully developed at a molecular mass of approximately 5 kDa. These results indicate that for sufficiently large dispersed microgels, the polyelectrolyte contribution will dominate over site-specific binding. Recent findings show that the molecular weight distribution of humic and fulvic acids have their maxima around 12000 and 600-700 Da, respectively.[27] To what extent association with primary HSs units or preferential complexation by larger units lead to the predicted large polyelectrolyte contribution will be the subject of further investigations.

3.2 Kinetics of Humate Interaction

Based on migration data in batch and column experiments, it was concluded that actinide migration is governed by the kinetics of binding and release of actinide ions by humic colloids. A kinetically controlled availability model (KICAM) was developed. The model describes how actinide ions bound to humic colloids distribute over HA binding environments with different accessibility to the bulk solution (Figure 6). With increasing contact time, actinide ions progressively transfer to HA sites (availability modes) that are kinetically more hindered from exchange with the bulk solution. Reasons for differences in the kinetic availability for exchange with the bulk solution can be 1) alteration of the local binding environment with increasing contact time, 2) transfer to binding sites with lower dissociation rates and 3) transfer to locations in the humate structure shielded from exchange with the bulk solution.

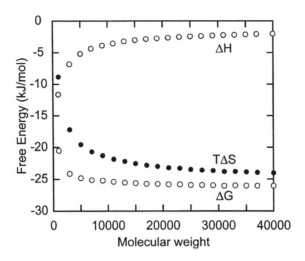

Figure 5 *Calculated contributions of different energies to the Ni(II) humate binding on dispersed polyelectrolyte microgels as a function of molecular weight*

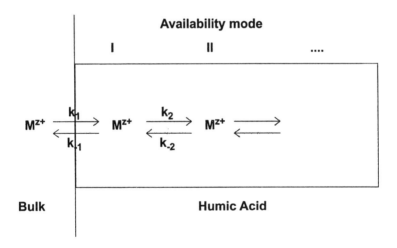

Figure 6 *Kinetic behavior of actinide humic colloid interaction as described by the kinetically controlled availability model (KICAM). Metal ions are bound to humic colloids by different availability modes that have different rates of exchange with the bulk solution*

Batch and column experiments were evaluated to compare the two metal ion humate binding modes in Figure 6 with dissociation rate constants on the order of hours and on the order of hundred of hours. The outcome is that the transition between the "inner" and "outer" binding modes follows first order kinetics. The entropy of activation is −230 to −260 J/K mol and the enthalpy of activation is ≈ 0 kJ/mol at pH 4.5 and 40 kJ/mol at pH 6.5. The total height of the activation barrier is about 100 kJ/mol and the difference in energy between the two binding modes is very small. Thus, the reason for different kinetically governed accessibility of the two binding modes is not a large difference in binding strength.

3.2.1 Designed Systems. In order to verify the findings from near-natural system investigations, the association and dissociation kinetics were investigated with designed laboratory systems. The methodology is based on varying the contact time between a humic acid and actinide ions followed by measurement of the actinide ion desorption kinetics. Desorbed metal ions are scavenged with a cation exchanger. A variety of ion exchangers have been tested and used (Cellphos, Hyphan, Dowex and Chelex). Eu(III), Th(IV) and U(VI) bound by Aldrich HA and Derwent FA have been studied.

In all systems investigated the primary association is found to proceed within minutes. Metal dissociation from an FA or HA depends on a variety of parameters, including the nature and concentration of metal ions, the nature and concentration of humic material, pH and ionic strength, and especially the metal ion-humic acid contact time prior to desorption measurements. Consecutive dissociation is described from either two or three different availability modes. No indication of more than three availability modes has been found in designed laboratory system investigations. Some results are shown in Figures 7 and 8. The results verify that actinide ions progressively are transferred with increasing contact time to binding modes with lower dissociation rates. The results also show that the dissociation rates depend on the nature of the complexing ion. More Th(IV) is found to enter into the kinetically hindered sites than U(VI). Ionic strength and pH also influence the dissociation kinetics. The dissociation rate decreases with decreasing ionic strength and decreasing proton ion concentration. Furthermore, the fraction of Th(IV) dissociated within 24 h is higher for a fulvic acid than for a humic acid of the same concentration and the fraction decreases with increasing FA concentration.

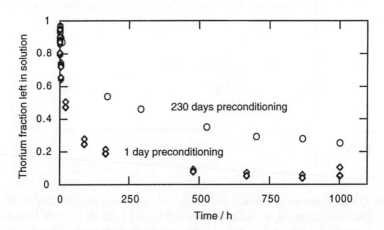

Figure 7 *Dissociation of Th from Derwent FA by exchange with Hyphan; 1 day preconditioning points are duplicates*

3.2.2 Designed Laboratory System versus Real Systems. Successful implementation of kinetic data for correct prediction of actinide transport in batch and column experiments is a major achievement of the project. The kinetic characteristics were verified by independent investigations of a large number of systems. The key question, however, is to what extent these findings are applicable to real systems?

Figure 9 shows results for Eu(III) dissociation from Aldrich HA at pH 8. As discussed above, metal dissociation becomes slower with increasing contact time between humic acid and the europium spike added. However, after some 2000 h more than 80 % of the europium has dissociated, irrespective of the contact time prior to addition of the Chelex 100 ion exchanger.

In parallel experiments, the dissociation of natural europium from humic colloids of the two Gorleben groundwaters Gohy-532 and -2227 was studied. Comparison of the behavior of Eu(III) added under laboratory conditions and natural europium constituents of groundwater humic colloids shows there are large differences in dissociation behavior. The natural europium shows a much lower dissociation rate constant and even after 2000 h about 40 % of the Eu(III) remains humic colloid bound. Continuation of the experiment

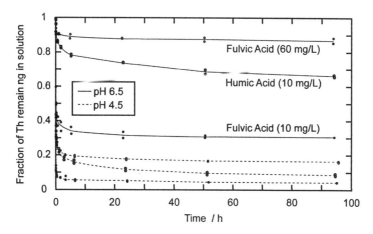

Figure 8 *Dissociation of Th(IV) from Aldrich HA and Derwent FA by contact with Dowex cation exchanger. The relative sequences of the curves are the same at both pH values*

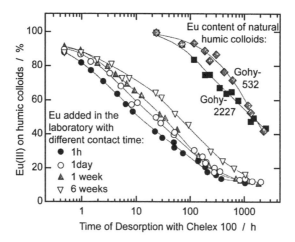

Figure 9 *Dissociation of Eu(III) from humic colloids by exchange with Chelex 100*

gives no indication for continued dissociation to a level comparable with the europium added to Aldrich HA in the laboratory. This shows that the kinetic data from laboratory systems, including column experiments with radionuclides equilibrated with humic colloids in the laboratory, are not representative of the real system.

Contrary to these laboratory results, much slower dissociation of multivalent trace metal ions may be expected in real systems. The key question for predictive modeling thus is to what extent the kinetic data from laboratory systems are appropriate. According to laboratory measurements, actinide ions will dissociate within a relatively short time and enhancement of actinide transport by humic colloids will be limited. On the other hand, if actinide ions under real conditions behave like the natural trace elements found in groundwater humic colloids, then a part will be transported with the mobile humic colloids.

3.3 Redox Reactions

The transport behavior of uranium, neptunium, plutonium and technetium is governed by their redox states. To allow interpretation of experimental results from column and batch experiments, the redox behavior of neptunium under relevant conditions had been investigated. The results for reduction of Np(V) to Np(IV) can be summarized as follows: 1) the reduction rate and reduction capacity in groundwater are pH dependent; 2) reduction is catalyzed by dissolved Fe(II) and Fe(III); 3) there may be more than one reaction route with different kinetics; and 4) reduction is significantly accelerated in the presence of sediment. The latter is demonstrated in Figure 10, where Np(V) reduction in the Gorleben groundwater Gohy-2227 under anoxic conditions (Ar + 1% CO_2) is shown. Without contact with sediment, approximately half of the Np(V) has been reduced after 25 days. However, in contact with sediment, almost quantitative Np(V)→Np(IV) reduction is found after one day. Therefore, in natural groundwater neptunium will be present as Np(IV).

The reduction of Tc(VII) also has been studied in the above Gorleben system. Experiments were conducted under an inert gas atmosphere (N_2/H_2, Eh = -200 mV) with initial Tc concentrations varying between 5×10^{-7} and 5×10^{-5} M in the pH range 7-10. The reduction of pertechnetate to Tc(IV) and subsequent association with the available HSs appears to occur in less than 6 hours. Thus, in real systems Tc(VII) is expected to be readily reduced to Tc(IV).

3.4 Actinide Transport

Actinide transport was investigated in the laboratory with natural ground water/sediment systems by batch and column experiments under near-natural conditions. More defined binary and ternary systems were investigated to better understand individual processes. This project component included designed systems with different minerals under controlled conditions, under a variety of controlled physicochemical conditions and investigations of natural minerals and sediments.

3.4.1 The Ternary System. As already well known, humic acid sorption behavior varies from one mineral to another. These variations cannot be accounted for merely by considering the respective mineral surfaces: specific properties and physicochemical conditions need to be accounted for. The same is true for the sorption of metal ions on different minerals. One important parameter is pH, which governs the charge of mineral surfaces and HA. The impact of pH on HA sorption on hematite and goethite is shown in Figure 11. HA sorption decreases with increasing pH. Furthermore, HA is more strongly

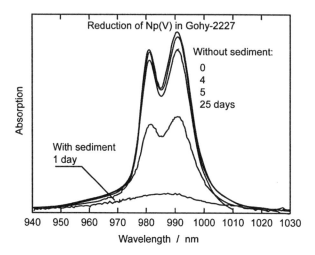

Figure 10 *Reduction of Np(V) to Np(IV) in the Gorleben groundwater Gohy-2227 under anoxic conditions (Ar + 1% CO_2) with and without the presence of sediment*

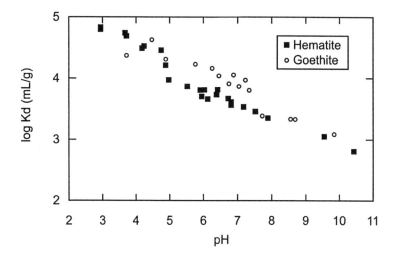

Figure 11 *Sorption of HA on hematite and goethite as a function of pH at ionic strength 0.1 M ($NaClO_4$) ([iron oxide] = 500 mg/L, [HA] = 10 mg/L)*

sorbed than FA. Ionic strength may be important, as is the case with goethite (Figure 12), or of less importance, as found, for example, with hematite.

A surface saturation value for HA sorption can be evaluated for many sediment surfaces. Figure 13 shows that sorption is not only depends on pH, but also on the degree of mineral surface coating. This issue must be addressed when evaluating the ternary system. We need to know to what extent the mineral surface is saturated with humic acid. Fractionation of HA also is observed during the sorption process, in that some HA fractions are preferentially sorbed and other fractions are enriched in solution. A sorption mechanism has been proposed to address these different parameters for HA sorption on minerals.

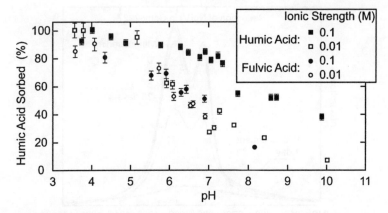

Figure 12 *Sorption of humic and fulvic acids on goethite as a function of pH ([α-FeOOH] = 500 mg/L, [HA] or [FA] = 10 mg/L)*

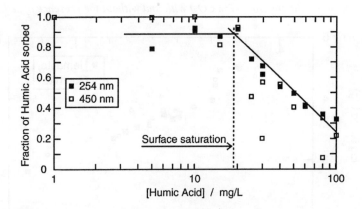

Figure 13 *Sorption of Aldrich HA on goethite (500 mg/L) at pH 5 and I = 0.1 M (NaClO$_4$). HA is measured by UV/Vis spectroscopy at 254 and 450 nm)*

Examples of the influence of HA on actinide ions sorption in ternary systems are shown in Figures 14 and 15. Figure 14 shows the sorption of Th(IV) on hematite as a function of pH in the absence and presence of HA (10 mg/L). In the latter case the mineral surface is saturated with humic acid. A rapid increase of sorbed Th(IV) is observed between pH 2 and 5, correlating with the progressive generation of Th hydroxo species. In the absence of HA, the sorption continues to increase between pH 5 and 10 and approaches 100 %. Addition of HA dramatically changes the sorption of Th on hematite. Sorption of Th is governed by its distribution between dissolved and sorbed humic acid in this ternary system.

A similar situation is found for the sorption of the U(VI) on phyllite (Figure 15). At a low humic acid concentration the sorption behavior is a combination of uranyl interaction with the phyllite mineral surface and the impact of its distribution between sorbed and dissolved humic acid. At higher humic acid concentration and above pH 5, the uranyl ion distribution closely resembles the humic acid distribution.

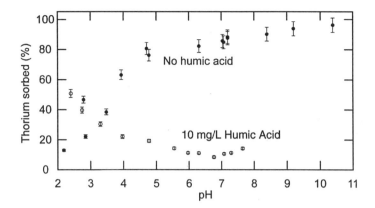

Figure 14 *Influence of Aldrich HA on the sorption of Th(IV) on hematite vs. pH. [Th] = 1.15×10^{-12} M, $[\alpha\text{-}Fe_2O_3]$ = 50 mg /L, [HA] = 10 mg/L, I = 0.1 M (NaClO$_4$)*

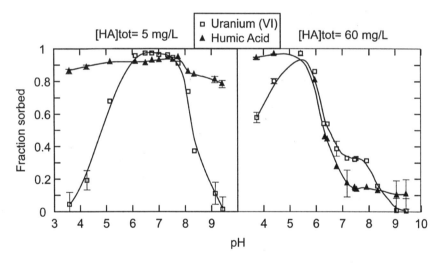

Figure 15 *Sorption of uranium(VI) and Kranichsee humic acid on phyllite as a function of pH and humic acid concentration*

3.4.2 Batch and Column Experiments under Near-natural Conditions. Batch and column experiments have been conducted on near-natural systems of thoroughly conditioned natural groundwater and sediments. Batch experiments were conducted with up to seven months of contact time. Column experiments were performed with both pulse and continuous addition of groundwater preconditioned with actinide ions for different times. The column lengths varied between 25 cm and 10 m. Figure 16 shows the most advanced column setup, in which migration experiments are conducted under controlled conditions. Samples can be measured after different migration lengths in one experiment. In most experiments, the same principle setup is used but with one column for groundwater conditioning immediately prior to the column for migration measurements.

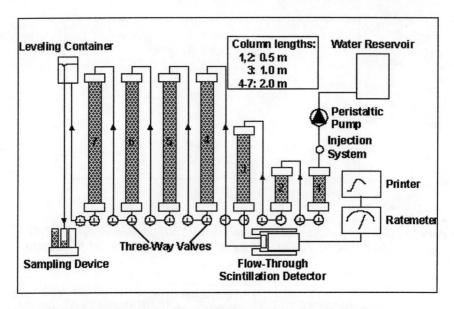

Figure 16 *Multicolumn system for migration experiments up to 10 m length. The system allows for measuring migration and breakthrough at various transport distances within a single experiment*

Typical results for the transport of multivalent metal ions in column experiments are shown in Figures 17 and 18. One fraction of multivalent metal ions is retarded in the

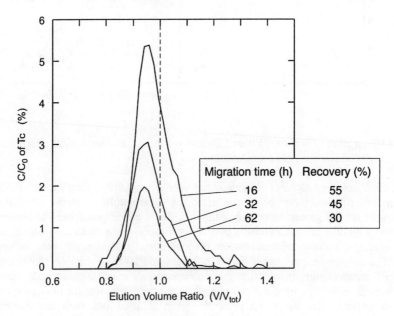

Figure 17 *Elution behavior of humic colloid borne thorium in the system Gorleben sediment and groundwater Gohy-532*

column and migrates so slowly that its elution cannot be monitored within practicable time scales. The other fraction is eluted slightly faster than the groundwater itself (as measured with a tritium tracer pulse). The recovery of metal ions in this second fraction increases with increasing equilibration time of the metal ion in groundwater prior to performing the column experiments. The recovery decreases with increasing residence time in the column ("migration time"). This behavior indicates that metal ion humic colloid interactions are governed by kinetic processes. A consistent set of kinetic parameters has been established based on batch and column experiments.

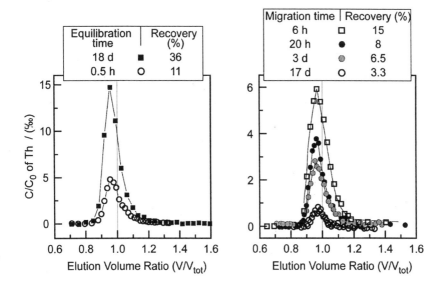

Figure 18 *Elution behavior of humic colloid borne technetium(IV) in the system Gorleben sediment and groundwater Gohy-532*

The kinetic approach was successfully tested by prediction of the results prior to a set of column experiments. Another example of the success of this approach can be seen in Figure 19. Eu transport experiments are shown from columns with length varying from 0.5 m to 10 m. The residence time in the columns also is varied by different groundwater flow velocities. The Eu groundwater pre-conditioning protocol is comparable between the different experiments. The Eu recoveries were found to depend only on the residence time in the columns, that is, the time for the Eu-humate complex to dissociate.

3.5 Origin, Stability and Mobility of Humic Colloids in Natural Aquifer Systems

The origin, stability and mobility of humic colloids in natural aquifer systems is of cardinal importance for the impact of humic colloids on the mobility of long-lived radionuclides in groundwater. A prerequisite for humic colloids to act as mediators of the transport of long-lived radionuclides is that they are stable and mobile. A prerequisite for answering the questions of their stability and mobility is their origin. Humic colloids, of course, have different origins. Once generated there is no indication of sorption or decomposition and the colloids migrate unhindered with groundwater flow. In other words, humic colloids migrate as ideal tracers until the groundwater leaves the aquifer system by discharge. Important results from two aquifer systems are discussed below. Detailed descriptions of aquifers studied and the results are given elsewhere.[28-32]

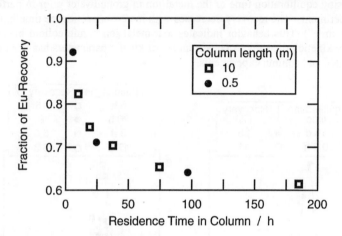

Figure 19 *Upscaling of humate mediated Eu transport in column experiments with 0.5 and 10 m column lengths*

A well known route for introduction of humic colloids into groundwater is generation in the soil zone and inflow with groundwater recharge. Another source is *in-situ* generation in groundwater, which requires a source of organic material and appropriate geochemical conditions. Other types of surfaces near groundwater sources are swamps and marshland, where relatively high humic colloid concentrations are found. The generation of humic colloids or humic colloid like material also can result from of anthropogenic activities, for example deposition of organic material in radioactive waste repositories. Another example is wood in abandoned mines, such as the "Königstein" uranium mine in Saxony, Germany.

3.5.1 Franconian Albvorland Aquifer System. The Franconian Albvorland aquifer system is typical of the introduction of humic colloids solely by recharge and relatively simple flow conditions. Figure 20 shows the ^{14}C age of fulvic acid as a function of the flow distance relative to Laibstadt I. FA in Greding I groundwater has a greater ^{14}C age than expected from the flow distance. This reflects more stagnant flow conditions in this part of the aquifer, i.e. higher residence time than expected solely from the distance between the wells. The other groundwaters show a very narrow correlation between the ^{14}C age of FA and the flow distance. The groundwater flow velocity resulting from this ^{14}C dating of FA falls in the range obtained from variation in the ^{18}O content along the flow-path as well as from hydrological modeling. This shows that fulvic acid introduced into the groundwater at a time where the soil layer started to develop (after the end of the latest Pleistocene temperature minimum, around 15,000 years ago) is still present. This is direct evidence that FA introduced by recharge into this aquifer system acts as an ideal tracer without retardation or decomposition over 15,000 years.

3.5.2 Gorleben Aquifer System. The Gorleben aquifer system is situated in Lower Saxony, Germany. One important feature of this aquifer system is the presence of deep groundwater with highly enhanced dissolved organic carbon (DOC) concentrations compared to recharge conditions.[28] Figure 21 shows the DOC concentrations for groundwaters from different sampling depths. Down to 27 m, DOC concentrations of 1 mg C/L or less are found and at 50 m, 2 mg C/L. These values are typical of DOC concentrations in young Gorleben recharge groundwaters.[28] Much larger DOC concentrations are found at depths between 50 and 200 m.

The strongly enhanced DOC concentrations are the result of *in-situ* generation of DOC through microbiologically mediated decomposition of sedimentary organic carbon (SOC). In this process, a portion of SOC is oxidized to dissolved inorganic carbon (DIC)

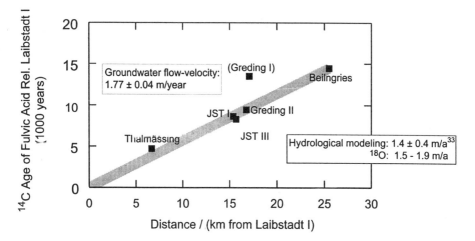

Figure 20 ^{14}C *dating of fulvic acid and the resulting groundwater flow velocity in the Franconian Albvorland aquifer system. For comparison the flow velocities from the Pleistocene to Holocene groundwater transition (^{18}O) and hydrological modeling are shown*

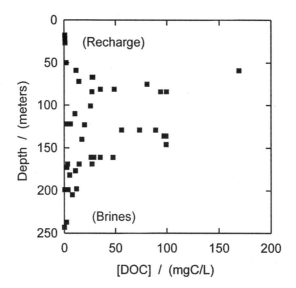

Figure 21 *DOC concentration of Gorleben groundwaters from the particular investigation area (see Fig. 20) as a function of depth*

and another portion is released as DOC. The process is driven by microbiological turnover of SOC that makes use of oxidizing agents and nutrients (see Figure 22). The DIC generated from the microbiologically mediated oxidation of SOC (mineralization) is released as carbonic acid. This carbonic acid dissolves an equal amount of DIC from sedimentary carbonate. The process is verified by analysis of the $^{13}C/^{12}C$ ratio of the dissolved inorganic carbon. The $^{34}S/^{32}S$ ratio also shows that sulfate is reduced by microbiological decomposition. Finally, enhanced DOC concentrations correlate with the concentration of phosphate, showing that the process takes place in a biofilm where SOC and phosphate are extracted from the sediment, sulfate from solution is reduced and DIC, DOC and phosphate are released into solution.

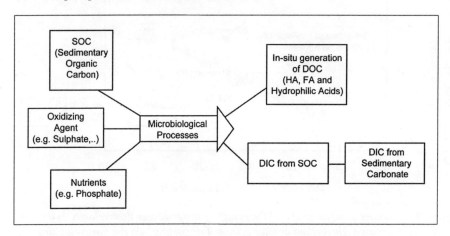

Figure 22 *Schematic overview of the in-situ generation of DOC in conjunction with microbiologically mediated mineralization of SOC*

One important question is to what extent decomposition of dissolved humic and fulvic acids can be expected. HSs decomposition is hindered because the concentration of microorganisms in solution is relatively low where access to required chemical compounds is limited. Furthermore, diffusion of negatively charged humic and fulvic acids to negatively charged sediment surfaces (where high populations of microorganisms can be found) is hindered by charge repulsion. As shown by the fact that humic and fulvic acids are released into solution, these substances also are less prone to microbiological decomposition. This could at least in part be the result of their enzymatic inhibition.[34]

The DOC and DIC concentrations of a large number of Gorleben groundwaters were plotted against each other (Figure 23). As expected from the above discussion, DOC and DIC correlate with each other. In addition to DIC from mineralization of SOC, DIC also is introduced from dissolution of sedimentary carbonate. The dissolution of sedimentary carbonate is enhanced in groundwaters where acid generating processes occur, i.e., the pH is lower than in other groundwaters (the dark shaded region in Figure 23, "pH: 6.9-7.7"). However there is no strong indication that DOC is decomposed or sorbed, which would give rise to excess DIC concentrations compared to the DOC. The situation is that DOC and DIC are co-generated and diluted away from the source.

In summary, in natural aquifer systems DOC (and thus humic colloids) originate 1) either from an overlaying soil-zone and are introduced with recharge or 2) are generated *in-situ* in conjunction with mineralization of SOC. Analogous generation from organic material in radioactive waste or engineered constructions is very likely. Humic colloids show no indication of decomposition or retardation, but can be expected to migrate as an ideal tracer until groundwater is discharged. Consequently, the humic colloid mediated transport of long-lived radionuclides will vary solely with the radionuclide-humic colloid interaction. These findings have important consequences for predictive modeling of humic colloid mediated radionuclide transport because the possible interaction between humic colloids and the sediments can be omitted.

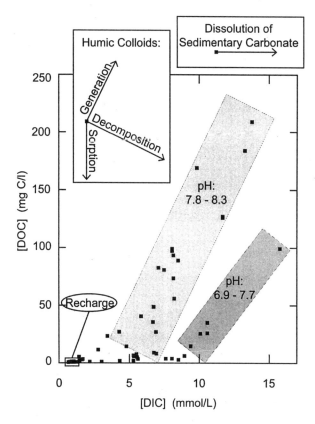

Figure 23 *Co-generation of DIC and DOC in Gorleben groundwater*

3.6 Migration Model Development and Testing

Previous models for radionuclides migration are based on a local equilibrium approach. These models fail to describe experimental observations from column experiments and also consistently fail to describe the outcome of batch and column experiments from comparable systems. This has led to a situation where no trustworthy prediction of the mobility of actinide ions in groundwater has been possible.

Based on the kinetic approach discussed above, an open transport code has been developed where equilibrium and kinetics can be implemented as appropriate.[34] This presently one-dimensional code is called k1D. The code has the possibility to include both equilibrium and kinetic processes. Wherever possible, equilibrium should be used because of the much lower numerical requirements. In principle, an optimization can be performed for each individual case. A considerable simplification and generalization is achieved by introduction of selection criteria for the need to apply either equilibrium or kinetics parameters. The selection criteria are based on the Damkohler number, which is the residence time along the migration path multiplied by the kinetic constant for the reaction concerned. In general terms, if the reaction rates are high enough then equilibrium can be assumed, and if reaction rates are low enough the particular reaction can be neglected. Kinetic processes must be accounted for in the intermediate region.

The improvement relative to equilibrium approaches on applying the k1D model is demonstrated in Figure 24. Application of a previous equilibrium based model making use of distribution constants from batch experiments leads to incorrect results. In contrast, the kinetic approach results in excellent agreement between experimental points and predictions of the model. The model was successfully tested by demonstrating correct predictions of results from a series of column experiments. It therefore can be concluded that the model and its underlying concepts are not issues for major further developments. However, the extension of the database and numerical refinement and optimization still is required.

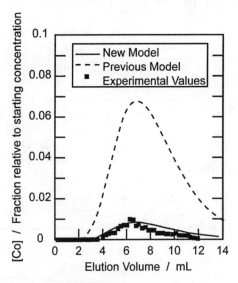

Figure 24 *Modeling of Co column transport*

3.7 Assessment of Impact on Actinide Transport

Three migration case studies are formulated for application of the transport code to real site conditions. This serves to visualize the impact of HSs on the predicted migration of tri-, tetra, penta- and hexavalent actinide ions under different conditions, making use of different input data. The three migration case studies are the above mentioned Gorleben aquifer system, the Dukovany sandy aquifer system and a uranium mining and milling rock pile. Results will be published in a separate report (in preparation). The most important findings for the transport prediction of plutonium at the Dukovany site are given below.

Basic features of the transport code are based on findings discussed above. Humic colloids are transported as ideal tracers along a one-dimensional groundwater flow. Humic colloids do not sorb or decompose. The somewhat faster transport velocity of humic colloids compared to groundwater flow is neglected. Actinide ions are found in three compartments, namely sorbed on the sediment surface, dissolved in the bulk solution (non-colloidal) and bound to humic colloids. The interaction between sediment surface and bulk solution is described by an equilibrium distribution. Total concentrations of non-colloidal actinides are used in the bulk solution. The humic colloids consist of sub-compartments where the actinide distribution between these sub-compartments and the bulk solution is described by kinetic rate constants or, if such simplifications are allowed, partly or entirely by equilibrium distributions. In some cases the humic colloid also contains an actinide containing sub-compartment that does not exchange with other humic colloid sub-compartments and thus also not with the bulk solution and the sediment surface. This is the case for irreversible actinide binding to humic colloids, as found for part of natural trace elements. Dispersion along with groundwater flow is accounted for.

The Dukovany migration case is in a valley near the Dukovany power plant, 300 m south of a shallow low-level waste repository in the Czech Republic. Exhaustive mineralogical, petrological and geochemical data are available for this shallow site in a sand deposit with some silt and clay. A continuous inflow of plutonium contaminated water from the near-surface radioactive waste repository is assumed for a period of 100 years at the Dukovany site. Calculations of the plutonium distribution in groundwater after these 100 years are shown in Figure 25. This Figure illustrates the dilemma of predictive actinide transport calculations at real sites. The tool for modeling has been developed and is ready for immediate application. However, the process understanding is insufficient with respect to selection of input data.

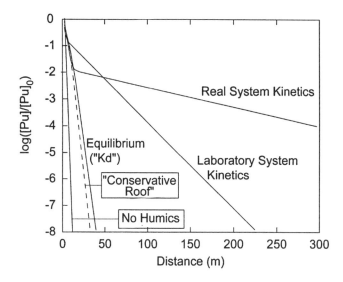

Figure 25 *Development of plutonium concentration in groundwater for a release according to the Dukovany migration case study and application of different approaches reflecting absence of humic substances, the "Conservative roof approach"[35] and differences in kinetics of relevant processes in the presence of humic substances*

The calculations demonstrate the very low plutonium mobility if humics are ignored ("no humics"). In this case the plutonium concentration is diluted by a factor of 10^4 at 6 m distance from the point of release. Application of equilibrium data (that is, the K_d concept) and the "conservative roof approach"[35] leads to a somewhat higher mobility of plutonium. In this case the same dilution (10^4) is obtained at around 20 m distance from the point of release. Contrary to this, application of kinetic data shows that the actual mobility of plutonium will be much higher. Applying the kinetic rate constants obtained from column experiments in the laboratory, a dilution of 10^4 will be found at approximately 100 m distance from the point of release. Dissociation of tetravalent trace metal ions from natural humic colloids indicates that a portion of approximately 80 % is practically irreversibly bound. Applying this "real system kinetics" therefore, 20 % of plutonium is retained almost immediately at the point of release and then migrates according to the "laboratory kinetics". The other 80 % of plutonium only shows a decrease in concentration reflecting dispersion of groundwater. The outcome is that a dilution by a factor of 10^4 is found at approximately 300 m distance from the point of release.

The Dukovany migration case study shows that a kinetic approach needs to be applied. However, the dilemma is 1) laboratory investigations where plutonium is contacted with groundwater humic colloids under conditions and time-frames that can be applied in the laboratory and 2) dissociation of tetravalent trace metal ions from natural humic colloids leads to very different results.

4 SUMMARY, CONCLUSIONS AND OUTLOOK

Progress within the project has been beyond our expectations when it was launched. The amount and quality of basic humate interaction data has improved. Basic process understanding of the actinide humate interaction process is emerging. One major breakthrough has been introduction of the kinetic concept based on findings from actinide transport batch and column experiments. For the first time, batch and column experiments can be consistently described and predictive modeling of such systems is possible. Real system analysis has brought insight concerning the geochemical behavior of humic colloids in natural aquifer systems. There is no indication of retention or decomposition of humic colloids in natural aquifer systems. Furthermore, the geochemical behavior of relevant humic colloid bound natural trace elements has been investigated. The results show large discrepancies between the geochemical behavior of natural actinide ions and natural trace elements on one hand, and actinide ions in laboratory investigations on the other hand. Demonstration of the impact of humic colloid mediated transport is performed through migration case studies. The outcome is that humic colloid mediated actinide transport is and will be much more extensive than previously assumed.

Despite this great progress, numerous issues still need considerable attention before we can obtain scientific acceptance of descriptions of humic colloid-mediated actinide and technetium transport. There still is a need for improvement of data, especially on the humate interaction with tetravalent actinide ions and the generation of mixed (ternary) complexes. More kinetic data are required for further development of the kinetic concept. With respect to demonstration of the impact of humic colloid mediated actinide transport via migration case studies, the hydrological and geochemical situation of complex systems needs to be more precisely described. The interaction of actinide ions with sediments needs a more precise description, including re-dissolution via sediment dissolution and retention via co-precipitation/solid solution generation. The most important issue, however, is continued work to generate the required actinide humic colloid interaction process understanding, which is necessary to resolve the reason for differences between results from laboratory investigations and the real system.

References

1. J. I. Kim, *Radiochim. Acta*, 1991, **52/53**, 71.
2. J. I. Kim, *MRS Bull.*, 1994, **19**, 47.
3. C. Degueldre, *Mat. Res. Soc. Symp. Proc.*, 1993, **294**, 817.
4. C. Degueldre, *Mat. Res. Soc. Symp. Proc.*, 1997, **465**, 835.
5. P. Vilks, H. Miller and D. C. Doern, *Applied Geochem.*, 1991, **6**, 565.
6. J. I. Kim, G. Buckau, F. Baumgärtner, Ch. Moon and D. Lux, *Mat. Res. Soc. Symp. Proc.*, 1984, **26**, 31.
7. J. I. Kim, G. Buckau and R. Klenze, in 'Natural Analogues in Radioactive Waste Disposal', B. Come and N. Chapman, (eds.), Graham &Trottman, London, 1987.
8. J. I. Kim, P. Zeh and B. Delakowitz, *Radiochim. Acta*, 1992, **58/59**, 147.
9. P. A. Smith and C. Degueldre, *J. Contam. Hydrol.*, 1993, **13**, 143.
10. P. Vilks and D. B. Bachinski, 'Natural Colloids in Groundwater from Granite and their Potential Impact on Radionuclide Transport', Report AECL-11635, 1997, 1.
11. M. J. Put, A. Dierckx, M. Aertens and P. De Canniere, *Radiochim. Acta*, 1998, **82**, 375.
12. P. Vilks, S. Stroes-Gascoyne, M. Goulard, S. A. Haveman and D. B. Bachinski, *Radiochim. Acta*, 1998, **82**, 385.
13. OECD-NEA "Binding Models Concerning Natural Organic Substances in Performance Assessment", Proceedings of an OECD-NEA Workshop organized in co-operation with the Paul Sherrer Institute, Bad Surzach, Switzerland, September 1994.
14. G. Buckau, (ed.), 'Effects of Humic Substances on the Migration of Radionuclides: Complexation and Transport of Actinides (First Technical Progress Report)', Report FZKA 6124, Research Center Karlsruhe, August 1998.
15. G. Buckau, (ed.), 'Effects of Humic Substances on the Migration of Radionuclides: Complexation and Transport of Actinides (Second Technical Progress Report)', Report FZKA 6324, Research Center Karlsruhe, June 1999.
16. G. Buckau, (ed.), 'Effects of Humic Substances on the Migration of Radionuclides: Complexation and Transport of Actinides (Third Technical Progress Report)', Research Center Karlsruhe, FZKA report, in preparation.
17. G. Buckau, (ed.), 'Effects of Humic Substances on the Migration of Radionuclides: Complexation and Transport of Actinides, Final Report', Research Center Karlsruhe, FZKA report, in preparation.
18. J. I. Kim, G. Buckau, R. Klenze, D. S. Rhee and H. Wimmer, 'Characterization and Complexation of Humic Acids', EC Report EUR 13181, Brussels, 1991.
19. K. Czerwinski, G. Buckau, F. Scherbaum, J. I. Kim, V. Moulin, P. Decambox et al., 'Effects of Humic Substances on the Migration of Radionuclides: Complexation of Actinides with Humic Substances', EC Report EUR-15914, 1995.
20. J. I. Kim and K. R. Czerwinski, *Radiochim. Acta*, 1996, **73**, 5.
21. K. R. Czerwinski, G. Buckau, F. Scherbaum and J. I. Kim, *Radiochim. Acta*, 1994, **65**, 111.
22. V. Moulin, I. Laszak, C. Moulin and C. Tondre, *Applied Spectrosc.*, 1998, **52**, 528.
23. C. Moulin, J. Wei, P. Van Isegem, I. Laszak, G. Plancque and V. Moulin, *Anal. Chim. Acta*, 1999, **396**, 253.

24. J. I. Kim and T. Sekine, *Radiochim. Acta*, 1991, **55**, 187.
25. G. Buckau, J. I. Kim, R. Klenze, D. S. Rhee and H. Wimmer, *Radiochim. Acta*, 1992, **57**, 105.
26. M. Wolf, G. Teichmann, E. Hoque, W. Szymczak and W. Schimmak, *Fresenius J. Anal. Chem.*, 1999, **363**, 596.
27. E. Hoque, M. Wolf, G. Teichmann, E. Peller, W. Szymszak, W. Schimmack and G. Buckau, *J. Chromatography A*, submitted for publication.
28. R. Artinger, G. Buckau, S. Geyer, M. Wolf, J. I. Kim and P. Fritz, *Applied Geochem.*, 2000, **15**, 97.
29. G. Buckau, R. Artinger, P. Fritz, S. Geyer, J. I. Kim and M. Wolf, *Applied Geochem.*, 2000, **15**, 171.
30. G. Buckau, R. Artinger, S. Geyer, M. Wolf, J. I. Kim and P. Fritz, *Applied Geochem.*, 2000, **15**, 583.
31. G. Buckau, R. Artinger, S. Geyer, M. Wolf, J. I. Kim and P. Fritz, *Applied Geochem.*, 2000, **15**, 819.
32. G. Buckau, R. Artinger, J. I. Kim, S. Geyer, P. Fritz, M. Wolf and B. Frenzel, *Applied Geochem.*, 2000, **15**, 1191.
33. L. Eichinger, 'Bestimmung des Alters von Grundwasser mit Kohlenstoff-14: Messung und Interpretation der Grundwässer des Fränkischen Albvorlandes', Dissertation, Ludwig-Maximilians-Universität München, 1981.
34. J. B. Jahnel, U. Schmiedel, G. Abbt-Braun and F. H. Frimmel, *Acta Hydrochim. Hydrobiol.*, 1993, **21**, 43.
35. W. Hummel, M. Glaus and L. R. Van Loon, in: 'Binding Models Concerning Natural Organic Substances in Performance Assessment', Proceedings of an OECD-NEA Workshop organized in co-operation with the Paul Sherrer Institute, Bad Surzach, Switzerland, September 1994.
34. P. Warwick, A. Hall, V. Pashley, N. D. Bryan and D. Griffin, *J. Contam. Hydrol.*, 2000, **42**, 19.

NATURAL ORGANIC MATTER FROM A NORWEGIAN LAKE: POSSIBLE STRUCTURAL CHANGES RESULTING FROM LAKE ACIDIFICATION

J. J. Alberts,[1] M. Takács[1] and M. Pattanayek[2]

[1] University of Georgia Marine Institute, Sapelo Island, GA 31327, USA
[2] Harding Lawson Associates, Novato, CA 94949, USA

1 INTRODUCTION

The Norwegian Humic Lake Acidification Experiment (HUMEX) was designed to study the impact of acidification on the characteristics and function of natural organic matter (NOM). Lake Skjervatjern, a dystrophic Norwegian lake situated in an area of Norway receiving very little acid rain, was bisected with a plastic curtain from its natural outlet to the far shore. One half of the lake and its watershed were artificially acidified with H_2SO_4 and fertilized with NH_4NO_3 for 5 years with sprinklers mounted in the highest trees in the catchment, while the other half of the lake and watershed remained an untreated control.[1,2]

As part of the assessment of the effects of this acidification, samples of NOM from both sides of the curtain were isolated by reverse osmosis techniques. In this report we present the results of potential compositional, fluorescence spectral and copper binding changes of the NOM and the effect upon metal toxicity of the natural organic chelators present in NOM as a result of lake and catchment acidification.

2 MATERIALS AND METHODS

The reverse osmosis method of concentrating NOM from the lakes has been reported elsewhere.[3] Therefore, only a brief summary of the isolation method will be presented here.

Between 10-12 October 1994, a group of scientists concentrated the organic matter from Lake Skjervatjern using a portable reverse osmosis system (RealSoft PROS/2S). In the field, 2.3 m^3 of water from each treatment were concentrated to 70 L. The two 70 L samples were then transported to the laboratory, where each was further concentrated to 5 L and then freeze-dried to obtain 70 - 75 g of solid sample from each treatment basin. Calculations indicate that 90% and 93% recoveries of NOM were obtained from the acidified and control treatments, respectively.

2.1 Pretreatment

Solid samples (150 mg) of NOM isolated from the acidified (A) and untreated (B) sections of the lake were re-dissolved in deionized water. Each solution was placed in dialysis tubing (MWCO 100, SpectroPor #7, Fisher Scientific) and dialyzed against deionized water for seven days with bath changes every 12 hours.

Following dialysis, aliquots of the samples were taken for fluorescence spectral analyses and dissolved organic carbon content (Shimadzu Model 500 High Temperature Carbon Analyzer). Part of each aliquot was freeze-dried for ash content and elemental C, H and N determination (Perkin-Elmer Model 2400 Elemental Analyzer). The remainders of the samples were used for the determination of copper binding capacity (CuBC).

In a separate experiment, a solution of phthalic acid (Supelco) was placed in the same types of dialysis bags and dialysed against deionized water for 48 hours with water changes every 12 hours. After 24 and 48 hours of dialysis, triplicate bags were removed and the internal solutions isolated. TOC contents of these solutions and an aliquot of undialysed starting solution were determined by high temperature combustion.

Aliquots of the freeze-dried solid materials from both the acidified and control sides of the lake were taken without dialysis for evaluation of their fluorescence spectral characteristics and their effect on metal toxicity.

2.2 Ultrafiltration

The undialysed solid materials were dissolved in deionized water and filtered through 0.45µ glass fiber filters. Aliquots of the samples were then ultrafiltered (Amicon Corp., Model 401 Stirred Cell) under nitrogen pressure to obtain nominal molecular weight (nmw) fractions of: >50, <50>10 and <10 kDa. Volumes of these solutions were recorded and samples were taken of them and the <0.45µ fraction for carbon analyses and % recovery calculations.

2.3 Fluorescence

Ultraviolet-visible spectra of the filtered and ultrafiltered samples from the acidified and control sides of the lake were collected (Perkin-Elmer Model Lambda 40). Samples were diluted with deionized water to the absorbance value at 254 nm of the lowest absorbance sample to normalize the fluorescence spectra. Ten mL of each sample was then treated with 1 mL of phosphate buffer to adjust all pH values to 6.8.

Total luminescence fluorescence spectra (λ_{ex} 220-400 nm, λ_{em} 350-600 nm, with a 5 nm increment between scans) were collected on a Perkin Elmer Model LS 50B Scanning Fluorescence Spectrophotometer in 1 cm quartz cuvettes. Daily scans of a standard quinine sulfate solution were made to assure comparability of spectra. Spectral data were collected with Perkin Elmer software and were analyzed using GRAMS32 (Galactic Industries).

2.4 Microtox®

Solid samples of the undialysed acidified and control NOM were dissolved in deionized water. Standard Microtox® basic serial dilution procedures[4] were used to determine the concentration at which the phosphorescence of the bacterium (*Vibrio fischeri*) was reduced by 50% (EC_{50}) for the organic matter, Cu^{+2}, Hg^{+2} and the metals in the presence of the organic matter. Results were corrected for internal light absorbance in the presence of the organic materials.

2.5 Proton and Copper Binding Capacities

Total exchangeable proton values were determined by potentiometric titration (Tanager Scientific).

Copper binding capacities were determined in duplicate (selective-ion electrode, Orion Corp.) by titration of the dialysed samples with $Cu(NO_3)_2$ (1000 ppm Cu Atomic Absorption Standard, Fisher Scientific). Fifty mL aliquots of sample were treated with 1 mL of 1 M $NaNO_3$ ionic strength adjustor and made to pH 5 by addition of NaOH. Titrations were conducted at pCu concentrations 3-7, pH was constantly readjusted (5.1 ± 0.1) and volume changes noted. Total copper complexing capacities were determined graphically.

3 RESULTS AND DISCUSSION

3.1 Changes in Basin Water Chemistry

The long-term changes of water chemistry in the acidified and control basins of Lake Skjervatjern over a seven year period have been discussed in detail.[5] Summary data are presented here (Table 1) for major cations, anions, organic carbon, organic nitrogen and acid neutralizing capacity (ANC). While standard deviations of the means of these data are large, randomized intervention analysis (RIA) of the data indicates that organic nitrogen and all the major ionic species listed, except Na^+, are significantly higher at the 99% confidence level in the acidified basin as a result of 5 years of sulfuric acid and NH_4NO_3

Table 1 *Mean water quality values of acidified and control basins of Lake Skjervatjern Norway after 5 years artificial acidification*[5]

Parameter	Acidified	Control
H^+	29 ± 8	26 ± 7
NH_4^+	4.7 ± 6.1	1.1 ± 0.7
NO_3^-	3.2 ± 2.8	0.7 ± 0.9
SO_4^{2-}	43 ± 13	27 ± 7
Al^{n+}	4.6 ± 1.6	4.5 ± 1.6
Ca^{2+}	13 ± 5	11 ± 4
Mg^{2+}	28 ± 12	25 ± 11
Na^+	5.7 ± 14	6.8 ± 15
ANC	-11 ± 16	-0.1 ± 16
TOC	6.1 ± 1.9	6.8 ± 2.2
Org-N	250 ± 118	177 ± 63

Values of TOC in mg C/L and Org-N in µg N/L, all other values are µeq/L; ANC is acid neutralizing capacity.

addition to the catchment. No significant differences occurred for TOC and Na^+ during that period, while the acidified basin has a significantly lower acid neutralizing capacity relative to the control during the same period.[5]

3.2 Chemical and Elemental Composition

Ash Content. Organic matter (NOM) isolated from the acidified portion of Lake Skjervatjern had an ash content of 23.5 ± 0.6 %, while the ash content of the NOM from the control was 15.8 ± 0.1 % (Table 2). This increased ash content may be a result of the increased Ca^{2+} and SO_4^{2-} contents of the acidified basin, which can be concentrated by the reverse osmosis technique.[3]

Elemental Composition. The elemental compositions of NOM isolated from both sides of the lake are very similar, with relatively low carbon and high oxygen contents (Table 2). Acidification of the lake and watershed appears to have resulted in very little change in the elemental composition of the NOM.

Table 2 *Elemental analyses and atomic ratios of NOM isolated from waters of Lake Skjervatjern, Norway*

	Acidified	Control
% AFDW	76.5 ± 0.6	84.2 ± 0.1
C	42.16 ± 0.03	42.56 ± 0.05
H	5.22 ± 0.03	4.98 ± 0.03
N	1.58 ± 0.02	1.21 ± 0.01
O	51.06 ± 0.02	51.26 ± 0.04
H:C	1.49	1.40
O:C	0.91	0.90
C:N	31.22	41.15

The H:C and O:C atomic ratios of the NOM vary little. The O:C values are relatively high compared to humic substances isolated from numerous sources. They are indicative of non-humic matter such as carbohydrates being associated with the organic isolates,[6] which is in agreement with NMR estimates of 25-28% of the carbon being in carbohydrates in these samples.[7] While being high for many types of humic substances, the H:C and O:C ratios are in good agreement with sedimentary humic and fulvic acids isolated from Lake Haruna in Japan.[8] This agreement may indicate that organic matter produced in lakes can be expected to have relatively high contents of non-humic-like compounds, which are successfully isolated by reverse osmosis.

The atomic C:N ratios do vary significantly as a result of in-lake acidification leading to an apparent nitrogen enrichment in NOM isolated from the acidified section of Lake Skjervatjern (Table 2).

Stoichiometry. The stoichiometry of the average, basic repeating units of the NOM isolates can be estimated for these samples by dividing the weight % of the element by its atomic weight, then dividing the results by the lowest resulting value for an element. In the case of these samples, nitrogen is always the smallest atomic percentage component of the isolates and thus is assigned a value of unity. With this method, the estimated stoichiometry of the basic structural unit for the NOM from the control portion of the lake is $C_{40}H_{57}NO_{35}$. Another sample of NOM was isolated from this lake in May 1996 using the same RO method. The latter material was dialysed in the same manner as the samples being discussed here. A similar calculation gave an estimated basic structural unit stoichiometry as $C_{51}H_{68}NO_{45}$ with atomic ratios of H:C = 1.33, O:C = 0.88, and C:N = 51.[9] The larger unit size and higher C:N ratio of this material may be a result of in-lake processes occurring during the two year period when the watershed was no longer being treated with acid and nutrients, or may be a result of spring/fall differences in lake NOM.

The NOM from the acidified section of the lake shows a significant decrease in the basic structural unit compared to the control. The material lost is relatively high in O and seems highly aromatic as evidenced by the low H:C ratios. The segment lost from the acidified section of the lake ($C_8H_{10}O_6$) has an elemental composition similar to vanillic acid ($C_8H_8O_4$), which can be obtained by the oxidation of vanillin, a component of lignin, or from coniferin via glucovanillin (Figure 1). Coniferin is the principal glucoside in conifers,[10] and would be expected to be present in the watershed of Lake Skjervatjern. Studies of lignin structure by CuO oxidation indicate that simple vanillyl phenols are the most abundant monomeric oxidation products for lignin derived from gymnosperms,[11,12] which would agree with the premise that the lost structural unit could be a vanillin derivative. In addition to the simple phenol products, 14 additional monomers and 30 dimers have been identified by CuO oxidation of lignin.[12] The monomers have elemental compositions ranging from $C_8H_8O_3$ to $C_{10}H_{12}O_6$ and include vanillic acid ($C_8H_8O_4$), syringic acid ($C_9H_{10}O_5$) and 6-carboxyvanillic acid ($C_9H_9O_6$) (Figure 1). Thus, it appears that the structural unit that was lost could have been lost by hydrolysis of gymnosperm removed by acidification of Lake Skjervatjern and its type lignin to remove substituted vanillic or syringic type compounds.

The absence of a structural unit similar to a vanillin derivative in NOM isolated from the acidified portion of Lake Skjervatjern may result from the dialysis procedure used in this study. The samples were dialysed across a membrane that is defined as a 100 Da cut-off. However, it is possible that small molecules (vanillic acid MW 202 Da) with a condensed structure might be able to penetrate the membrane's pores and be removed during the 7 day dialysis. Dialysis experiments with phthalic acid (1,2-benzenedicarboxylic acid, $C_8H_6O_4$, MW 166 Da), whose structure is very similar to 6-carboxyvanillic acid mentioned above, show a loss of 65.9 ± 1.3 %C after 48 hours of dialysis against deionized water. Further support for this mechanism is given by NMR analyses showing higher aromatic carbon contents in the NOM from the acidified section of the lake compared to the control and by pyrolysis GC-MS analyses, which also indicated higher polyhydroxy aromatic contents in the acidified samples.[7] At first this appears to argue against the proposed mechanism, but the NMR and pyrolysis GC-MS analyses were conducted on samples that had not been dialysed but rather were the freezed-dried RO concentrates. Thus, the small, highly oxygenated aromatic molecules would appear as part of the whole in those studies.

The estimates are not precise, and it cannot be said conclusively that the units lost are vanillic acid or 6-carboxyvanillic acid. Other potential compounds would be coumaric acid, ferulic acid or syringic acid and their derivatives (Fig. 1). All are known components of lignin. However, the estimates do indicate that significant structural units are being lost during acidification of NOM in the lake. Furthermore, the unit size of the lost material is consistent with groups of compounds that are major components of lignin and would be expected to be present in the NOM of the lake. If this hypothesis is true, relatively long-

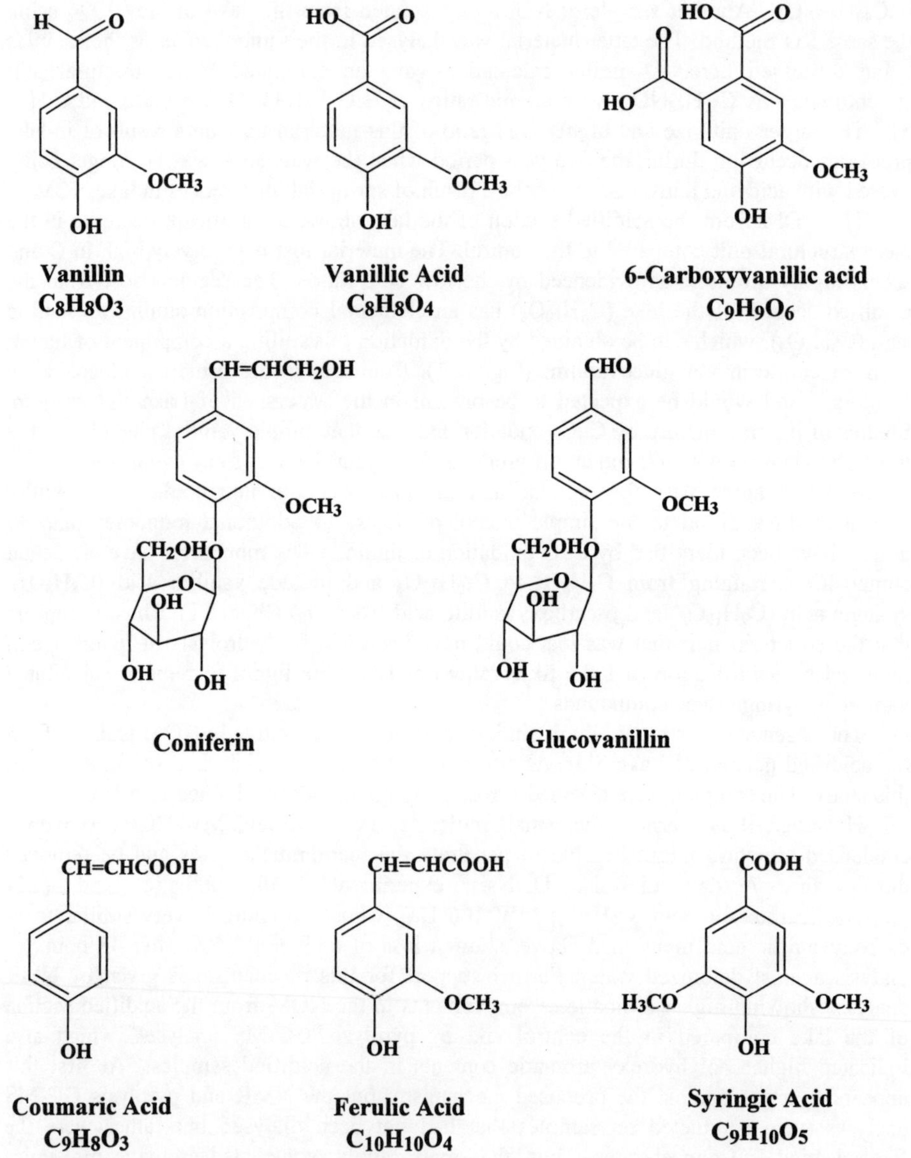

Figure 1 *Some components of lignin*

term acidification of the Lake Skjervatjern watershed has led to the cleavage of the larger molecular structure of NOM.

3.3 Copper, Proton and Metal Binding Characteristics

NOM from the acidified section of the lake appears to have lower CuBC and lower total acidity content (as determined by potentiometric titration) than that of the control (Table 3). This observation is consistent with the earlier discussion of the loss of small carboxylic acids from the NOM in the acidified section of Lake Skjervatjern as a result of catchment acidification.

Table 3 *Various parameters of NOM isolated from Lake Skjervatjern, Norway*

	CuBC µeq Cu/mg C	Total Acidity µeq/mg C	Ultrafil % (recovery)	EC_{50} ppm
Acid.	1.32	12.2	(96.7)	98.3
+ Cu^{+2}				0.39 ± 0.01
+ Hg^{+2}				0.23 ± 0.02
>50K			62.9 ± 11.7	
<50K>10K			6.7 ± 1.3	
<10K			27.9 ± 2.3	
Cont.	1.79	14.4	(100.5)	68.1
+ Cu^{+2}				0.32 ± 0.02
+ Hg^{+2}				0.24 ± 0.02
>50K			58.2 ± 10.8	
<50K>10K			14.1 ± 2.8	
<10K			28.2 ± 2.4	
Metals				
Cu^{+2}				0.39 ± 0.01
Hg^{+2}				0.13 ± 0.01

3.4 Ultrafiltration and Toxicity

Dissolved organic carbon in both sections of Lake Sjkervatjern are predominantly large molecular size molecules with greater than 50% of the DOC being in the fraction that is nominally >50 kDa. Approximately 30% of the remaining DOC is in the smallest size fraction, with the remaining (<15%) being in the middle size class as defined by ultrafiltration (Table 3). Acidification of the lake appeared to have little effect on the distribution of DOC among fractions. Although there did seem to be a slight decrease in the middle size fraction as a result of acidification, the errors inherent in the method do not support this as a significant change.

It is known that the free form of most metals is the most toxic to organisms and that NOM, in particular humic and fulvic acids, can bind metals often resulting in a decrease in that toxicity. However, the results of the acidification of Lake Skjervatjern and the associated influence of the NOM upon the toxicity of Cu^{+2} and Hg^{+2} are not straightforward (Table 3).

Complexation of Hg^{+2} by NOM from both the acidified and control treatments reduce the EC_{50} of that metal by approximately 45% and show no apparent difference as a result of the acid treatment (Table 3). However, complexation of copper with NOM from the acidified section of Lake Skjervatjern has no apparent effect on Cu^{+2} toxicity, while complexation of NOM from the control seems to slightly increase Cu^{+2} toxicity to the test organisms. These observations regarding Cu^{+2} are in opposition to the copper binding capacity and total acidity trends observed earlier. Furthermore, the organic matter by itself has a slightly negative effect on the test organisms (albeit two orders of magnitude less toxic than the metals), but again the NOM from the control site has a slightly greater effect on the organisms than the NOM from the acid treatment site.

3.5 Fluorescence Spectra

The total luminescence fluorescence spectra of NOM from both the control and acidified sides of the lake have a very similar bimodal peak distribution (Figure 2, only NOM from acidified site shown), characteristic of aquatic organic matter isolated from open ocean, estuaries and rivers, with the two peaks centered at approximately 330-5 nm λ_{ex} and 437-80 λ_{em} and 225 nm λ_{ex} and 426-8 nm λ_{em}, respectively[19-23] (Table 4). There is very little difference in either the peak positions, relative peak intensities (RFI) or ratio of peak 1 to peak 2 RFIs in the spectra of the <0.45µ samples for the control and acidified portions of the lake (Table 4). However, the spectra of the ultrafiltration fractions (Figure 2, only NOM from acidified site shown) show a definite trend of increasing RFI for both peaks with decreasing nominal molecular weight of separated fractions (Table 4). Thus, smaller sized molecules fluoresce more strongly per mass of carbon, perhaps as a result of more efficient energy transfer through internal conversions in the large sized molecules.

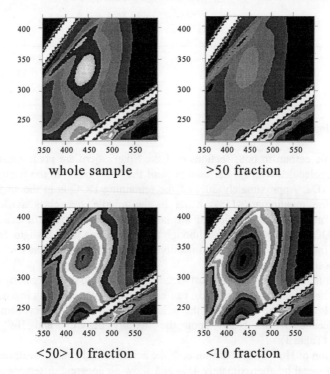

Figure 2 *Total luminescence spectra of whole water (<0.45µ) and 3 ultrafiltration fractions of NOM isolated from the acidified side of Lake Skjervatjern, Norway*

Although there are no obvious peak shifts in the total luminescence spectra of the ultrafiltration fractions, the spectra can be deconvoluted by taking the second derivative of the entire spectrum (Figure 3). When this analysis is performed we observe four peaks in the spectra at $\lambda_{ex}:\lambda_{em}$ 310:400, 330:430, 340:458 and 380:512, (peaks A, B, C and D respectively). Again we observe changes in the spectra of the ultrafilter fractions with respect to these peaks. As nominal molecular size decreases, peaks A and D increase in the acidified NOM samples relative to the control, while peaks B and C remain relatively unchanged (Table 5).

Table 4 *Peak positions and relative fluorescence intensities (RFI) of NOM and ultrafiltered size fractions isolated from Lake Skjervatjern, Norway*

	Peak 1			Peak 2			
	λ_{ex}	λ_{em}	RFI1	λ_{ex}	λ_{em}	RFI2	RFI1:2
Acidified							
<0.45μ	330	437	17.73	225	428	23.91	0.74
>50 kDa	335	447	11.28	225	430	18.01	0.63
<50>10 kDa	330	444	34.67	225	427	43.33	0.80
<10 kDa	330	434	48.37	225	428	57.23	0.85
Control							
<0.45μ	335	438	17.59	225	426	22.70	0.78
>50 kDa	335	447	13.26	225	426	16.46	0.81
<50>10 kDa	335	442	24.64	225	427	31.20	0.79
<10 kDa	330	438	43.74	225	428	52.63	0.83

Table 5 *Percentage change in peak intensity of NOM from acidified section of Lake Skjervatjern relative to NOM from control section*

Peak identification ($\lambda_{ex}:\lambda_{em}$)	Nominal Molecular Size	
	<0.45μ	<10 kDa
Peak A (310:400)	29.7	16.3
Peak B (330:430)	3.3	3.1
Peak C (340:458)	4.1	8.6
Peak D (280:512)	27.0	14.5

Examination of the literature[24,25] shows that a wide variety of fluorescent compounds can occur over the range of wavelengths covered by the four peaks (Fig. 4). However, closer inspection indicates that peak A is a region of fluorescence for carboxylic acids (Fig. 4), while peaks A and D are representative of nitrogen containing compounds (Fig. 4). The latter group contains larger molecules such as natural products like vitamins, metabolic products, etc. (Figure 5), while peak A is represented by smaller nitrogen containing compounds (Figure 6).

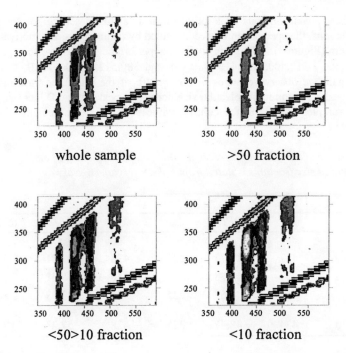

Figure 3 *Second derivative total luminescence spectra of whole water (<0.45μ) and 3 ultrafiltration fractions of NOM isolated from the acidified side of Lake Skjervatjern, Norway*

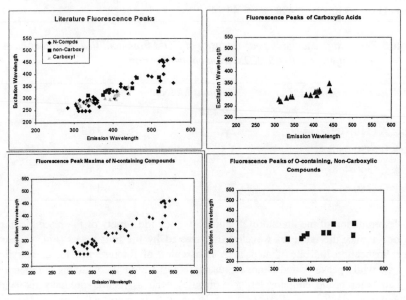

Figure 4 *Excitation and emission peaks of various organic compounds (pH 5-9) reported in the literature*[24,25]

Literature Values of Fluorescence Excitation and Emission Peaks for Various Compounds

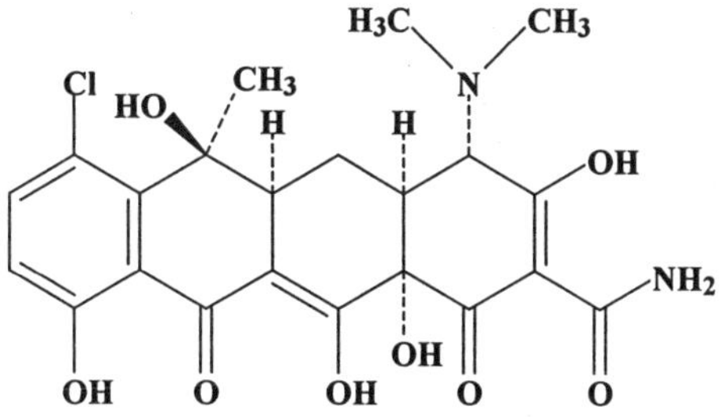

Chlortetracycline
ex:em 345:520   400:530
$C_{22}H_{23}ClN_2O_8$ mol. wt. 479

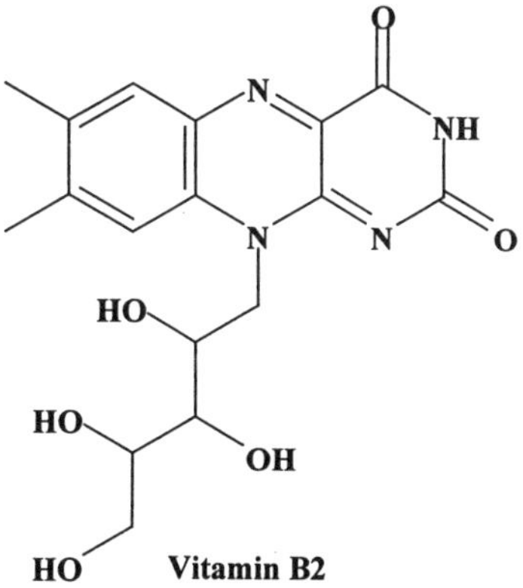

Vitamin B2
ex:em 373:523
$C_{17}H_{17}N_4O_6$ mol. wt. 396

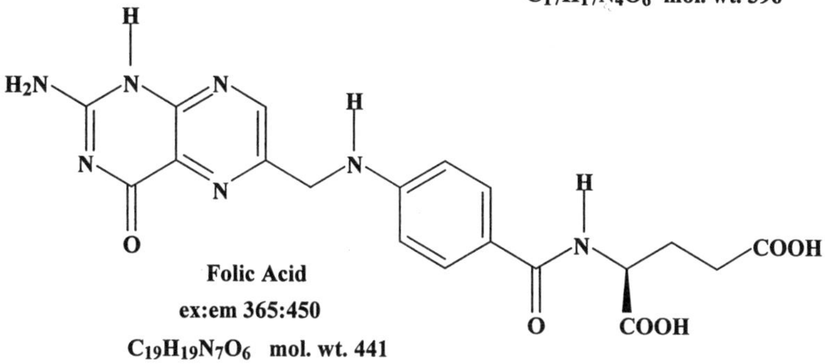

Folic Acid
ex:em 365:450
$C_{19}H_{19}N_7O_6$ mol. wt. 441

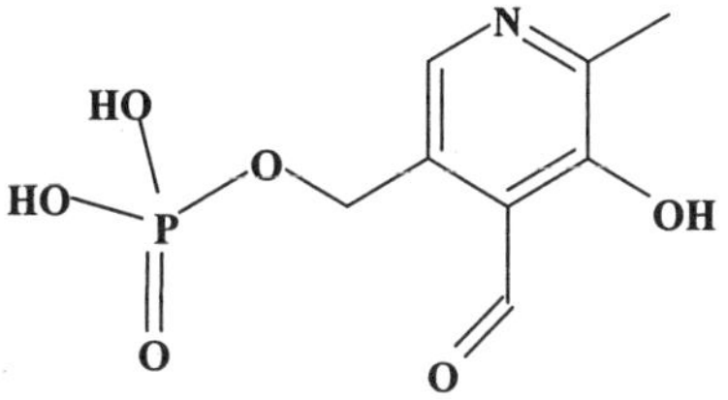

pyridoxal-5'-phosphate
ex:em 390:500
$C_8H_{10}NO_6P$ mol. wt. 247

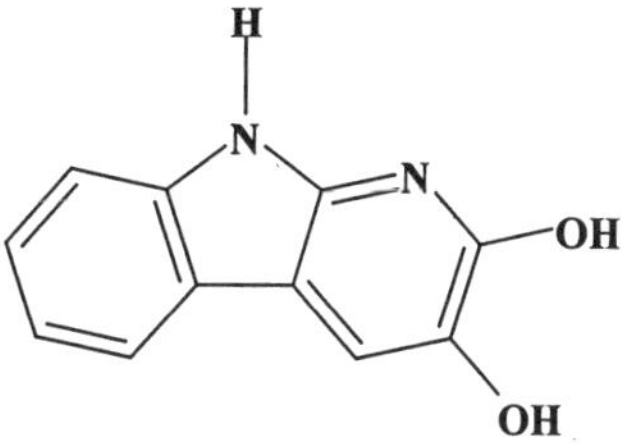

Fluorodaturatine (3,4-dihydroxycarboline)
ex:em 395:491
$C_{11}H_7N_2O_2$ mol. wt. 199

Figure 5 *Compounds reported to have excitation and emission peaks in the region of peak D of NOM isolated from Lake Skjervatjern, Norway*

Literature Values of Fluorescence Excitation and Emission Peaks for Various Compounds

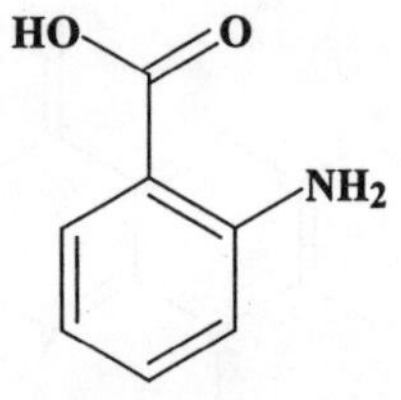

2-aminobenzoic acid
ex:em 300:405
$C_7H_6NO_2$ mol. wt. 136

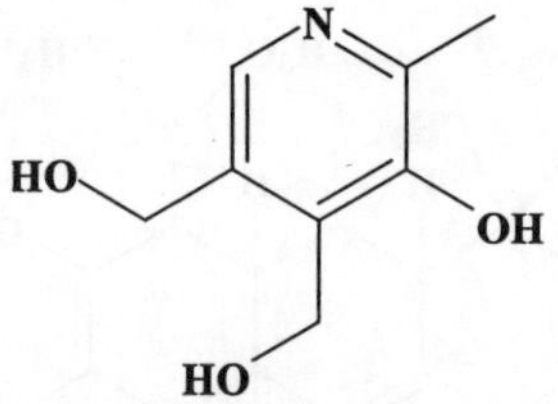

pyridoxine
ex:em 324:390
$C_8H_{11}NO_3$ mol. wt. 169

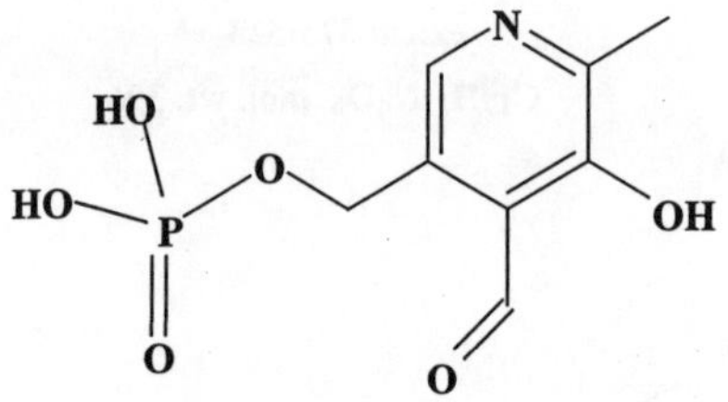

pyridoxal-5'-phosphate
ex:em 330:400, 330:410
$C_8H_{10}NO_6P$ mol. wt. 247

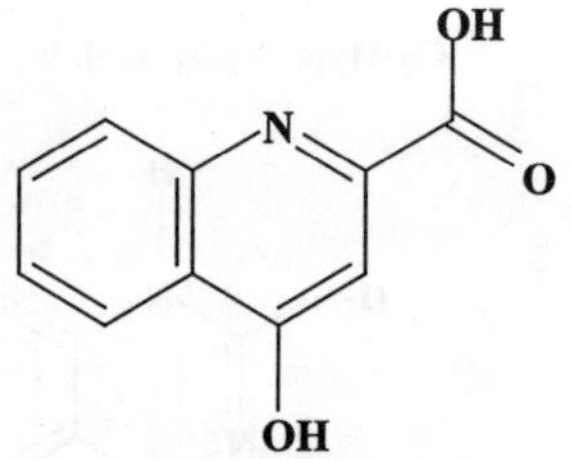

4-hydroxy-2-quinolinecarboxylic acid (kynurenic acid)
ex:em 335:385
$C_{10}H_7NO_3$ mol. wt. 189

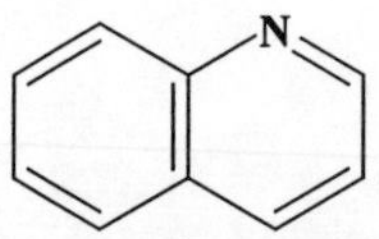

quinoline
ex:em 331:382
C_9H_7N mol. wt. 129

Figure 6 *Compounds reported to have excitation and emission peaks in the region of Peak A of NOM isolated from Lake Skjervatjern, Norway*

If we consider again the lost carbon structure discussed above and examine the total luminescence spectra of some of the lignin degradation products we observe that vanillic, syringic and phthalic acid do not fluoresce in the region of peak A, but ferulic, coumaric and salicylic acids do (Figure 7). Since the fluorescence spectra were determined on samples which had not been dialysed, the presence of peaks A and D and the relative increase of these peaks in the acidified NOM relative to the control NOM samples support the suggestion of loss of small, highly oxygenated organic molecules related to lignin oxidation products as a result of long-term acidification of the lake and its watershed.

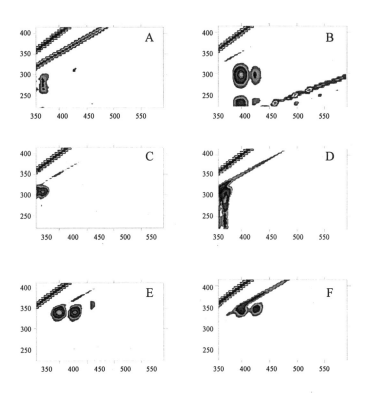

Figure 7 *Second derivative total luminescence spectra of selected carboxylic acids: A. phthalic acid, B. salicylic acid, C. vanillic acid, D. syringic acid, E. ferulic acid, F. coumaric acid*

4 SUMMARY

Acidification and fertilization of Lake Skjervatjern and its watershed continued for 5 years. Changes in overall lake water quality during that period were statistically significant but subtle. For example, the occurrence of "sea-salt" events[26] increased the acid neutralizing capacity (ANC) of the control basin while the ANC of the acidified basin remained unchanged compared to its value two years prior to the beginning of treatment. Yet concentrations of major cations and anions, labile and total reactive Al and organic nitrogen were all significantly increased in the treated basin relative to the untreated basin.[5] These subtle and often unanticipated changes can result from "scaling" effects that occur when moving from controlled laboratory experiments to catchment-wide manipulations. This phenomenon is accepted in engineering disciplines but often is ignored in environmental studies.

The acidification of a section of Lake Skjervatjern and its watershed appears to have reduced the size of the average basic structural unit of the NOM by removing small, highly oxygenated aromatic molecules. In addition, copper binding capacity and total acidity of the acidified NOM sample was reduced relative to the control. However, changes of these parameters did not affect the ability of the NOM from both treatments to reduce apparent toxicity of Hg^{+2}. Thus, subtle environmental changes appear to be occurring in the NOM of the large scale lake experiment that may go unnoticed in studies not including the entire ecosystem.

Acidification of the lake did not have large effects on the qualitative fluorescence spectra of the total NOM. However, fluorescence of fractions separated by ultrafiltration demonstrate a trend of increasing relative fluorescence intensity with decreasing molecular size. These changes support the hypothesis of reduced molecular size of the basic structural unit of NOM as a result of in-lake basin acidification.

Acidification of Lake Skjervatjern and its watershed occurred over a relatively short period of time on a catchment-wide scale, but was an extremely long-term study by most standards for laboratory and field systems. The results indicate that changes are occurring in the chemistry of the NOM of the lake system. Considering that the lake and its watershed represent a potential buffer to acidification of unknown capacity, the subtle changes observed in this study beg the question of whether the rates of change will increase with time and further acidification, indicating that the natural buffering capacity of the system has been exceeded, or that dynamic processes not previously considered, e.g., "sea-salt" events or secession of acid inputs, will act to moderate and/or negate the observed changes.

ACKNOWLEDGEMENTS

The authors wish to thank Drs. Egil Gjessing (Agder College) and Espen Lydersen (NIVA, Norwegian Institute for Water Research) for providing samples of the Lake Skjervatjern organic matter. This work was supported in part by the University of Georgia Marine Institute's Visiting Scientist Program and from U.S. Environmental Protection Agency award R825147-01-0 (NCERQA). This is contribution #855 of the University of Georgia's Marine Institute.

References

1. E. T. Gjessing, *Environ. Int.*, 1992, **18,** 535.
2. E. T. Gjessing, *Environ. Int.*, 1994, **20,** 267.
3. S. M. Serkiz and E. M. Perdue, *Water Res.*, 1990, **24,** 911.
4. Azur Corp., 'Microtox® Manual A Toxicity Testing Handbook', Vol. II. Detailed Protocols, Carlsbad, CA, 1992, p. 178.
5. E. Lydersen, E. Fjeld and E. T. Gjessing, *Environ. Int.*, 1996, **22,** 591.
6. C. Steelink, in 'Humic Substances in Soil, Sediment and Water: Geochemistry, Isolation and Characterization', G. R. Aiken, D. M. McKnight, R. L. Wershaw and P. MacCarthy, (eds.), Wiley, New York, 1985, p. 457.
7. E. T. Gjessing, J. J. Alberts, A. Bruchet, P. K. Egeberg, E. Lydersen, L. B. McGown, J. J. Mobed, U. Munster, J. Pemkowiak, E. M. Perdue, H. Ratnawarra,

D. Rybacki, M. Takács and G. Abbt-Braun, *Wat. Res.*, 1998, **32,** 3108.
8. R. Ishiwatari, in 'Humic Substances: Their Structure and Function in the Biosphere', D. Povoledo and H. L. Golterman, (eds.), Centre for Agricultural Publishing and Documentation, Wagenigen, The Netherlands, 1975, p. 87.
9. J. J. Alberts and M. Takács, *Environ. Internat.*, 1999, **25,** 237.
10. M. Windholz, S. Budavari, L. Y. Stroumtsos and M. N. Fertig, 'The Merck Index', 9th Edn., Merck, Rahway, NJ, 1976.
11. J. I. Hedges and D. C. Mann, *Geochim. Cosmochim. Acta*, 1979, **43,** 1803.
12. M. A. Goñi and J. I. Hedges, *Geochim Cosmochim. Acta*, 1992, **56,** 4025.
13. J. J. Alberts, J. P. Giesy and D. W. Evans, *Environ. Geol., Water Sci.*, 1984, **6,** 91.
14. J. J. Alberts and Z. Filip, in 'Trends in Chemical Geology', Trivandrum, India, 1994, p. 143.
15. J. J. Alberts and J. P. Giesy, in 'Aquatic and Terrestrial Humic Materials', R. F. Christman and E. T. Gjessing, (eds.), Ann Arbor Science, Michigan, 1983, p. 333.
16. J. J. Alberts and Z. Filip, *Environ. Technol.*, 1998, **19,** 923.
17. N. Senesi, G. Sposito and J. P. Martin, *Sci. Total Environ.*, 1986, **55,** 351.
18. N. Senesi, G. Sposito and J. P. Martin, *Sci. Total Environ.*, 1987, **62,** 241.
19. P. G. Coble, *Mar. Chem.*, 1996, **51,** 325.
20. P. G. Coble, S. A. Green, N. V. Blough and R. Gagosian, *Nature*, 1990, **348,** 432.
21. P. G. Coble, C. A. Schultz and K. Mopper, *Mar. Chem.*, 1993, **41,** 173.
22. S. K. Hawes, 'Quantum Fluorescence Efficiencies of Marine Fulvic and Humic Acids', MS Thesis, University of South Florida, 1992, p. 92.
23. J. J. Alberts, T. M. Miano, J. Nelson and J. R. Ertel, 'Humic Substances in the Environment: New Challenges and Approaches', Georgia Institute of Technology, Atlanta, GA, Aug. 27-Sept. 1, 1995.
24. O. S. Wolfbeis, in 'Chemical Analysis', P. J. Elving, J. P. Winefordner and I. M. Koltoff, (eds.), Wiley, New York, 1985, p. 167.
25. N. Senesi, *Anal. Chim. Acta*, 1990, **232,** 77.
26. D. O. Anderson and H. M. Seip, *J. Hydrology,* 2000, **224,** 64.

ORGANOCLAYS REMOVE HUMIC SUBSTANCES FROM WATER

George R. Alther

Biomin, Inc., Ferndale, MI 48220, USA

1 INTRODUCTION

Groundwater and surface waters throughout the world are often colored with humic materials. Such brown colored waters usually are more of an aesthetic nuisance than a health hazard. The sources of color in waters are peat moss, coal beds, peat soil, grass and grass clippings, leaves, needles from trees, algae and other natural organic matter (NOM).

The methods of removing these colors from water are mainly based on the use of anion and cation exchange resins, coal and lignite derived activated carbon, reverse osmosis and ozonation. Anion exchange resins are expensive ($5/lb). Coal based activated carbon is not very efficient at removing colors; its pores are too small to accommodate the large humic and fulvic acid molecules. Lignite based carbon is much more efficient at removing HSs and is inexpensive. However, none of these sorbents work well in all situations. The colored waters in some areas such as Saskatchewan resist all of the commercial sorbents and only expensive reverse osmosis (RO) methods work.

One reason for unsuccessful decolorization of water is the influence of pH. A pH of 12 or higher is detrimental to activated carbon and resins. Second, temperatures near the freezing point seem to cause HAs to curl up, thus exposing fewer charged sites for adsorption or ion exchange by the sorbent and resulting in much lower removal efficiency.[1,2] Humic solubility is a third factor. Fulvic acids are far more soluble than HAs and thus they are more difficult to remove. It stands to reason that the ratio of fulvic to humic acids is a determining factor in the ease of HSs removal by sorption. pH also is a factor. Fulvic acids are soluble below pH 2, where many HAs precipitate.

Humic acids are more soluble at higher pH. Models such as that shown by Kunin[3] indicate that HAs can have both positive and negative charges. This means that 1) they are polar and 2) they also are of medium to low solubility if the less soluble fractions predominate.

Organically modified clays called organoclays remove organic compounds by two mechanisms depending on their character and composition. The two mechanisms are partition and ionic bonding. It is postulated that removal of humic and fulvic acids from water can be achieved by a combination of ionic bonding and partitioning. Organoclays are capable of adsorption by both mechanisms.[3]

Organoclays are made from clays like bentonite that have been modified with quaternary amines. Organoclays are used as thickeners in paints, inks, greases and drilling muds, as binders for foundry sand and as filtration media to remove oils, greases, PAHs and other low solubility organic compounds from water. When prepared in a specific manner, organoclays also can be used to remove humic and fulvic acids from water. This paper summarizes the chemistry of organoclays and illustrates a number of laboratory tests demonstrating the ability of organoclays to remove a variety of humics from water under various conditions. The data show that organoclays are just as effective as synthetic resins and much more effective than activated carbon for the removal of humic colors from water.

2 ORGANOCLAYS

Organoclays are manufactured by modifying bentonites with quaternary amines. Bentonites originate from volcanic ash that has been chemically altered after deposition in fresh or salt water. A good bentonite from Wyoming or Mississippi contains 90% montmorillonite clay. These clays are swelling clays if the ash was deposited in salt water or non-swelling clays if it was deposited in fresh water. Their ion exchange capacities range from 70 to 90 meq/gram.

After modification, some 40% of the bentonite surface is covered by the quaternary amines and the remainder still is available for ion exchanging heavy metals. Sodium, calcium and magnesium are the primary exchangeable ions. Sodium becomes hydrated when exposed to water, which is why sodium bentonites swell. The sodium cation has twice the hydrated radius of calcium.[4] The quaternary amines consist of a positively charged hydrophilic nitrogen ion and a carbon tail. The nitrogen exchanges onto the clay surface by replacing a sodium or calcium ion. The quaternary amines in use are cationic surfactants. The chloride ion that is loosely attached to the amine combines with the exchanged sodium ion and becomes a salt that is washed out once the clay is introduced into water.

Figure 1 shows this system. The organoclay is a non-ionic surfactant. Upon introduction into water, the hydrocarbon chains extend perpendicularly into the water, which allows them to capture humic and fulvic acids either by partitioning or by ionic bonding (Figure 1). If the surface of the clay is covered by the stoichiometrically correct amount of amine chains, then partition is the removal mechanism. If there is an excess of amine chains they will invert and attach themselves to the attached chains via a coulombic tail-tail interaction in a bilayer,[5] which probably also is a partition mechanism. But the positively charged nitrogen ion now extends into the water, allowing removal of negatively charged humic and fulvic acids by ionic bonding. This is in contrast with anion exchange resins, where the negatively charged humic or fulvic acids simply replace a chloride or sulfate anion attached to the divinylbenzene backbone of the resin.

The advantage of the organoclay over the activated carbons and resins is that the molecular size of the HA is irrelevant because sorption takes place on the clay surface rather than inside a pore. Thus, blinding of pores does not occur in sorption by organoclays.

3 MATERIALS AND METHODS

Samples of humic acid (Quantum H) containing 6%w/w HA with pH 9.6 and a dark brown color, and of fulvic acid (Fulvic 6000) containing 6%w/w FA with a pH of 1.9 and

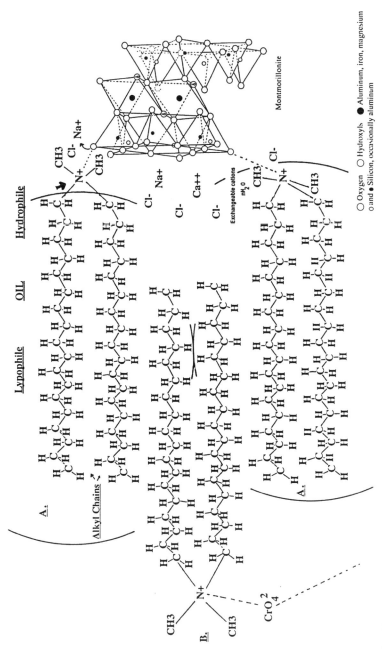

Figure 1 *Model of an organoclay indicating pertition and ion exchange adsorption mechanisms*

A. Partition reaction that removes Non-Polar Organic coumpounds.   B. Ionic bonding that removes Anionic Inorganic and Organic compounds.

a dark yellow color were obtained from Horizon Ag Products (Kennewick, WA), which supplies these products as soil conditioners and fertilizers. Several blends of the two solutions where prepared with a total organic carbon (TOC) concentration of 2000 ppm. The original solutions were tested with both the cationic and non-ionic organoclays to observe if both partition and ionic bonding mechanisms operate. Three types of laboratory tests were conducted. Preliminary trials were conducted with jar tests. Water was spiked with the concentration of the contaminant desired or the raw water from a site was used.

One gram of sorbent was introduced into a vial filled with 30 mL of the water to be tested and placed on a shaking platform for 48 hr at room temperature. After this time the mixture was centrifuged to separate the solids and the supernatant was filtered through a 0.45 μ Millipore filter. Then the total organic carbon (TOC) in the supernatant was measured.

The second test was with a micro-column technique. A small vial converted to make a micro-column was filled with 1 gram of sorbent and the prepared solution was passed through this column by means of a small pump at a rate of 4-5 mL/min. The effluent was continuously monitored for its TOC content. Breakthrough and equilibrium concentrations were determined and the amounts of fluid that passed through the column when these points were reached were recorded. The third test was the standard ASTM approved isotherm method. The isotherms were of the Freundlich type.

4 RESULTS

The observations summarized in Table 1 were made.

Table 1 *Adsorption tests with commercial humic and fulvic acid samples*[a]

Solution	Sorbent	pH	Final TOC mg/L	% Removal TOC	TOC mg/L
Humic Acid (100%)	Control	8.42	2000		
	PT-1E	7.31	1800	10	6
	PC-1	7.63	1880	6	3.6
Fulvic Acid (100%)	Control	2.35	2000		
	PT-1E	3.08	1890	5.5	3.3
	PC-1	3.15	1850	7.5	4.5
Humic Acid (75%)	Control	4.08	2000		
Fulvic Acid (25%)	PC-1	5.66	190	90.5	54.3
Humic Acid (50%)	Control	2.53	2000		
Fulvic Acid (50%)	PC-1	5.26	110	94.5	56.7
Humic Acid (25%)	Control	2.47	2000		
Fulvic Acid (75%)	PC-1	4.29	80	96.0	57.6

[a] Initial TOC in each test was 2000 mg/L

Introduction of the organoclays decreases the pH of the humic acid solutions and increases it for the FAs. This probably is because the organoclay contains 10-15% moisture, which may be extracted and dilutes the original solution. Both sorbents showed poor removal capacity at the extreme pHs and the highest concentrations of the FA and HA samples. At the highest HA concentration the non-ionic PT-IE sorbent performed better than cationic PC-1, while at low pH and the highest FA concentration the opposite was the case. This indicates that HAs are less soluble than FAs. As a result, partition as exhibited by the PT-IE sorbent predominantly occurs. At lower pH, the opposite is the case and the cationic organoclay works better with the more polar FA.

Experiments with FA-HA blends showed the following. Figure 2 indicates that the performance of the cationic organoclay improves as the ratio of FA to HA increases. An anionic organoclay should have been included in this study, at least at the extreme pHs. The original, undiluted Horizon extracts had pH 12.0 for the HA and pH 1.0 for the FA. This simply is the result of the use of sodium and potassium hydroxide for extraction of the HA and phosphoric acid for extraction of the fulvic acid.

To further prove that PC-1 is a cationic organoclay, an isotherm was constructed to test its ability to remove chloride from water (Figure 3). The isotherm clearly shows that PC-1 does remove chloride and therefore is cationic.

Table 2 shows jar test results with a variety of FAs and HAs and compares the performance of organoclays with other sorbents.

Table 2 *Jar tests employing a variety of sources of humic and fulvic acids and sorbents*

pH	Control mg/L	Non-ionic Organoclay	Cationic Organoclay	Anion Exchange Resin	Bituminous Activated Carbon	Lignite based Carbon
IHSS Standard Suwannee River FA						
6.91	1950					
% TOC removed			73.8	11.8		
IHSS Standard Nordic Aquatic FA						
6.91	1930					
% TOC removed			75.1	13.5		
Sodium humate extracted from soil (Aldrich)						
7.67	2100					
%TOC removed			99.0	6.0	33.0	
FAs derived from grass clippings soaked in water						
7.18	2060					
%TOC removed		61.8	98.5	55.8	51.0	24.0
Tannic acid derived from Pine Needles soaked in water						
6.89	2040					
%TOC removed			83.0	56.6	67.9	69.2

These tests showed that the cationic organoclay performed better than all the other sorbents. Also worth noting is that in groundwater and surface waters the pH tends to be close to neutral. In the author's experience there is only one location where large amounts of water are contaminated with humics at pH 12.5. This is in southern Wyoming near a trona deposit. The results of the tests with water extracted from grass clippings showed that soluble FAs and insoluble HAs were present in the extract since the nonionic organoclay performed quite well. Here the main sorption mechanism is partition.

The next set of results shows the performance of the different sorbents in micro-column experiments (Tables 3 and 4).

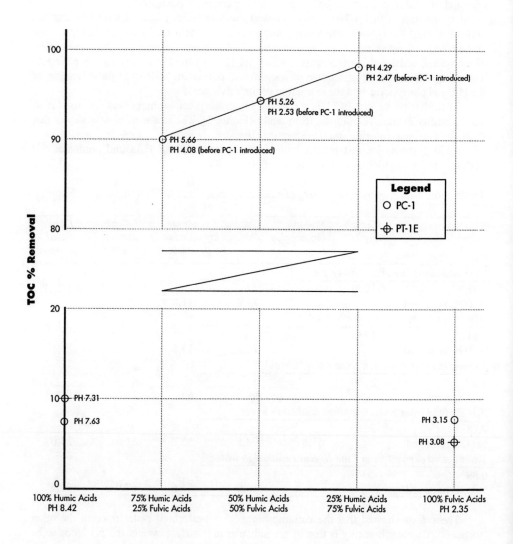

Figure 2 *Removal capacity of a non-ionic organoclay (PT-1E) and a cationic organoclay (PC-1) for a commercial HA and a commercial FA*

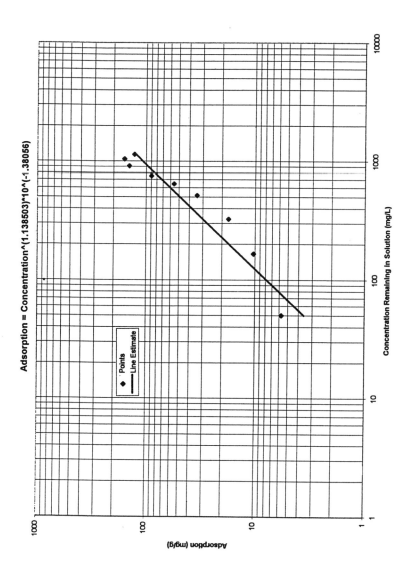

Figure 3 *Isotherm for chloride binding by cationic organoclay PC-1*

Table 3 *Micro-column test data for water extracted from grass clippings and pine needles*

Water source	Medium	Sorbent loading at breakthrough, mg/L	Equilibrium sorbent loading, mg/L
Grass Clippings FA[a]	Cationic Organoclay	448	546
	Bituminous Activated Carbon	175	234
	Anion Exchange Resin	78	195
Tannic Acid Pine Needles[b]	Cationic Organoclay	280	
	Bituminous Activated Carbon	100	
	Anion Exchange Resin	120	

[a] The TOC of each test solution was adjusted to 1950 mg/L; [b] The TOC of each test solution was adjusted to 2040 mg/L.

Table 4 *Breakthrough data for Saskatchewan water in micro-column experiments with different solid media*

Sorbent Medium	Breakthrough quantity, mL
Cationic organoclay	510
Anionic Organoclay	460
Blended anion/cation	390

Figure 4 is an illustration of these data. Figure 5 displays two Freundlich isotherms that compare the performance of bituminous activated carbon with that of the cationic organoclay. The difference is dramatic, the organoclay being much more efficient than the carbon. The HA used to obtain the data for Figure 5 was purchased from Aldrich.

5 TESTING OF WATER FROM SASKATCHEWAN AT VARIOUS TEMPERATURES

The results in this section were obtained from micro-column experiments. The HA content of the water was 0.65 mg/L and it was stored at 60°F. Table 5 shows the effect of temperature on the amounts of TOC remaining after contact with different media in 30 min and 8 hr in jar tests.

The first conclusion from these data is that the humic substances in the Saskatchewan water are dipolar, with anionic and cationic charges on the molecules. This is concluded because anionic and cationic organoclays or a blended resin perform well at removing the HSs from water. This conclusion is reasonable if we accept the fact that HAs and FAs are at least partly hydrophobic, that they consist of many different organic compounds and that they contain some some metals such as iron.

Secondly, because the initial concentration of humics is so low (0.65 mg/L), the quantity of TOC removed is only 30-40%.

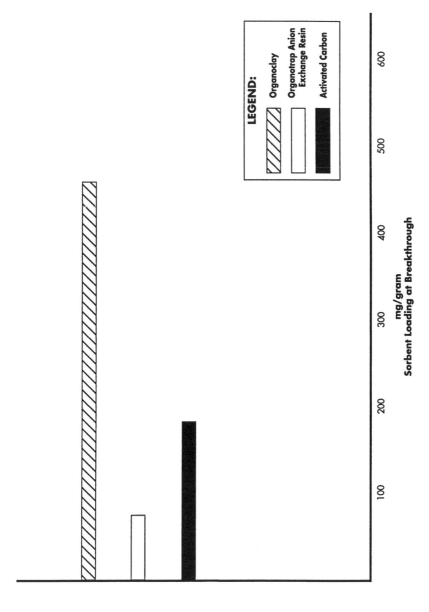

Figure 4 Micro-column data comparing the performance of organoclay and activated carbon for the removal of FA from water

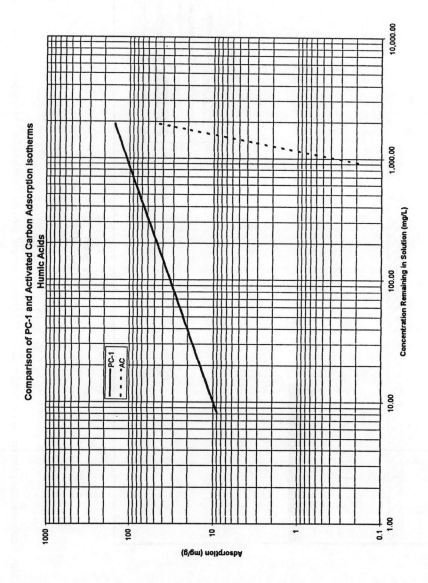

Figure 5 *Isotherms for adsorption of Aldrich HA from water by cationic organoclay PC-1 and activated carbon*

Table 5 *Effect of temperature on sorption capacity of various sorbents*[a]

Sorbent Medium	Temperature, °C	30 min. contact mg/L	8 hr contact mg/L
Cationic Organoclay	4	0.49	0.24
	25	0.50	0.27
Anionic Organoclay	4	0.52	0.30
	25	0.53	0.32
Blended Anion	4	0.57	0.39
Cation Exchange Resin	25	0.48	0.37

[a] the initial TOC concentration in each test was 0.65 mg/L

Thirdly, the effect of temperature is apparent with the organoclays, which remove very slightly less of the HSs from the water at 4°C than at 25°C. This is not the case with the cation exchange resin, which actually removes more HSs at the lower temperature.

It also was found in a field test with one ft^3 of each material that the cation exchange resin lasted longer in continuous application than the cationic organoclay, although neither medium lasted longer than 3 days even with regeneration. It is possible that the freezing phenomenon has more of an effect on the hydrated organoclay than on the resin.

The organoclay needs to be treated with water to be activated. This allows the amine chains to stand up and extend into the water. The viscosity of water is higher at lower temperatures, which may prevent full extension of the amine chains into the water. The cation exchange resin operates by ion exchange inside a pore and thus is not subject to that phenomenon. Also, the resin contains a fair amount of moisture inside its pores that is warmer than the ground water.

The decrease in performance with decreasing temperature also supports the idea that HA chains curl up on lowering the temperature. If so, the HA charges may be less accessible to the amine chains, both in the organoclay and the resin (anion exchange resins contain attached quaternary amines).[3]

6 REGENERATION OF ORGANOCLAYS

It is well known that ion exchange resins and activated carbon can be regenerated. Carbon is regenerated by burning the organics off in a furnace at about 1800°F. Cation exchange resins are regenerated with dilute hydrochloric or sulfuric acid solution, and anion exchange resins with dilute sodium or potassium chloride solution or sodium or potassium hydroxide solution. In attempting to regenerate an organoclay, the first question to be addressed is can the sorbed HA and FA be leached off with distilled water? Second, could the quaternary amines be recovered from the organoclay by leaching with a regenerant solution? The answers to these questions are as follows.

One gram of cationic organoclay (PC-1) was extracted with 50 mL of regenerant solution consisting of 3% NaCl and 1% NaOH for 8 hr and then was washed with deionized water. The solid PC-1 then was separated from the liquid by centrifugation and the supernatant was filtered through a 0.45 μ Millipore filter before TOC analysis.

The PC-1 solid was resuspended in purified water and leached for another 8 hr. TOC then was determined after following the separation procedures described above. The results are shown in Table 6.

Table 6 *Results of PC-1 regeneration experiments*

Extractant	TOC, mg/L	Total extracted, mg
3%NaCl/1% NaOH	7.9	0.4
Water Wash	ND[a]	ND

[a] ND = none detected.

One gram of PC-1 organoclay consists of about 400 mg quaternary amine and 600 mg of clay. Thus, 0.4 mg is a negligible amount of organic carbon from the quaternary amine that was leached out.

A micro-column test was next performed with a PC-1 sample that was saturated with Aldrich HA. The same micro-column test was performed after the PC-1 was regenerated with brine as described above. The results are as follows: Control, 1020 mg HA/L; breakthrough sorbent loading, 393 mg HA/g; equilibrium sorbent loading, 460 mg HA/g. The PC-1 was washed again after regeneration and only 4.19 mg/l TOC was detected.

The conclusion is that a cationic organoclay can be regenerated in the same manner as anion exchange resins but with one difference. The brine removes the HAs and FAs by competing with them for the charged amine site, while in the case of anion exchange resins the NaOH simply exchanges for the HA and FA acid on the exchange site. Therefore, the regeneration mechanism for cationic organoclays is not the same as those for resins.

7 CONCLUSIONS

The results of this study show that cationic organoclays are very good sorbents for removal of HAs and FAs from water. Field data show that in locations such as southern Ireland, the concentration of humics is high enough to warrant the use of a coarse grained cationic organoclay (8 x 30 US mesh size) with excellent results. The amount of humus there can be as high as 50 ppm, with pH $\approx$ 7.5.[6] A smaller grain size (16 x 30 US mesh size) works well in North America for efficient HSs removal from water. The data in this paper also show that this cationic organoclay can be regenerated at least once and probably several times.

References

1. G. R. Alther, *Appl. Clay Sci.*, 1986, **1**, 273.
2. S. Borman, *Chem. & Engineer. News*, April 5, 1999, p. 7.
3. R. Kunin, 'Ion Exchange Resins', Krieger, Boca Raton, FL, 1990.
4. G. Sheng and S. A. Boyd, *Clays and Clay Minerals*, 2000, **48**, 43.
5. J. A. Smith, P. Jaffe and G. Chiou, *Environ. Sci. Technol.*, 1990, **24**, 1167.
6. M. H. B. Hayes, personal communication.

MASS SPECTROMETRY AND CAPILLARY ELECTROPHORESIS ANALYSIS OF COAL - DERIVED HUMIC ACIDS PRODUCED FROM OXIHUMOLITE. A COMPARISON STUDY

D. Gajdošová,[1] L. Pokorná,[1] A. Kotz[2] and J. Havel[1]

[1] Department of Analytical Chemistry, Faculty of Science, Masaryk University, 611 37 Brno, Czech Republic
[2] TEHUM s.r.o., 415 01 Teplice, Czech Republic

1 INTRODUCTION

Humic substances (HSs) are natural organic compounds found in coals, compost, soils, sediments and waters.[1] HSs traditionally are divided into three main parts: humic acids (HAs), the fraction soluble at high pH and insoluble in acids, fulvic acids (FAs), which are soluble at all pH values, and humin, which is insoluble in water at any pH value. HAs are the major extractable component of coal and soil humic substances.

In the Czech Republic (Northern Bohemia) there is an extraordinary source of humic compounds called oxihumolite. It is a part of brown coal that was formed by oxidation some million years ago. For its light brown color it also is called "capuccine." Its chemical composition (content of water, ash and humic acids) depends on the different formation conditions. The average content of humic acids in oxihumolite is high (usually about 70 % w/w) and therefore HAs are produced on an industrial scale by alkali leaching of oxihumolite.

The aims of this work were to analyze oxihumolite raw material and the sodium and potassium humates derived from it by MALDI-TOF mass spectrometry and capillary zone electrophoresis (CZE), and to compare the results of analysis with data for humic and fulvic acids from other sources.

2 MATERIALS AND INSTRUMENTATION

2.1 Chemicals

All reagents used were of analytical grade. NaOH and $Na_2B_4O_7$ were from Lachema (Brno, Czech Republic). H_3PO_4 for conditioning of the capillary and mesityl oxide used as an EOF marker were from Merck (Darmstadt, Germany). Deionized water used to prepare all solutions was doubly-distilled from a quartz apparatus (Heraeus Quartzschmelze, Hanau, Germany).

2.2 Humic Acids

The raw oxihumolite material and the oxihumolite-derived sodium and potassium humates were obtained from TEHUM s.r.o. (Teplice, Czech Republic). The HAs from Fluka were Fluka HA No. 53680 (analysis No. 38537/1 293, Fluka I) and Fluka HA No. 53680 (analysis No. 38537/1 594, Fluka II). HAs from the International Humic Substances Society (IHSS) were the Peat HA standard (1S103H) and the Leonardite HA standard (1S104H). HA samples from the People's Republic of China and from Australia were supplied by TEHUM. A fulvic acid from Química Foliar, Mexico also was investigated.

2.3 Instrumentation

Mass spectra were measured with a Shimadzu Kompact MALDI III mass spectrometer. The instrument was equipped with a nitrogen laser (wavelength 337 nm, pulse duration $\tau =$ 10 ns, energy pulse 200 µJ). The energy of the laser applied was in the range of 0-180 units. One hundred laser shots were used for each sample, the signals of which were averaged and smoothed. Mass spectra were measured in linear positive mode and insulin was used for mass calibration.

A Beckman CZE (Model P/ACE, System 5500) equipped with a diode array detection (DAD) system, automatic injector, a fluid-cooled column cartridge and a System Gold Data station was utilized for all CZE experiments. Fused silica capillary tubing of 47 cm (40.3 cm length to the detector) x 75 µm I.D. was used. The normal polarity mode of the CZE system (cathode at the side of detection) was applied. pH was measured with a glass G202C electrode, a Radiometer K401 standard calomel electrode and a Radelkis Precision Digital OP 208/1 pH-meter. Standard buffer solutions from Radiometer and/or Radelkis were used for pH calibration.

3 RESULTS AND DISCUSSION

3.1 Dissolution of HS Samples

Solid HA samples were dissolved in diluted aqueous sodium hydroxide as it was found that use of NaOH solutions is the best way to dissolve the samples. Each HA sample (1 mg) was dissolved in 36 µL of 1 M NaOH and the solution was completed with doubly distilled water to 1 mL. For mass spectrometry, 1 µL of the sample solution was pipeted onto a sample slide. After drying in a stream of air at room temperature, the slide was introduced into the MALDI III instrument and the HAs were analyzed.

3.2 MALDI – TOF Mass Spectrometry

Mass spectra of HAs and FAs were measured in Laser Desorption/Ionization (LDI) mode, i.e., no matrix was used. A suitable matrix for good ionization of HA was not found. The ionization of HAs was poor with gentisic, sinapinic, caffeic, 5-chlorosalicylic and α-cyano-4-hydroxycinnamic acids as matrices. The mass spectra of oxihumolite, potassium humate and sodium humate are presented in Figure 1. It is evident that all spectra are very similar. We can observe a group of peaks between 200 and 1200 m/z.

Because the mass spectra of sodium humates were better developed, the sodium humate product from oxihumolite was studied in detail. The mass spectra of sodium humate for different laser energies are given in Figure 2a. For energies in the range 100-130 (arbitrary units) a family of regularly distributed peaks was observed. The differences in m/z values of about a multiple of 14 in the group of peaks around 700 – 1500 m/z were observed. However, with increasing laser energy another group of poorly resolved peaks of lower intensity is formed around 558.9 m/z. This might be caused by the formation of doubly charged ions at higher laser energy.

The mass spectrum of sodium humate at optimal laser energy (E = 120) is presented in Figure 2b. In this mass spectrum we can observe good resolution of peaks between 700 and 1200 m/z with differences of 14 m/z, 16 m/z and sometimes also 28 m/z. These values can be accounted for as $-CH_2-$, $-O-$ or $>C=O$ groups in humic acids.

Mass spectra of HAs from other producers were measured for comparison. The mass spectrum of coal derived Fluka I HA is presented in Figure 3. Also in this case we can observe several peaks with regular differences between the peaks equal to 74 m/z.

The mass spectra of coal derived HA from Australia and the People's Republic of China are given in Figures 4 and 5. Also in this case we can observe a difference in m/z values of about a multiple of 14 in the group of peaks around 800 m/z.

Comparison of mass spectra of two natural humic acids derived from peat is given below. The mass spectra of IHSS Peat standard humic acid and fulvic acid are presented in Figure 6. There is no marked difference between the mass spectra of these humic and fulvic acids. To confirm this result, the spectrum of FA (Química Foliar) is presented in Figure 7. It is evident by comparison with Figure 6 that there really is no significant difference between mass spectra of humic and fulvic acids. There are similar features in the spectra and even peaks with the same m/z values are observed.

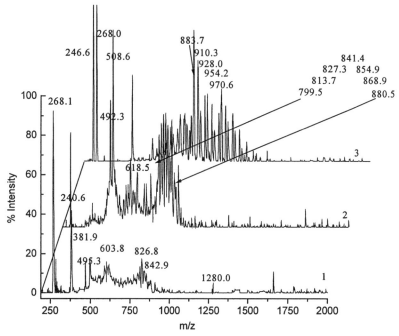

Figure 1 *Mass spectra of oxihumolite (1) and humic acids derived from this raw material as potassium humate (2) and sodium humate (3)*

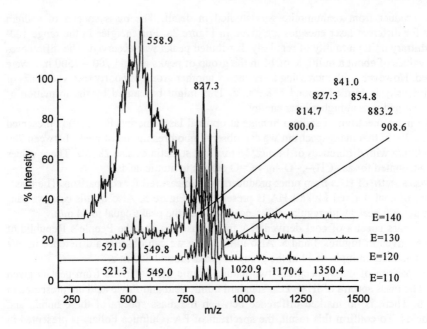

Figure 2a *Mass spectra of sodium humate as a function of laser energy*

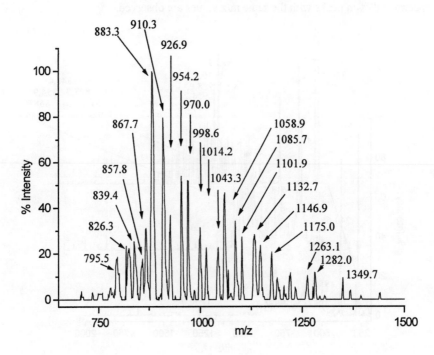

Figure 2b *Mass spectrum of sodium humate measured at optimal laser energy (E = 120 arbitrary units)*

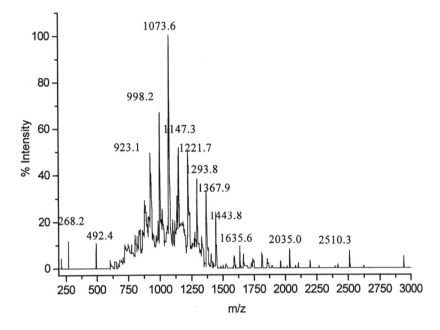

Figure 3 *Mass spectrum of Fluka I measured at laser energy E = 120 (arbitrary units)*

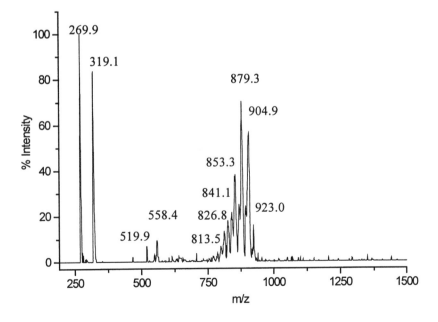

Figure 4 *Mass spectrum of HA from Australia*

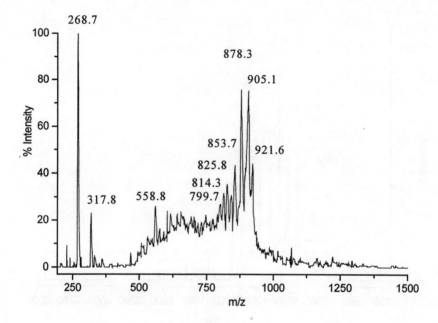

Figure 5 *Mass spectrum of HA from the People's Republic of China*

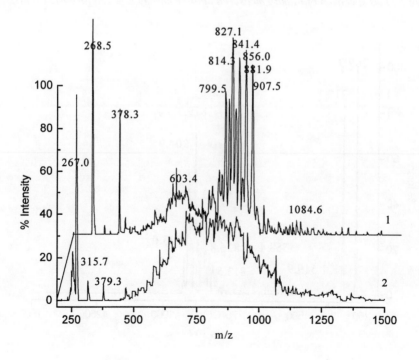

Figure 6 *Mass spectra of IHSS Peat HA standard (1) and IHSS Peat FA standard (2)*

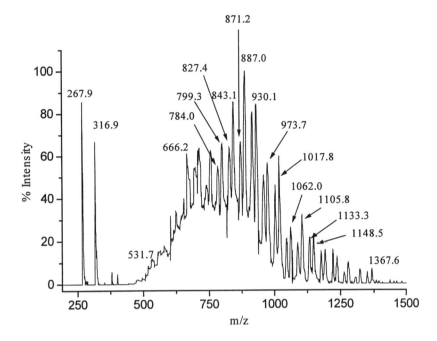

Figure 7 *Mass spectrum of HA from Química Foliar, Mexico, measured at optimal laser energy (E = 140, arbitrary units)*

3.2.1 Effect of aging of HA solutions. In this work we also studied the effect of HA solutions aging. The index of humification E_4/E_6 was measured, where E_4 and E_6 are the absorbances of an HA in solution at 465 and 665 nm, respectively. The E_4/E_6 ratio for a humic acid usually is below 5 and decreases with increasing molecular weight and condensation of aromatic constituents.[10] We have observed an interesting phenomenon: the E_4/E_6 ratio varies with time. Table 1 gives the results of E_4/E_6 measurements in 10 mM NaOH for some HAs. Sodium hydroxide solution was used because the humic acids studied are quite soluble in this medium. It was found that from the initial value of approximately 5 (for the IHSS Peat HA standard), the value increased to 5.4 after 6 days and to 6.4 after 39 days. It was observed by MALDI TOF mass spectrometry that several fractions, both for HAs or FAs, are the same. We found that a group of peaks appears between 200 and 2000 m/z, depending on the age of solution, with the difference between the peaks usually 14 m/z, 16 m/z or 28 m/z. Sometimes peaks at 267 and 317 m/z values were observed, but with increasing laser energy the intensities diminished or the peaks were not observed at all.

3.3 Capillary Zone Electrophoresis

Capillary zone electrophoresis was found to be the most efficient method of separatiing humic acids. Recently, different background electrolytes and conditions for capillary electrophoresis analysis of humic acids were examined and proposed.[4,7-9] Considering these results and our experience, sodium tetraborate was selected as a background electrolyte for CZE measurements and the following conditions were found to be optimal for the analysis: 50 mM $Na_2B_4O_7$; pH 9; hydrodynamic injection 22 s; separation voltage 15 kV; wavelength 210 nm; temperature 40°C.

Table 1 *Index of humification E_4/E_6 measured as a function of HS solution age*

Origin of HS	Solution age, days	E_4/E_6 ratio
IHSS Peat HA standard	0	5.0
	6	5.4
	39	6.4
Sodium humate (Tehum)	0	5.3
	6	6.5
	39	7.5
Potassium humate (Tehum)	0	5.4
	6	6.7
Oxihumolite (Tehum)	0	6.5
	6	7.3

Electropherograms for raw oxihumolite material and TEHUM's sodium or potassium humate products are given in Figure 8. We can find similar parts of the electropherograms for all three TEHUM products. The sodium humate electropherogram is very similar to that of the oxihumolite raw material. In potassium humate electropherograms there is no significant peak for migration time about 11 min. This is in contrast to electropherograms for sodium humate or oxihumolite. Also, for potassium humate there is no significant peak with migration time about 19 min, whereas this feature is observed for the oxihumolite raw material and the sodium humate.

Electropherograms for comparison of TEHUM's sodium humate with Fluka II, IHSS Peat HA and FA standards (Figure 9) demonstrate that the electropherogram for TEHUM's sodium humate has similar features to those of Fluka and IHSS Peat HA and FA standards.

4 CONCLUSION

Oxihumolite is a raw material that is rich in humic acids. Sodium and potassium humates produced from it are similar or equivalent to other industrially produced humates. By MALDI-TOF-MS it was found that peaks with the same m/z values are observed for these products and also for natural humic acids and for coal derived humic acids from other producers. Structured groups of peaks usually are observed in the range 200 – 2000 m/z. The differences between the peaks usually are multiples of 14, 16 or 28 m/z values. We found that the index of humification E_4/E_6 changes with the age of aqueous HA solutions. From capillary zone electrophoresis experiments it also was possible to conclude that TEHUM's humic acids[12] are of similar composition to the humic products from other companies. Similar peaks are observed in all samples of humic acids. The peaks differ

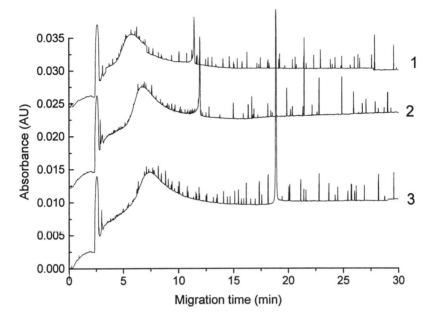

Figure 8 *Comparison of electropherograms for potassium humate (1), sodium humate (2) and oxihumolite (3)*

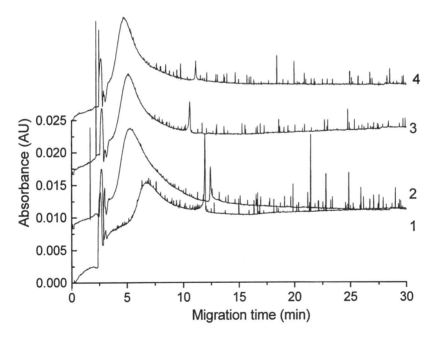

Figure 9 *Electropherograms of sodium humate (1), Fluka II (2), IHSS Peat HA standard (3) and IHSS Peat FA standard (4)*

mostly only in the height and sometimes in the migration time, confirming that humates derived from oxihumolite are comparable in composition and behavior with those of other producers.

ACKNOWLEDGEMENTS

Shimadzu GmbH, Korneuburg, Austria and Shimadzu GmbH Prague, Czech Republic are gratefully acknowledged for supporting this work via sponsoring the Shimadzu Demonstration Laboratory at the Department of Analytical Chemistry, Faculty of Science, Masaryk University, Brno, Czech Republic.

References

1. G. Davies, A. Fataftah, A. Radwan, R. F. Raffauf, E. A. Ghabbour and S. A. Jansen, *Sci Total Environ.*, 1997, **201**, 79.
2. M. Klavins, L. Eglite and J. Serzane, *Crit. Rev. Anal. Chem.*, 1999, **29**, 197.
3. J. C. Rocha, I. A. S. Toscano and P. Burba, *Talanta*, 1997, **44**, 69.
4. S. Pompe, K. - H. Heise and H. Nitsche, *J. Chromatogr. A*, 1996, **723**, 215.
5. D. Fetsch, M. Hradilová, E. M. Peña-Méndez and J. Havel, *J. Chromatogr. A*, 1998, **817**, 313.
6. D. Fetsch, A. M. Albrecht-Gary, E. M. Pena-Méndez and J. Havel, *Scripta Fac. Sci. Nat.*, Univ. Masaryk Brun. Chemistry, 1997, **27**, 3.
7. D. Fetsch and J. Havel, *J. Chromatogr. A*, 1998, **802**, 189.
8. L. Pokorná, D. Gajdošová and J. Havel in 'Understanding Humic Substances: Advanced Methods, Properties and Applications', E. A. Ghabbour and G. Davies (eds.), Royal Society of Chemistry, Cambridge, 1999, p. 107.
9. A. Rigol, J. F. López-Sánchez and G. Rauret, *J. Chromatogr. A*, 1994, **664**, 301.
10. W. Garrison, P. Schmitt and A. Kettrup, *Wat. Res.*, 1995, **29**, 2149.
11. F. J. Stevenson, 'Humus Chemistry', Wiley, New York, 1982, p. 267.
12. http://www.tehum.cz

ANALYSIS AND CHARACTERIZATION OF A "STANDARD" COAL DERIVED HUMIC ACID

L. Pokorná,[1] D. Gajdošová,[1] S. Mikeska[2] and J. Havel[1]

[1] Department of Analytical Chemistry, Faculty of Science, Masaryk University, 611 37 Brno, Czech Republic
[2] Chemapex s.r.o., 430 03 Chomutov, Czech Republic

1 INTRODUCTION

Humic substances (HSs) are natural compounds that are widely distributed in nature but still of unknown structure. The properties of HSs are determined by their composition, which results from the humification of living matter. HSs are intermediate phases in the humification process and traditionally are divided into several groups: fulvic acids (FAs), humic acids (HAs) and humin. HAs are mixtures of complex organic polyelectrolytes.[1] Their acid-base properties are determined by their functional groups. HSs also are found in coal.

So called "capuccine" is a special form of oxidized bohemian brown coal from the western Bohemia region (Czech Republic), where coal is mined on a large scale from open-cast mines. Capuccine is the waste in coal production but it is very rich in HSs. Capuccine can be used for large scale industrial manufacture of alkali metal humates for production of new types of combined fertilizers, as additives in concrete or ceramics and for other purposes.[2]

Raw humates extracted from capuccine were used as a starting material for the production of so called Chemapex HA standard (labeled as "Standard" in the text). The objective of this work is to characterize this coal-derived HA "Standard" by elemental analysis and ash content, and to investigate it with capillary zone electrophoresis (CZE), matrix assisted laser desorption time of flight mass spectrometry (MALDI-TOF-MS), ICPAES and other methods. Details of the use of these methods can be found elsewhere.[3-5] The aim also was to make a comparison of "Standard" with IHSS standard HAs and other humic products.

2 MATERIALS AND METHODS

2.1 Chemicals

All reagents used were of analytical grade purity. NaOH and $Na_2B_4O_7$ were from Lachema (Brno, Czech Republic). H_3PO_4 for conditioning of capillaries and mesityl oxide used as an EOF marker for CZE were from Merck (Darmstadt, Germany). Deionized water used to prepare all solutions was double-distilled from a quartz apparatus (Heraeus

Quartzschmelze, Hanau, Germany).

2.2 Humic Acids

The products analyzed were 1) raw capuccine material, 2) the intermediate product of the capuccine purification process, and 3) "Standard" HAs used for the comparison study were Chinese humic acid supplied by Chemapex, HA from Aldrich (sodium salt, Cat. No H1, 675-2) and International Humic Substances Society (IHSS) standard Peat (1S103H) and Leonardite (1S104H) HAs.

2.3 Instrumentation

Mass spectra were measured with a Kompact MALDI III mass spectrometer (Kratos Manchester, U.K.) in linear mode. The instrument was equipped with a nitrogen laser (wavelength 337 nm, pulse duration $\tau = 10$ ns, pulse energy 200 µJ). The energy of the laser applied was in the range 0-180 units. One hundred laser shots were used for each sample (positive mode), the signals of which were averaged and smoothed. Insulin was used for mass calibration.

A Beckman CZE (Model P/ACE, System 5500) equipped with a diode array detection (DAD) system, automatic injector, a fluid-cooled column cartridge and a System Gold Data station was utilized for all CZE experiments. Fused silica capillary tubing of 47 cm (40.3 cm length to the detector) x 75 µm I.D. was used. The normal polarity mode of the CZE system (cathode at the side of detection) was applied. ICP-AES measurements were performed with a Jobin-Yvon JY 170 ULTRACE ICP spectrometer. pH was measured with a glass G202C electrode, a Radiometer K401standard calomel electrode and a Radelkis Precision Digital OP 208/1 pH-meter. Standard buffer solutions from Radiometer and/or Radelkis were used for pH calibration.

3 RESULTS AND DISCUSSION

3.1 Capillary Zone Electrophoresis

Capillary zone electrophoresis is a suitable method for analysis and characterization of humic acids. Recently, different background electrolytes and conditions for capillary electrophoresis analysis of HAs were examined.[6-8] In consideration of these results and our other experiments, sodium tetraborate was selected as the background electrolyte and the following conditions were found to be optimal: background electrolyte 50 mM $Na_2B_4O_7$; pH 9; hydrodynamic injection 22 s; separation voltage 15 kV; detection wavelength 210 nm; temperature 40°C.

CZE analysis of capuccine raw material and of the intermediate and final products of its purification were performed. The electropherograms are summarized in Figure 1.

During the purification of capuccine by repeated precipitation of HAs from aqueous alkaline extracts, minor peaks are diminished or disappear and in the final product the minor peaks are suppressed. This means that some impurities were removed to give the Chemapex "Standard." Some distinct peaks are still present, but many other peaks were eliminated during the purification process. The final so-called "Standard" product seems to be of high purity.

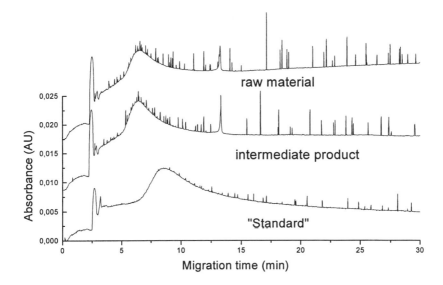

Figure 1 *Electropherograms for products of different steps of the capuccine purification process. Conditions for CZE analysis are given above*

Electropherograms comparing Chemapex "Standard" HA with other products, such as Aldrich and Chinese HAs and IHSS Peat and Leonardite standard HAs are given in Figure 2. The electropherogram for Chemapex "Standard" is similar to that of the IHSS Peat HA standard. Small peaks with practically the same migration times are observed and distinct peaks present in electropherograms for Aldrich and Chinese products are absent. Also, the peak areas and shapes of the "humps" are very similar for IHSS Peat HA and the "Standard" and different from data for the other samples analyzed. Electropherograms for Chinese HA and Aldrich HA have many peaks that are similar to those observed for raw capuccine material before the purification procedure.

The electropherogram of the Chemapex "Standard" has some small peaks that cannot be identified. This is a reason why different modifications of background electrolytes for CZE were tested. The first modification used derivatized polysaccharides in the BGE. The electropherograms for these analyses are given in Figure 3.

Although several other polysaccharides were examined, (including ethylcellulose and some oligosaccharides such as maltose and saccharose), the best results were obtained for hydroxyethylcellulose modified BGE and good results were observed also for dextran sulfate modified BGE. For dextran sulfate practically the same number of peaks as for unmodified BGE was found, but the peaks were more intense. Many peaks also are observed on the electropherogram with hydroxyethylcellulose in the BGE. However, in contrast to electropherograms with unmodified BGE, these peaks show greater height and area and some new peaks also were observed.

Comparison of electropherograms for "Standard" and HA products from other sources using hydroxyethylcellulose modified BGE are given in Figure 4.

Another possibility of modification of BGE with cyclodextrins (CDs) also was investigated. Addition of individual α, β, γ–CDs and their mixtures was studied. Electropherograms for these analyses are given in Figure 5.

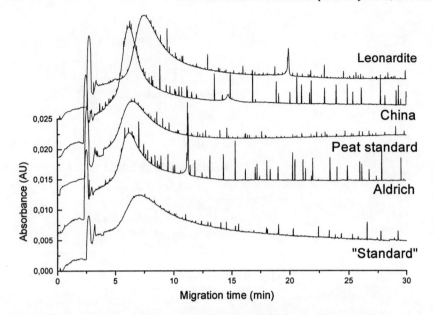

Figure 2 *Electropherograms for Chemapex "Standard" HA and products from other sources. Conditions of CZE analysis are as given in Figure 1*

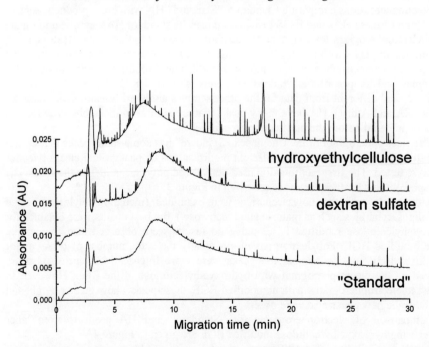

Figure 3 *Electropherograms of Chemapex "Standard" HA in unmodified and modified BGE (7 mM hydroxyethylcellulose, 5 mM dextran sulfate). Conditions the same as in Figure 1*

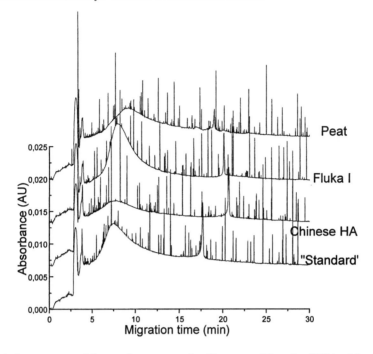

Figure 4 *Comparison of electropherograms for Chemapex "Standard" HA with other products in BGE modified with 7 mM hydroxyethylcellulose. Conditions the same as in Figure 1*

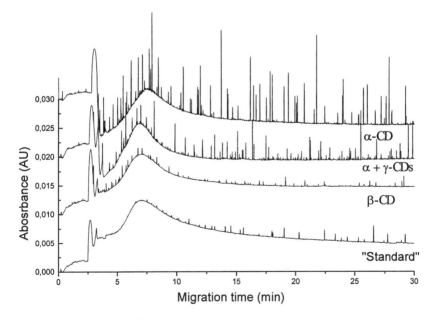

Figure 5 *Electropherograms for analysis of "Standard" using cyclodextrins modified BGE (30 mM + 50 mM α+γ –CDs, 30 mM α-CD, 15 mM β-CD). Conditions the same as in Figure 1*

The best results for cyclodextrins as modifiers of BGE were obtained using α-CD, β-CD or the mixture of α+γ-CDs. When using β-CD as BGE modifier, electropherograms with practically the same peaks as with the unmodified BGE and only a "hump" with some more peaks was observed. This effect is more evident when using α-CD or the mixture of α+γ-CDs. It was observed that addition of α-CD can cause the appearance of several unidentified and sometimes not well-separated peaks.

Comparisons of electropherograms of Chemapex "Standard" and other companies' products when using β-CD modified BGE are given in Figures 6 and 7.

3.2 Matrix Assisted Laser Desorption/Ionization Time of Flight Mass Spectrometry

All samples were dissolved in a small amount of 0.036 M NaOH. After dissolution, the solution obtained was diluted with distilled water. A small amount (~1 μL) was deposited on a metallic support. After drying in a stream of air at room temperature, the platform was placed inside the instrument and the spectrum was measured. The mass spectra were obtained in the laser desorption/ionization (LDI) mode, i.e. no matrix was used.

Mass spectra of "Standard" were compared with those of HAs from other producers. Figure 8 shows mass spectra of raw capuccine material, the intermediate product and the "Standard" HA. There is a group of peaks in the range 500-900 m/z. We observed a difference in m/z values of about a multiple of 14 in the group of peaks around 800 m/z. Figure 9 compares the mass spectra of the Chemapex "Standard" and HAs from other producers. Very similar spectra are observed. The peaks are distributed in two groups of peaks at around 500 and 800 m/z. There are peak differences corresponding to multiples of 14 m/z.

3.3 ICP-AES

The ICP-AES method was used to analyze "Standard" and other HA samples. This method was used for determination of trace levels of some elements in the raw material, the intermediate product and in the "Standard" HA. A comparison of these analyses with those of other products was made. The results are given in Table 1. It was observed that the "Standard" HA has a three times lower ash content than the other products.

Table 1 *ICP-AES analyses (ppm) for "Standard" and other HAs*

Sample	As	PO_4^3	Zn	Cd	Ni	Cr	Mg	V	Cu	Ti	Al	Ash, %
Chinese	<LD[a]	300	21	<LD	34	9	220	16	31	97	8600	20.87
Raw material	0.048	900	29	0.004	21	26	645	30	17	820	12620	24.72
Intermediate product	<LD	700	22	0.003	14	29	598	21	24	1080	10440	12.52
"Standard"	<LD	500	31	0.004	14	24	330	14	28	986	9140	7.3
Aldrich	<LD	900	24	<LD	12	10	3810	9	12	427	5610	22.89

[a] LD = less than the detection limit.

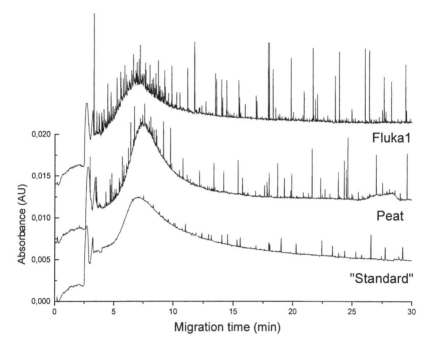

Figure 6 *Electropherograms of "Standard" HA, Fluka HA and IHSS Peat standard HA with 15 mM β-CD modified BGE. Conditions the same as in Figure 1*

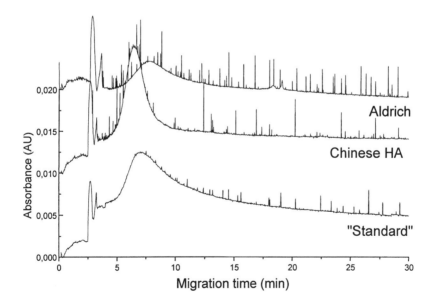

Figure 7 *Electropherograms of "Standard" HA, Aldrich HA and Chinese HA using β-CD (15mM) modified BGE. Conditions the same as in Figure 1*

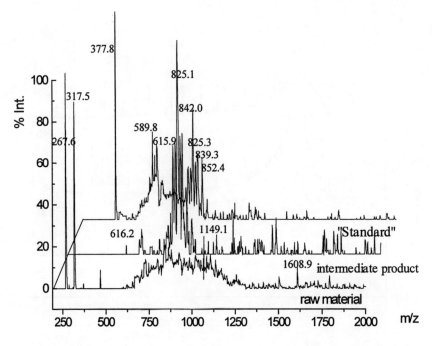

Figure 8 *Mass spectra of raw capuccine material, the intermediate product and the "Standard" HA product*

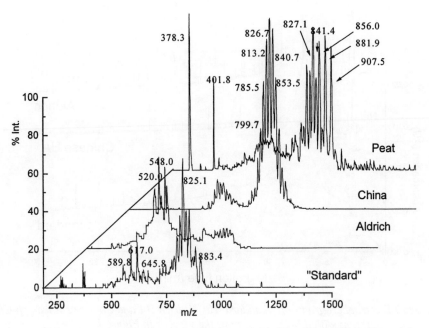

Figure 9 *Comparison of mass spectra of Chemapex "Standard" and other HA products*

4 CONCLUSION

Based on the results obtained by capillary zone electrophoresis it is possible to say that the Chemapex product labeled "Standard" obtained by purification of capuccine raw material is a sufficiently pure and cheap humic acid standard to be useful for comparative studies. Capillary zone electrophoresis and MALDI-TOF mass spectrometry showed that this product is very similar to the IHSS Peat and Leonardite HA standards. The IHSS Leonardite standard HA also is a coal-derived humic acid.

References

1. J. C. M. de Witt, W. H. van Riemsdijk, M. M. Nederlof, D. G. Kinniburgh and L. K. Koopal, *Anal. Chim. Acta,* 1990, **232**, 189.
2. CHEMAPEX s.r.o., http://www.chemapex.cz.
3. M. Remmler, A. Georgi and F. -D. Kopinke, *Eur. Mass Spectrom.*, 1995, **1**, 403
4. D. Fetch, M. Hradilová, E. M. Peña-Méndez and J. Havel, *J. Chromatogr. A*, 1998, **817**, 313.
5. L. Pokorná, D. Gajdošová and J. Havel, in 'Understanding Humic Substances: Advanced Methods, Properties and Applications', E. A. Ghabbour and G. Davies, (eds.), Royal Society of Chemistry, Cambridge, 1999, p. 107.
6. D. Fetsch, M. Fetsch, E. M. Peña-Méndez and J. Havel, *Electrophoresis*, 1998, **19**, 465.
7. D. Fetsch and J. Havel, *J. Chromatogr. A*, 1998, **802**, 189.
8. L. Pokorná, M. L. Pacheco and J. Havel, *J. Chromatogr. A,* in press.

PERFORMANCE IMPROVEMENT AND APPLICATIONS OF HUMASORB-CS™: A HUMIC ACID-BASED ADSORBENT FOR CONTAMINATED WATER CLEAN UP

A. K. Fataftah, H. G. Sanjay and D. S. Walia

ARCTECH, Inc., Chantilly, VA 20151, USA

1 INTRODUCTION

The number of hazardous waste sites requiring treatment for soil and groundwater remediation under current federal and state regulations is estimated to be about 217,000 sites in the United States. These sites include those that fall under the National Priorities List (NPL, Superfund), Resource Conservation and Recovery Act (RCRA) Corrective Action, Department of Defense (DOD) and Department of Energy (DOE) installations. The soil and groundwater at these sites are contaminated with various toxic metals (about 50-70% of the sites) and with organic contaminants (40-70% of the sites).[1] In addition, radioactive contamination is found at 90% of the DOE installations. The DOE estimates that more than 5,700 groundwater plumes have contaminated over 600 billion gallons of water and 50 million cubic meters of soil throughout the DOE complex.[2] Mixed waste containing multiple hazardous and radioactive contaminants is a problem at a number of installations. The types of contaminants present at the sites include toxic metals such as lead, chromium, arsenic, cadmium, nickel, mercury and others, organic chemicals such as benzene, toluene, xylenes and chlorinated hydrocarbons such as trichloroethylene (TCE), perchloroethylene (PCE), energetic chemicals such as nitroesters and others, and radioactive contaminants such as uranium, plutonium, cesium, strontium, tritium and others.

The remediation of contaminated surface and groundwater typically is attempted with treatments such as precipitation, ion exchange, membrane separation and activated carbon adsorption. The method used most frequently to treat groundwater is the conventional pump-and-treat technology. The groundwater is pumped to the surface and treated using various technologies. At sites having mixed contaminants, two different processes are required to remediate a site, an approach that results in complex and costly processing steps. A typical approach is to remove organics using activated carbon followed by ion exchange to remove metals. However, this method is not very effective in meeting the desired cleanup criteria for sites with various types of contaminants, especially when the aquifers are contaminated with non-aqueous phase liquids (NAPLs). Pump-and-treat effects are expected to last 30-70 years at a number of sites that contain NAPLs, thus increasing the treatment costs.

The limitation of present treatment approaches can be overcome by the use of HUMASORB-CS™, a humic acid based product. HUMASORB-CS™ was developed by Arctech as a single step process to remediate water that contains mixed waste contaminants. HUMASORB-CS™ was developed by cross-linking and immobilizing a coal-derived humic acid.[3] HUMASORB-CS™ is a stable and water insoluble adsorbent that retains the properties of humic acid to bind metals, radionuclides and organic contaminants from aqueous waste streams in a single processing step.

The results presented in earlier studies[4] clearly show that HUMASORB-CS™ is effective for removal of organic and inorganic contaminants. In this paper, efforts made to improve HUMASORB-CS™ performance and enhance cost-effectiveness are documented. The approach used to improve HUMASORB-CS™ included modification of HUMASORB-CS™ to enhance the removal of metals present in anionic forms, use of various drying techniques during HUMASORB-CS™ production and also the use of wet beads after storage for extended time periods for removal of multiple contaminants.

2 MATERIALS AND METHODS

HUMASORB-CS™ was produced by cross-linking and immobilizing a liquid humic acid extracted from coal with proprietary methods.[3] HUMASORB-CS™ also was made with another proprietary method to enhance removal of anionic toxic metals.

HUMASORB-CS™ was tested for removal of contaminants in both batch and column modes. In these tests, HUMASORB-CS™ was evaluated with actual and simulated waste streams. Simulated waste solutions were prepared by dissolving the desired contaminants in water. The batch tests were conducted by shaking the simulated waste solution (25 mL) and HUMASORB-CS™ (one gram) at 300 rpm and 25°C for the desired contact time. The mixture then was centrifuged to separate the solid and liquid phases. The liquid phase was analyzed by inductively coupled plasma spectroscopy (ICP) for the metals and with gas chromatography (GC-ECD and/or GC-MS) for organics.

Column tests were conducted in glass columns having different dimensions. The columns were slurry packed with HUMASORB-CS™. Actual and simulated waste streams were passed through the columns with a peristaltic pump. The flow rate was adjusted by controlling the pump speed and with a valve at the column outlet. Column tests were conducted at relatively similar rates defined as Empty Bed Contact Time (EBCT). EBCT is the time required for the fluid to pass through the volume occupied by the adsorbent bed. The Empty Bed Contact Time (EBCT) and the bed volumes used in these tests are based on the volume occupied by dry HUMASORB-CS™ in the column. The amount of HUMASORB-CS™ in the column and the bulk density (~ 1gm/mL) was used to estimate the volume to calculate EBCT. A control sample was collected at the beginning of the test and samples were collected at the column effluent after every 5 to 10 bed volumes, based on the amount of dry HUMASORB-CS™ in the column.

Barrier application experiments were conducted with two large columns having an internal diameter of two inches and a length of 36 inches. The first column (A) was slurry packed with 100% HUMASORB-CS™ (870 gm) and was subjected to a pressure of 10 psig using nitrogen. The second column (B) was packed with a mixture of 50% sand and 50% HUMASORB-CS™ (750 gm) on a weight basis and was subjected to a pressure of 100 psig. Each column was connected to a pressurized tank containing the simulated contaminated water under the desired pressure, as shown in the experimental setup of

Figure 1. A simulated waste stream containing chromium(VI), cerium (surrogate for plutonium), copper, trichloroethylene, and tetrachloroethylene was prepared and passed at flow rate of approximately 1.0 mL/min and 0.5 mL/min through columns A and B, respectively. The flow rate was maintained by controlling the output flow with a needle valve (column A) and a metering valve (column B). The flow rates were monitored and measured twice a day. Two samples were collected at the outlet and two at the inlet of each column every ten days for the metal and organic analyses, with the samples at the inlet acting as control. The samples were analyzed for organics and metals using calibrated GC-ECD and ICP, respectively.

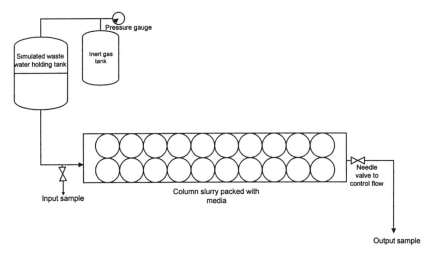

Figure 1 *Experimental setup to evaluate HUMASORB-CS™ under simulated barriers conditions*

3 RESULTS AND DISCUSSION

As reported earlier, HUMASORB-CS™ is effective for removal of organic and inorganic contaminants.[4] This paper summarizes efforts made to improve HUMASORB-CS™ performance and reports some bench scale and field applications of HUMASORB-CS™. The method of preparation of HUMASORB-CS™ was modified to enhance its removal of anionic toxic metals such as chromate and perrhenate. Various drying methods were evaluated during HUMASORB-CS™ production to determine their effect on surface area and contaminants removal. The potential to use wet beads of HUMASORB-CS™ after storing for extended time periods for contaminants removal also was evaluated.

3.1 Modification of HUMASORB-CS™

As a cation exchange material humic acid has little capacity for metals present in anionic form such as chromate, perrhenate and arsenate. The mechanism reported for removal of such metals is based on the ability of humic acid to reduce the metals species to lower oxidation states, where they become available in cationic forms.[5,6] HUMASORB-CS™ is capable of removing chromium(VI) from contaminated water by reducing it first to chromium(III) under acidic conditions.

The capacity of HUMASORB-CS™ and the rate of removal of toxic metals present as anions were improved by modification of HUMASORB-CS™. The new, modified HUMASORB-CS™ was tested for the removal of chromium(VI) (present as chromate) and rhenium(VII) (present as perrhenate) from different simulated waste streams containing 5 ppm of each metal. The tests also were conducted with non-modified HUMASORB-CS™. The tests were conducted in a batch mode as described in the materials and methods section and the contact time was only two hours. The results shown in Figure 1 clearly indicate the big improvement in the removal of these toxic metals in a short contact time. The removal of both toxic metals was increased to about 90% with the use of modified HUMASORB-CS™ as compared to only 5% removal with the non-modified HUMASORB-CS™.

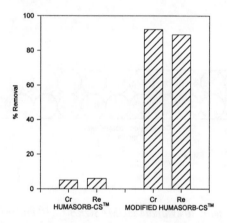

Figure 2 *Removal of Cr(VI) and Re(VII) with the modified HUMASORB-CS™*

3.2 Drying Techniques for Surface Area Improvement

The surface area of humic acid can be increased significantly depending on the drying method used, as reported by Radwan et al.[7] The use of various drying methods was evaluated to increase the surface area of HUMASORB-CS™ and enhance the removal of various contaminants. The drying methods used included air-drying (usually used in the production of HUMASORB-CS™), oven drying, freeze-drying and supercritical fluid CO_2 drying using acetone to replace water in the matrix. The samples lost approximately 90% of their mass with the different drying methods and the products were uniform materials with spherical shape. The products were hard and not brittle, except for the freeze-dried product, which was very brittle and could be easily crushed.

The surface areas of the dried samples were measured at 77 K with liquid nitrogen as the adsorbate using the Brunauer-Emmett-Teller (BET) method. The surface area of HUMASORB-CS™ produced by supercritical fluid drying was the highest among the materials tested. The surface area of the freeze-dried sample was also relatively higher, but as indicated earlier the beads are very brittle.

Samples produced by different methods of drying were evaluated for chromium and TCE removal in batch mode. The batch tests for chromium removal were conducted using simulated water with chromium concentration levels of 10,000, 5,000, 2,000, 1,000 and 500 ppm and were contacted for 24 hours. The objectives of the tests were to evaluate chromium removal at different concentrations and to estimate the maximum capacity based on the isotherms.

The results from the tests were used to develop isotherms for chromium removal using HUMASORB-CS™ produced by the different methods. The isotherms were then used to estimate the maximum capacity for uptake of chromium. Maximum capacities for chromium and the corresponding surface areas for the different products are tabulated in Table 1. The maximum capacity for chromium removal as estimated from the isotherms indicates that the capacity for contaminant removal with oven dried and supercritical dried samples is similar and is approximately twice that with the air-dried samples. The freeze-dried samples have the highest capacity as shown in Table 1. However, the freeze-dried samples are extremely brittle and are essentially in a powder form. The difference in the particle size of the freeze-dried samples and the other samples in part could be responsible for the higher capacity of the freeze-dried sample.

The results in Table 1 also show that there is no correlation between the surface area as measured by the BET method and the capacity for chromium removal. However, when the air-dried sample was ground (increasing the external surface area), the capacity for chromium removal increased significantly. This indicates that an increase in the external surface area has a significant effect on the capacity for contaminant removal compared to the internal surface area. It is believed that this is because of easier and better access to the functional groups and sites on the external surface when the particle size is decreased.

Table 1 *Chromium removal capacity for HUMASORB-CS™ samples*

Drying Method	Maximum Capacity, mg/gm	Surface Area, m^2/g
Freeze Drying	73.5	5.0
Supercritical Fluid Drying	36.4	15.8
Oven Drying	35.6	0.8
Air Drying	19.9	2.0
Air dried-Ground	74.6	1.3

Batch tests also were conducted with simulated water contaminated with TCE (6 ppm) in 25-mL vials with zero headspace. The contact times were two and 24 hr. The tests were conducted in duplicate and appropriate controls were used in the tests. The results in terms of TCE removal in Figure 3 show that the results are similar to that with metals (chromium). The removal of TCE is better with the samples dried by supercritical fluid drying and oven drying. The freeze-dried sample had the highest removal, but as indicated earlier the particles are brittle and are in powder form.

3.3 Comparison of Wet and Dry HUMASORB-CS™ Beads

In order to reduce the overall costs, the wet beads of HUMASORB-CS™ were evaluated for contaminant removal (immediately after they were produced). The objective of the tests was to determine the effect of storing the wet beads on the physical handling properties and the ability to remove contaminants. If the physical properties are not affected by extended storage periods of the wet beads and if the metal removal is not affected, it is possible to either reduce or eliminate the drying required to produce the beads, thereby enhancing the cost-effectiveness significantly. About 80 lbs of wet HUMASORB-CS™ beads were made and stored in two five-gallon sealed buckets for four months. Comparison of HUMASORB-CS™ beads before and after storage showed that the beads maintain their shape and strength on storage.

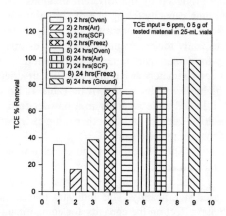

Figure 3 *TCE removal using HUMASORB-CS™ samples*

The HUMASORB-CS™ beads were tested for the removal of metals as wet beads and were compared to dry beads before and after storage. The tests were conducted in both batch and column modes.

The batch tests were conducted as described earlier and the contact time was two hours. The amount of HUMASORB-CS™ used was approximately 10 gm for wet beads and one gram for the dry beads. The 10 gm of wet beads contain approximately one gram of the active component after drying. The results from the batch tests for both the dry and wet beads (before and after storage) are shown in Table 2. Wet beads were still effective for chromium removal and are not affected by storage for at least four months as wet beads.

Table 2 *Chromium(III) removal with wet and dry HUMASORB-CS™ before and after storage*

Sample	Input, ppm	Output, ppm
	Before Storage	
Dry HUMASORB-CS™	479	11.6
Wet HUMASORB-CS™	479	0.57
	After four months storage	
Dry HUMASORB-CS™	464	5.35
Wet HUMASORB-CS™	464	0.68

The column tests were performed by slurry packing 150 gm of wet beads in one column and 15 g of dried beads in another column. The simulated wastewater containing 25 ppm of chromium(III) was passed through the column at a flow rate of 2 mL/min. The column tests were conducted with the fresh wet beads (before storage) and also with wet beads that were stored for an extended time period (four months). The results are shown as breakthrough curves in Figures 4 and 5.

The column tests results clearly show that the wet HUMASORB-CS™ beads are effective in removing chromium. The wet HUMASORB-CS™ is effective for chromium removal and can treat more than 400-500 bed volumes before breakthrough. The wet beads were physically intact during the course of the tests, but the water treated with the wet beads had significantly higher levels of humic substances, which might indicate that cross-linking was not complete. In addition, the tests showed that wet HUMASORB-CS™ could be stored for at least four months without any impact on the physical handling properties and without losing the capacity to remove various contaminants.

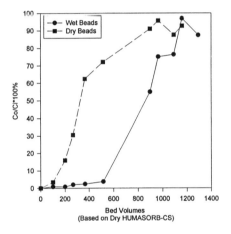

Figure 4 *Chromium breakthrough curves in columns packed with fresh wet and dry HUMASORB-CS™*

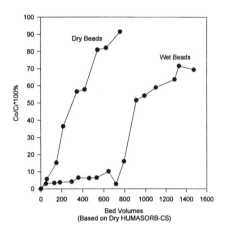

Figure 5 *Chromium breakthrough curves in columns packed with stored wet and dry HUMASORB-CS™*

4 APPLICATIONS OF HUMASORB-CS™

HUMASORB-CS™ has been deployed for demonstration and treatment of industrial waste streams at many different locations. HUMASORB-CS™ has been tested and deployed for removal of toxic metals from brine solution (high sodium) from Johnston Island. Also, as was reported earlier,[4] HUMASORB-CS™ was pilot tested for recovery of valuable resources such as metals and micronutrients from the Berkeley Pit, a superfund site in Butte, Montana. HUMASORB-CS™ can be deployed in existing pump and treat systems or for in-situ groundwater remediation in shallow and deep barriers. In this section the partially completed data from the ongoing barriers experiment will be discussed. The applications of HUMASORB-CS™ in the treatment of brines at Johnston Island also will be discussed.

4.1 Barrier Applications

The simulated barrier tests were conducted to evaluate the effectiveness of HUMASORB-CS™ to treat groundwater in-situ. The rate of groundwater flow is very low and the residence time in a permeable barrier typically is 1-2 days. The residence time in the simulated barrier tests based on the amount of HUMASORB-CS™ present in the column was approximately 14 hours for Column A and 25 hours for Column B. Column A was stopped after 21 months, but column B is still running (the reported data are for 24 months). The breakthrough curves for the metals in columns A and B are shown in Figures 6 and 7, respectively. In this discussion, the breakthrough of the contaminants is assumed when the output concentration is more than 5% of the input. The simulated water was prepared on a regular basis throughout the nearly two-year study and the concentration of the contaminants in the input varied during the course of the study. The average input concentrations of the contaminants are shown in Figures 6 and 7. The breakthrough curves clearly indicate that there is no breakthrough of copper and cerium (a surrogate for plutonium). The chromium(VI) concentration in the output approaches breakthrough after 350 bed volumes but does not increase until at least 700 bed volumes in Column A. The chromium(VI) output data is a little scattered, but the output concentration is very close to the breakthrough limits. For Column B, the concentration increased above the breakthrough limit after approximately 300 bed volumes, but gradually decreased after 350 bed volumes and remained lower than the 5% limit after more than 700 bed volumes. The anomaly in the chromium(VI) breakthrough between 300-350 bed volumes is primarily because of analytical issues and also is reflected to a certain extent in the breakthrough curves for Column A.

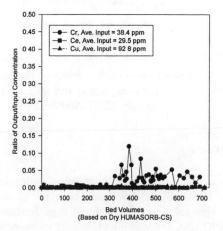
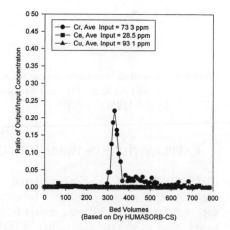

Figure 6 *Column A breakthrough curves for inorganic contaminants*

Figure 7 *Column B breakthrough curves for inorganic contaminants*

The breakthrough curves for PCE and TCE in columns A and B are shown in Figures 8 and 9. There is no breakthrough of PCE in Column A until approximately 400 bed volumes and the column was 30-40% saturated at 700 bed volumes. In Column B, PCE breakthrough occurs after 600 bed volumes and the column was nearly 60% saturated at 700 bed volumes. The higher number of bed volumes passed through Column B before breakthrough is due to the longer residence time compared to that in Column A. TCE breakthrough occurs in both the columns at approximately 150 bed volumes and the output concentration continued to increase above the input concentration. This increase in the TCE concentration at the column output is believed to be the result of a number of factors. HUMASORB-CS™ has a higher affinity for PCE than for TCE, resulting in possible preferential adsorption of PCE. In addition, there is some evidence to suggest the possible degradation of chlorinated organic compounds after adsorption by HUMASORB-CS™ as reported earlier.[4] TCE is a product of PCE degradation and the steady increase in TCE concentration could be a result of PCE degradation by HUMASORB-CS™.

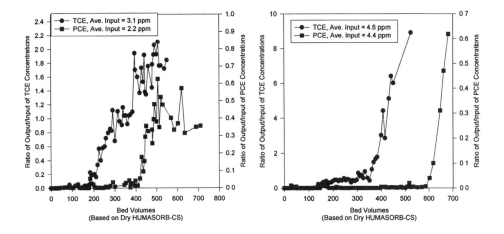

Figure 8 *Column A breakthrough curves for organic contaminants*

Figure 9 *Column B breakthrough curves for organic contaminants*

4.2 Treatment of Waste Brines

The successful treatment of Spent Decontamination Solution (SDS) at Johnston Island using a HUMASORB-CS™ based treatment system[4] showed that the process can be implemented easily at various chemical agent disposal facilities to not only treat SDS, but also as a potentially effective and viable option for in-line treatment of waste brines generated at chemical agent disposal facilities. The HUMASORB™ technology was selected by the US Army for a treatability study to evaluate its effectiveness for the treatment of brines produced from the chemical weapons destruction process at Johnston Island.

The off-gases produced from the chemical weapons destruction process are treated by washing down with a caustic solution, which results in the formation of a brine solution. The brines generated during the process contain multiple toxic metals contaminants in cationic and anionic forms. In addition, the high concentration of sodium coupled with the suspended solids content of the brines present a challenge selectively to remove the metals using conventional treatment systems such as ion-exchange, membrane treatment or chemical precipitation.

Tests were conducted at Arctech using actual brine samples collected from the chemical weapon destruction facilities at Johnston Island. Due to the high sodium concentration, which might affect the stability of the HUMASORB-CS™, the liquid form of the HUMASORB™ technology termed HUMASORB-L™[4] was used instead. The tests with HUMASORB-L™ using actual brine samples were conducted with a factorial experimental design and the results were analyzed statistically. The objective of the tests was to identify the optimum values of the process parameters such as pH, contact time and HUMASORB-L™ dosage. The analysis included development of surface and contour plots for various values of the parameters and regression analysis to help predict HUMASORB™ system performance as a function of the various parameter values. There was good agreement between the observed and predicted values from the regression analysis. The optimum conditions for the treatment of brines were identified based on the statistical analysis as pH 4.5-5.0, contact time five minutes and HUMASORB-L™ dosage of 1% by wt. of the brine.

The confirmation tests were then completed using the optimum values of the parameters. The results show that the concentration of the contaminants was reduced below the identified target limits in the tests.

5 CONCLUSIONS

The objective of this work was to improve the performance of HUMASORB-CS™ for the removal of different types of contaminants from waste streams. HUMASORB-CS™ has been modified by a proprietary method to remove anionic species such as chromate and perrhenate. The removal of Cr(VI) and Re(VII) present in anionic forms was increased from about 5% removal with the non-modified HUMASORB-CS™ to about 90% removal with the modified HUMASORB-CS™.

Using different drying techniques produces different HUMASORB-CS™ batches with different physical properties, such as surface area. The supercritical fluid CO_2 drying method gave the highest surface area. Increasing the internal surface area of HUMASORB-CS™ did not impact the removal capacity, but increasing the external surface area (by grinding) had a significant positive effect on contaminants removal. This indicates that the capacity of HUMASORB-CS™ can be improved by producing smaller particle size beads.

The column tests conducted with wet and dry beads confirm that the capacity of the wet HUMASORB-CS™ beads is higher than that of the dry ones. Also, these tests showed that the wet beads could be stored for a long time without affecting their chemical and physical properties.

The barriers tests results showed that HUMASORB-CS™ is effective for removing various contaminants under simulated barrier conditions and can be used as a Permeable Reactive Barrier (PRB) material for in-situ treatment of groundwater. The HUMASORB-CS™ treatment systems have been demonstrated to be effective for various applications such as treatment of acid mine waters and removal of metals from waste brines.

ACKNOWLEDGEMENT

The authors acknowledge, with thanks, the support and technical guidance provided by the

Department of Energy over the course of this study. We also acknowledge the Humic Acid Research Group at Northeastern University for drying the HUMASORB-CS™ sample and for surface area measurements.

References

1. US EPA Report EPA 542-R-96-005, 'Clean Up the Nation's Waste Sites: Markets and Technology Trends', April, 1997
2. US DOE, Office of Science and Technology, 'Technology Summary Reports', August, 1996.
3. H. G. Sanjay, K. C. Srivastava and D. S. Walia, 'ADSORBENT' US Patent 5, 906, 960.
4. H. G. Sanjay, A. K. Fataftah, D. S. Walia and K. C. Srivastava, in 'Understanding Humic Substances: Advanced Methods, Properties and Applications', E. A. Ghabbour and G. Davies, (eds.), Royal Society of Chemistry, Cambridge, 1999, p. 241.
5. A. Szalay and M. Szilagyi, *Geochim. Cosmochim. Acta*, 1967, **1,** 31.
6. G. G. Choudhary, *Toxicol. Environ. Chem.*, 1983, **6,** 127.
7. A. Radwan, R. J. Willey and G. Davies, *J. Appl. Phycol.,* 1997, **9,** 481.

HUMIC ACID PRODUCTS FOR IMPROVED PHOSPHORUS FERTILIZER MANAGEMENT

K. S. Day,[1] R. Thornton[2] and Harry Kreeft[3]

[1] Kenneth Day Consulting, Fresno, CA 93755, USA
[2] Thornton Consulting, Inc., Benton City, WA 99320, USA
[3] Western Laboratories, Inc., Parma, ID 83660, USA

1 INTRODUCTION

Humic acid products are being used on farms throughout the world. The most accepted use of humic acid products is in combination with phosphorus fertilizers to improve uptake efficiency. Humic acids have been shown to improve phosphate availability and uptake by crop plants.[1]

1.1 Band Applied Humic Acids Affect Potato Yield

Replicated field trials conducted by the University of Idaho showed higher potato yields when liquid humic acid products were added to phosphorus fertilizers.[2] The liquid humic acid products were derived from Leonardite. In each treatment the humic acids were added to banded applications of ammonium polyphosphate fertilizers. Only the humic acid rates varied.

The experiment was conducted at two locations in 1989[2] and was repeated at one of the locations the following year. Yields increased relative to the control for the 1.5 through 6.1 Liter per hectare (ha) humic acid applications. Average yields increased for the 6.1 L humic acid rate from 500 to 700 kg/ha. There was no further advantage to increases in the humic acid rate beyond 6.0 L/ha. In fact, the 24.4 L/ha treatment yielded less than the 6.1 or 12.2 L/ha rate. Increased potato yields were attributed to improved early plant growth from an increased supply of available phosphorus and other nutrients.

1.2 Dry and Liquid Products Affect Potato Yield and Quality

In 1987 a large farming operation in the Columbia Basin of Washington State began experimenting with humic acid products on their research farm.[3] Broadcast applications of 45 kg/ha of 70% dry granular humic acids from Leonardite (Horizon Agri-Plus) were made to the treated plots prior to planting the potatoes. An additional 6.0 L/ha of 3% liquid humic acids (Horizon Quantum-H) were applied to the same treated plots in 1.9 Liter increments over an eight-week period during the growing season through the sprinkler irrigation system. The plots were replicated 20 times.

There was no difference in the watering or fertilization schedule for the treated or untreated plots. Significant yield and quality differences were noted at harvest time.[2] The humic acid treated plots yielded 5% higher than the control. There were substantial improvements in tuber size and specific gravity. There were also measurable reductions of internal quality, including hollow heart, brown center and heat necrosis.

1.3 Effects on Fertilizer Inputs

In order to investigate the possibility for decreasing farm inputs with humic acids use,[2] a second year of experiments was moved to the commercial farming operation so that varying the fertilizer and irrigation inputs was possible. Ten pairs of fields were carefully selected for their similarity in soil type, slope, crop rotation and yield history. Once again, the humic acids were applied in two forms. The dry granular Leonardite (Horizon Agri-Plus 70% humic acids) was broadcast applied prior to planting at 45 kg/ha. The liquid humic acid product (Horizon Quantum-H; 3%) was applied weekly through the irrigation system in combination with phosphorus and nitrogen fertilizers. A total of 6.1 L/ha of Quantum-H was applied through the sprinkler system in 1.9 Liter increments.

Nitrogen and phosphorus fertilizers were applied based on weekly petiole analyses.[2] By the end of the season, the humic acid treated fields increased yield by 3.6%, yet required less nitrogen and phosphorus. Five percent less nitrogen and 15.1% less phosphorus were applied to the humic acid treated fields. Even though less fertilizer was applied, humic acid treated fields had higher tissue phosphorus values 100% of the time. Nitrate petiole levels were higher 90% of the time as well. A summary of the increased yields and reduced inputs is shown in Figure 1.

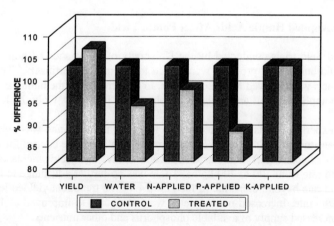

Figure 1 *Changes in potato yield and inputs for humic acid treated fields*[2]

2 MECHANISMS FOR IMPROVED PHOSPHORUS UPTAKE

Nardi reviewed the literature and reported that phosphorus uptake by plants was enhanced by humic acids in nine out of ten studies.[4] There are several proposed mechanisms by which humic acids improve phosphorus uptake. Actual improvements in the field may result from a combination of the various proposed modes of action.

2.1 Competing Ions Are Complexed by Humic Acids

In high pH soils, especially those with an abundance of free lime (calcium carbonate), calcium reacts with various forms of phosphates in the soil and forms insoluble precipitates. Phosphorus moves to plants by diffusion in the soil solution. Humic acids, having several acidic functional groups capable of complexing (chelating) calcium,[1] can prevent calcium from reacting with phosphorus in the soil solution. The result is a higher concentration of soluble phosphorus in the soil solution. This leads to greater phosphorus uptake by plants in the presence of humic acids.

Iron and aluminum ions can also form precipitates with phosphates in the soil solution when soil pH levels are low (below 6.0). Even at high soil pH ranges, oxides of iron and aluminum contribute to phosphorus tie-up.[5] Humic acids help reduce tie-up problems by complexing species of iron and aluminum.[5]

Numerous field studies have demonstrated the conflict between zinc and phosphorus uptake. For example, high phosphorus rates have been shown to induce zinc deficiency in corn. Humic acids can help resolve this problem by chelating zinc and phosphorus ions.

2.2 Amine Groups Complex Phosphates

Much of the nitrogen contained in humic acids has been shown to exist in the form of amine functional groups. In acidic soil conditions, these positively charged amine groups are capable of complexing the negatively charged phosphate ion. Being complexed, the phosphorus would be less likely to react with iron and aluminum to form insoluble precipitates.

2.3 Enhanced Respiration Rates Cause Faster Active Nutrient Uptake

At the root surface, phosphorus enters the root by active uptake (rather than by passive flow). Active uptake requires energy from respiration. Studies show that humic acids enhance the rate of respiration in roots[6] and lead to substantially faster active uptake processes.[7]

2.4 Root Interception is Increased Due to Greater Root Area

Reports in the literature of how humic acids improve root initiation, elongation and surface area abound.[1,4,6] Improved zinc uptake has been proposed as the primary mode of action,[1,4] though hormone-like effects have not been ruled out. Regardless of the mode of action, the increased surface area of the roots leads to increased root interception and uptake of most nutrients, including phosphorus.[4]

2.5 Anti-transpirant Effects of Fulvic Acid Leading to Better Phosphorus Uptake

Xudan reported a 41% increase in phosphorus uptake for moisture stressed wheat plants when fulvic acid was applied to the foliage.[8] Table 1 shows his findings. His extensive research documented that fulvic acid caused partial closure of stomates during periods of moisture stress. Fulvic acid sprays allowed drought stricken plants to conserve water and avoid most of the yield loss associated with the lack of rain and the high environmental demand. Since phosphorus uptake depends on the presence of soil moisture, conservation of soil moisture apparently led to improved phosphorus uptake.

Table 1 *Counts from uptake of radioactive labeled phosphorus fertilizer in drought stressed wheat plants*[8]

	Ear	Stem	Flag Leaf	2nd Leaf	3rd Leaf	Total	Percent
Control	7,514	6,879	1,203	954	173	16,723	100
Fulvic Acid	9,076	10,366	1,800	1,474	827	23,543	141

2.6 Humic Acids Encourage Mycorhizal Fungi Symbiosis

Western Laboratories, Inc. in Parma, Idaho, conducted replicated field trials with onions that demonstrated a strong synergy between humic acids and mycorhizal fungi inoculates. In previously fumigated soils, the combination of humic acid applications at 56 Liters/hectare and mycorhizal fungi inoculation resulted in significantly improved phosphorus uptake and much greater sized onions. The combination of humic acids and mycorhizae was far more effective than either treatment alone. Table 2 shows the marked improvements in tissue nutrient values for the average of each of the treatments. Phosphorus levels were nearly double in the combined treatment of humic acids and mycorhizae.

Table 2 *Average level of elements in onion roots on July 28 (midseason) projected in PPM*[3]

Treatment	N*	P**	K**	S**	Ca*	Mg*	Mn*	Zn**	Cu*	B*
Check	0.40	21.5 B	177 B	21.8 AB	208	71.3	1.41	0.28 B	0.12	0.45
Humic acid	0.55	21.8 B	177 B	17.4 B	206	70.0	1.51	0.31 AB	0.14	0.44
Mycorrhiza	0.61	22.5 B	181 B	21.2 AB	210	73.6	1.77	0.33 AB	0.14	0.49
Humic acid + mycorrhiza	0.47	40.5 A	254 A	30.0 A	267	91.4	2.20	0.41 A	0.20	0.67

*No statistical differences between treatments (LSD, P<0.1); **Numbers in the same column followed by different letters means statistically different (LSD, P<0.1)

3 METHODS OF APPLICATION IN THE FIELD

Methods of applying humic acid products with phosphorus fertilizers vary depending on the form of phosphate applied and the particular farming system in question. Both liquid and dry forms of humic acid products are being applied with phosphorus fertilizers.

3.1 Broadcast Applied Dry Products

Dry forms of humic acids are generally broadcast applied with dry phosphorus fertilizers before planting. Rates of the granular humic acid can vary from 45 to 500 kg/ha depending on the manufacturer's recommendations. Applications are made by airplane or ground spreaders. In most cases the fertilizer and humic acids are applied just before formation of the beds for vegetable crops.

3.2 Band Applied Liquid Products

Liquid phosphorus fertilizers are usually banded to lessen tie-up problems. Humic acid liquid products are often mixed with the liquid phosphate fertilizers in a ratio of 1:10 to 1:20 Liters of humic acid per Liter of liquid fertilizer.

3.3 Sprinkler Applied Liquid Products

Many growers apply phosphate fertilizers in irrigation water. This practice is especially popular where the calcium and bicarbonate levels in the irrigation water are relatively low. Each week, 5 to 20 Liters per hectare of liquid humic acids are injected into the irrigation water. Rates of 40 to 100 Liters per hectare are applied throughout the growing season. Increased petiole levels for nitrogen, phosphorus, potassium and micronutrient levels have been observed. Improved water penetration and reduced runoff also have been reported.[2,4]

References

1. Y. Chen and T. Aviad, in 'Humic Substances in Crop and Soil Sciences: Selected Readings', P. MacCarthy, C. E. Clapp, R. L. Malcolm and P. R. Bloom, (eds.), ASA, SSA, Madison, WI, 1990, Chapter 7, p. 161.
2. M. Thornton, M. Seyedbagheri and R. Thornton, *Proceedings of the 1993 Washington State Potato Conference & Trade Fair,* Use of humic acid in potato production.
3. J. Taberna, *Proceedings of the 1999 Oregon State University Fresh Market Onion Commission Report*, Effect of humic acids and mycorhizal innoculants on onion size and nutrient status.
4. S. Nardi, G. Concheri and G. Dell'Agnola, in 'Humic Substances in Terrestial Ecosystems', A. Piccolo, (ed.), Elsevier, Amsterdam. 1996, Chapter 9, p. 361.
5. D. Vaughan and R. E. Malcolm, 'Soil Organic Matter and Biological Activity', Martinus Njihoff/Dr. Junk Publishers, Dordrecht, The Netherlands, 1985.
6. Z. Sladky, *Biol. Plant*, 1959, **1**, 142.
7. A. Maggioni, *Sci. Total Environ.*, 1987, **62**, 355.
8. X. Xudan, *Austr. J. Agric. Res.*, 1986, **37**, 343.

Subject Index

Abiotic oxidation, 47
Absorbance, 16, 39, 42, 57, 67, 99, 120, 126, 129, 155, 156, 165, 179, 207, 211, 219, 221, 262, 263
Absorption
  band, 39, 45, 46
  K edge, 228
  peak maximum, 228
  spectra, 39, 126, 129, 131
Abundances, 91
Accumulation wall, 216
Acer campestre L., 5
Acetic acid, 99, 100, 112, 114-122
Acetone, 13, 16, 95, 194, 312
Acetonitrile, 13, 99, 145, 146, 193, 227
Acid mine waters, 318
Acidic groups, 93, 97
Acidification, 14, 101, 112, 125, 261, 264-267, 273, 274
Acquisition
  parameters, 55, 148, 150
  time, 13, 113, 127
Actinide-humate, 238, 239, 258
Actinide
  ions, 235, 237, 238, 241, 242, 244, 246, 248, 249, 255, 256, 258
  migration, 235, 242
Activated carbon, 277, 278, 284-287, 309
Activation barrier, 243
Active
  hydrogens, 144
  uptake, 323
Adsorbed water, 45
Adsorption, 11, 112, 154, 157, 165, 192-196, 199, 200, 202, 215, 218, 277, 279, 280, 286, 309, 317
  capacity, 194-196, 200
  isotherm, 193, 195, 197, 198
Aesthetic nuisance, 277

Agglomeration, 83
Aggregation, 41, 42, 135-140, 143, 149, 156, 158, 173, 218, 221, 223, 224,
  dynamics, 223
Aging, 43, 49, 156, 158, 295
Agricultural
  practices, 53, 191
  soils, 96
Air-drying, 40-42, 312
Albite, 237
Alcohols, 4, 44, 46, 68, 77
Aldehydes, 68, 76, 77, 186
Aldohexose, 26
Aldrich, 11, 68, 153-157, 159-162, 169, 178, 180, 181, 183-185, 192, 206, 237-240, 244-246, 248, 249, 281, 284, 286, 288, 300, 301, 304
  HA, 153-157, 159-162, 169, 178, 180, 181, 183, 184, 237-239, 240, 244-246, 248, 249, 286, 288, 301, 305
Aliphatic
  carbons, 32, 39, 45, 56, 71, 181
  C-H, 39, 45
Alkaline extracts, 106, 112, 300
Alkyl carbons, 44, 87
Aluminum, 37, 48, 154, 323
  oxides, 102, 153, 237
Amides, 32
Amines, 98, 205, 206, 278, 287, 288, 323
Amino acids, 2, 10, 23, 32, 56, 68, 77
Ammonium polyphosphate, 321
Angiotensin I, 146
Animal residues, 1-3, 177, 205
Anion exchange chromatography, 238, 284
Anionic forms, 310, 317, 318
Anomeric carbons, 68, 76, 77, 87, 113
Anthracene, 169

Aqueous
  colloids, 218
  HSs, 13, 66, 153, 154
  origins, 11
Aquifer, 153-155, 159, 161, 162, 172, 236, 252, 253
Aromatic
  amines. 205, 206
  carbons, 32, 40, 44, 46, 56, 68, 77, 87, 127, 130, 132, 180, 181, 185, 265
  ethers, 39
  hydrocarbons, 10, 11
  structure, 130, 132
Aromatic/amide protons, 85
Aromaticity, 32, 35, 56, 130, 165, 180, 181
Arsenate, 311
Artifacts, 2, 39, 218, 238
Ash content, 39, 112, 219, 264
Atomic force microscopy (AFM), 38, 43, 46-48, 237
Automatic titrator, 114
Autoxidation, 9, 11, 15, 40, 76
Availability modes, 242-244
Azo dyes, 205

Ba^{2+}, 224
Background electrolytes, 295, 300, 301
Barriers, 166, 311, 315, 316, 318
Base parameter (pK_{HB}), 94, 95, 99, 100
Bases, 2, 5, 93, 97, 98, 102, 113, 195
Basic structural unit, 265, 274
Batch and column experiments, 236, 237, 242, 244, 246, 251, 255, 258
Bed volumes, 310, 315, 316
Bentonite, 277, 278
1,2-Benzenedicarboxylic acid, $C_8H_6O_4$, 265
Bicarbonate, 68, 325
Bimodal peak distribution, 268
Binding sites, 183, 185, 227, 242
Bioavailability, 53, 57, 153, 177
Bioconcentration, 166, 167, 170, 171, 173
Biodegradation, 21, 92, 155, 156
Biofilms, 254
Biological concentration factors (BCFS), 167, 170
Biomaturity, 22, 32, 35

BIONAPL/3D, 155, 159-161
Bioremediation, 11, 19, 162
Biosynthesis, 9
Bisphenol A (BPA), 192, 195-197, 199
Bituminous activated carbon, 284
Bohemian brown coal, 299
BPA, 192-196, 199-202
BR-24 multiple-pulse homonuclear dipolar decoupling, 84
Breakthrough curves, 154, 315-317
Brunauer-Emmett-Teller (BET), 312
Buffers, 9, 206
Building blocks, 9, 93
Butanol, 68

C/N ratio, 125, 126
$C_8H_{10}O_6$, 265
$C_{40}H_{57}NO_{35}$, 265
Ca^{2+}, 96, 100, 154, 223, 263, 264
Caffeic acid, 290
Calcium, 223, 224, 241, 278, 323, 325
Calibration, 111, 113-115, 120-122
Capillary electrophoresis (CE), 215, 237, 289, 295, 300
Capillary zone electrophoresis (CZE), 289, 290, 295, 296, 299-302, 307
Capuccine, 289, 299-301, 304, 306, 307
Carbohydrates, 4, 5, 10, 32, 44, 56, 68, 71, 77, 87, 91, 181, 186, 264
Carbon dioxide, 4, 55
Carbonaceous material, 101, 169
Carbonyl carbons, 56, 68, 71, 77
Carboxyl groups, 4, 5, 40, 45-48, 65, 77, 93, 100, 205, 208, 212, 213, 223, 239
Carboxylic
  acids, 68, 77, 267, 269, 273
  carbons, 44, 56, 87, 88, 113
6-Carboxyvanillic acid ($C_9H_9O_6$), 265
Catchment acidification, 261
Catechol, 37-40, 42-49
Cation
  bridging, 96
  exchange capacity (CEC), 93
  exchange resin, 103, 112
Cationic organoclay, 281-284, 286-288
Cationic surfactants, 278
Cattle manure, 21
Cellphos, 244
Cellulose, 26, 32, 35, 101, 106, 167, 220
Cereal crops, 53

Subject Index

Cerium, 311, 316
Cesium, 309
Chain shortening, 4
Channel flow, 216, 218, 219
Chaotic genesis, 9
Charcoal, 4, 169
Charge, 9, 47, 96-98, 102-104, 106, 114, 115, 117, 121, 173, 208, 212, 213, 216, 232, 246, 254, 277, 284, 287
   density differences, 97
   neutralization model, 238-242
   transfer, 144, 232
Chelex, 244, 245
Chemical
   agent disposal facilities, 317
   compositions, 91, 187
   constituents, 3
   shifts, 32, 44, 45, 68, 76, 77, 79, 85, 87, 88, 127
   structure, 3, 5, 26, 57, 63, 68, 111, 177, 196
   weapons destruction, 317
Chicken manure, 21
Chloride ion, 278
Chlorinated organic compounds, 317
Chlorination, 63
Chlorite, 237
Chloroform, 22, 146
5-Chlorosalicylic acid, 290
Cholesterol, 23, 26
Chromate, 311, 312, 318
Chromism, 118
Chromium, 309, 311-316
   Cr(VI), 311, 312, 316, 318
   Cr(III), 312, 314, 315
Citric acid, 5, 12
Clay
   formations, 235
   minerals, 215, 218
Closed genesis, 9
Clusters, 143, 148, 149
   analysis, 146, 148, 149
   formation, 15, 148
   ions, 144, 148-150
Coal, 64, 65, 83, 112, 169, 277, 289, 291, 296, 299, 307, 310
Cohesive energy densities, 95, 166
Coiled conformations, 111
Colloid mediated actinide transport, 236, 258

Colloids, 106, 138, 216, 219, 257
Column calibration, 115, 122
Commercial
   Aldrich HA, 180
   FA, 282
   HA, 153, 161, 162, 282
   HSs, 153, 162, 185
Complexation constants, 238-241
Composition, 26, 53, 57, 59, 67, 84, 93, 100, 111, 115, 117, 120, 122, 143, 145, 148, 149, 151, 165, 177-180, 183, 193, 227, 264, 265, 277, 289, 296, 298, 299
Compositional changes, 54
Compost, 21-23, 26, 32, 34, 35, 66, 216, 289
Condensation, 10, 26, 68, 76, 78, 221, 223, 295
Conditioning, 237, 249, 251, 289, 299
Conductivity, 11, 102, 103
Configuration, 3, 183, 184, 187, 219
Conformation, 96, 101, 102, 113, 120-122, 215, 223
   changes, 112
   energies, 120
   stability, 120, 122
Coniferin, 265
Conjugate acid, 98
Conjugate base, 97, 98, 100
Conservative roof approach, 258
Contact time, 55, 64, 71, 75, 77, 84-86, 127, 223, 242, 244, 245, 249, 310, 312, 314, 318
Contaminants, 153-155, 162, 185, 192, 205, 215, 223, 309-313, 315-318
Contin analysis, 136
Coordination shell, 227
Copper, 227, 230, 241, 261-263, 267, 274, 311, 316
   binding capacity (CuBC), 262, 267, 268, 274
   ^{63}Cu, 220, 221
   $[Cu(H_2O)_6]^{2+}$, 228
   $[Cu(H_2O)_6]^{2+}$, 227, 229-232
   $[Cu(OH)_4(H_2O)_2]^{2-}$, 227-232
   Cu metal foil, 228
   Cu(II), 3
   Cu/C molar ratio, 227
   Cu^{2+}, 227-232
   Cu-O distances, 228

Copper, *continued*
 Cu-PSM, 228, 230, 232
Co-precipitation, 92, 102, 258
Core structures, 9
Correlation, 165
 coefficients, 186, 196, 199, 200
Covalent binding, 205
Cover crop systems, 53, 54, 57, 59
Critical micelle concentration, 138
Crop rotation, 53, 322
Cross
 flow, 216, 219
 polarization dynamics, 71, 77
 linking, 9, 33, 34, 35, 310, 315
Crystalline $(CH_2)_n$ chains, 87
Cutins, 4, 5
Cyano group, 185
α-Cyano-4-hydroxycinnamic acid, 290
Cyclodextrins, 301, 303, 304
Cytidine, 146
Cytochrome C, 146

Damkohler number, 256
DDT (2,2-bis(4-chlorophenyl)-1,1,1-trichloroethane), 136, 138, 169, 184
Dead plant tissue, 1
Dealkylation, 37
Decarboxylation, 37, 48, 76, 93, 106
Decomposition, 2, 37, 177, 183, 251-255, 258
Deep soil HAs, 193, 196, 202
Degradation, 1, 2, 4, 18, 63, 71, 76, 92, 205, 272, 317
 processes, 15
Dehydration, 241
δ_{HOM}, 165, 173
δ-values, 169
Dendritic structure, 137
Department of Defense (DOD), 309
Department of Energy (DOE), 232, 309, 319
Depolymerization, 4
Deprotonation, 47
Desorption, 13, 144, 148, 150, 152, 154, 157, 158, 160, 161, 165, 192, 194, 196, 199-202, 244
 coefficients, 196, 199
 parameters, 195
Detection limit, 172, 194, 304

Detector, 13, 113, 114, 116-122, 194, 219, 223, 290, 300
Dextran, 3, 301
Dextran sulfate, 301, 302
Dialysis, 98, 103, 112, 178, 215, 238, 262, 265
Diethyl ether, 22
Diethylaminoethyl (DEAE)-cellulose, 178
Difference in binding strength, 244
Diffuse reflectance Fourier transform infrared spectroscopy (DRIFTS), 53-55, 57-59, 219
Diffusion, 84-86, 157, 215, 216, 218, 219, 221, 254, 323
 coefficients, 216, 221
 rates, 215, 218
Dilemma, 257, 258
Dimethyl sulfoxide (DMSO), 92, 95, 97-101, 103, 104, 106
Dimyristoylphosphatidylcholine, 171
Dioxane, 99
Dioxins, 191
Dipolar, 84-87, 95, 98, 284
 aprotic solvents, 95, 98
 dephasing, 84-87
 interaction, 84
Dipole moment, 94
Direct polarization, 65, 67
Direct polarization magic-angle spinning (DP/MAS) ^{13}C NMR spectroscopy, 64
Dispersion forces, 95, 100
Dissociation rates, 242-244
Dissolution, 13, 94, 97, 100, 101, 112, 125, 146, 155, 166, 193, 254, 258, 304
Dissolved inorganic carbon (DIC), 253-255
Dissolved organic carbon (DOC), 1, 12, 14, 67, 68, 102, 178, 235, 252-255, 262, 267
Dissolved organic matter (DOM), 63, 65, 66, 68, 71, 76, 77, 91, 102, 103, 177
Distribution of radionuclides, 235
Divinylbenzene, 278
DNA, 14
Donor-acceptor, 10
Double layer relaxation, 241
Dowex, 112, 244, 245
DP/MAS spectra, 67, 69, 70
Drainage waters, 97

Subject Index

Drought, 323, 324
Drying techniques, 310, 312, 318
Duck excreta, 21, 23, 24, 29-34
Duck manure, 26
Dukovany sandy aquifer, 256
Dynamic light scattering (DLS), 136
Dynamic transport, 153

E_2/E_3 ratio, 179-181
E_4/E_6 ratio, 101, 179, 180, 182, 186, 295, 296
EED, 192-196, 200-202
Effective ionic charge, 241
Electronegativity, 49
Electronic properties, 232
Electropherograms, 296, 297, 300-305
Electrophoretic ion focusing, 238
Electrospray ionization, 143
Electrostatic repulsion, 47
Elemental
    analysis, 12, 93, 98, 113, 178-182, 218, 237, 264, 299
    compositions, 66-68, 83, 264, 265
    distributions, 221
Elution
    profile, 127, 128, 130
    volumes, 115, 117, 119, 120
Emergence times, 216
Emission, 68, 125, 126, 128, 129, 194, 206-213, 270-272
    spectra, 208, 209, 211, 212
    wavelength, 126, 128, 209-211, 213
Empty bed contact time (EBCT), 310
Endocrine disruptors, 191, 202
Endothermic, 240
Energy, 9, 95, 97, 144, 146, 166, 232, 241, 243, 290, 291, 295, 300, 323
    transfer, 268
Enols, 93, 97
Enthalpy, 166, 242, 243
    of evaporation, 166
Entropy, 9, 11, 240, 242, 243
Environmental pollution, 21
Environments, 3, 49, 68, 77, 88, 91-93, 177, 187, 215, 218, 222, 223, 242
Enzymatic inhibition, 254
Enzyme activities, 34
Esfenvalerate, 177, 179, 183-187
Ester, 5, 15, 26, 32, 34, 44, 91-93, 102, 105, 181, 185-187

    groups, 32, 44, 186
    linkages, 5
Estrogenic compounds, 191
Ethanol, 12, 13, 15-18, 99, 100, 102, 193, 194
Ethers, 44
Ethyl acetate, 179
Ethylcellulose, 301
Ethylenediamine (EDA), 98, 99
17-α-Ethynilestradiol (EED), 192
Eu(III), 238-240, 244, 245
    humate, 240, 251
EXAFS, 227-230, 232, 239, 240
Exchange resins, 277, 278, 287, 288
Excitation, 67, 125, 126, 128, 129, 171, 194, 207-213, 271, 272
    wavelength, 126, 128, 207-212
Exothermic, 240
Extraction, 2, 4, 11, 15, 54, 84, 96-98, 100-102, 104, 106, 136, 149, 193, 194, 219, 239, 281
    procedures, 2, 104, 136

Fast atomic bombardment, 143
Fatty acids, 23, 26, 34, 91-93, 105
FEFF6 computer code, 228
Fertilizers, 54, 56, 57, 59, 280, 299, 321, 322, 324, 325
Field applications, 161, 162, 311
Field-flow fractionation (FFF), 15, 215-221, 223, 237
    inductively coupled plasma-mass spectrometry (FFF-ICP-MS), 218-223, 227
Filtration media, 278
Fingerprint characterization, 149, 150
First derivatives, 231
First order kinetics, 243
Flory-Huggins theory, 166, 169
Flow field-flow fractionation (FFFF), 15, 216-219, 221, 223, 224, 237
Flow rates, 218, 219, 311
Fluka HA, 290, 305
Fluorene, 173
Fluorescence
    fractions, 126-130, 132
    mode, 228
    quenching, 165, 169, 171, 173, 213
    spectra, 262, 272, 274
    spectroscopy, 206

Fluorescent
  fractions, 130, 132
  peak, 127, 128
  substances-rich fraction, 125
Fluorophore, 206, 208
Fourier transform infrared spectroscopy (FTIR), 13, 17, 38, 39, 42, 43, 46, 53, 57
  spectra, 13, 17, 39, 42, 45, 46
Fourier transform ion cyclotron resonance mass spectrometry (ESI FT-ICR MS), 143, 144
Fractal
  dimension, 137, 138
  geometry, 135, 137
Fractionation, 3, 54, 63, 91-93, 101-106, 216, 218-221, 247
Fractions, 1-4, 10, 11, 53-55, 57, 58, 66, 71, 92, 93, 96, 97, 100, 102, 104, 105, 112, 118, 126-132, 135, 143, 150, 154, 180, 181, 185, 187, 192, 202, 215, 218, 221, 223, 247, 262, 267-270, 274, 277, 295
Fractograms, 221-223
Fragmentation, 14, 143
Free radicals, 48, 64, 65, 69, 71, 98
Freeze dried, 12-15, 38, 55, 99, 103, 112, 114, 125, 126, 178, 261, 262, 312, 313
Freeze drying, 12, 13, 102, 103, 193, 312
Freezing point depression, 111
Fresh waters, 181
Freundlich
  adsorption, 196, 199
  equation, 194-196, 200
  isotherms, 194-196, 198-200, 280, 284
  model, 196
Fulvic acids, 1-3, 5, 15, 37, 54, 55, 57, 65, 66, 69, 70, 92, 135, 136, 143-146, 149, 150, 169, 177, 178, 181, 183, 184, 206-213, 215, 237-239, 241, 242, 244, 248, 252-254, 264, 267, 277-281, 289- 291, 299, 323
Functional groups, 11, 15, 19, 46, 47, 53, 56-59, 64, 68, 93, 143, 177, 187, 196, 202, 205, 237, 240, 299, 313, 323
Fungicides, 191, 192

Gallotannins, 5
Gas chromatography (GC), 105, 310
Gated decoupling, 65, 67

Gel permeation chromatography (GPC), 105, 237
Gentisic acid, 146, 290
Geochemical
  behavior, 258
  data, 257
Glass fiber filters, 262
Globular
  protein standards, 115
  structures, 15
Glucovanillin, 265
Goethite, 237, 246-248
Gorleben aquifer, 237, 252, 256
Greenhouse gas emissions, 21
Groundwater, 11, 159, 162, 172, 178, 180, 236, 237, 239, 245-247, 249-259, 277, 281, 287, 309, 315, 316, 318
  recharge, 252
Guaiacyl structures, 26

H/C atomic ratio, 264
H_2SO_4, 126, 261
H_3PO_4, 101, 289, 299
Hairy vetch, 54
Heat necrosis, 322
Heavy metals, 218, 278
Hematite, 237, 246-249
Herbicides, 191
Heterogeneity, 5, 53, 63, 64, 83, 111, 135, 139, 146
Heteronuclear J coupling, 84
Hexachlorobenzene, 171
Hexacoordinated complex, 47
High performance
  liquid chromatography (HPLC), 13, 113, 146, 193, 194, 219
  size exclusion chromatography (HPSEC), 13, 14, 105, 111-122, 125-127, 129, 130, 215, 218
High pH soils, 323
Hollow heart, 322
Homogeneity, 105
Homonuclear dipolar interaction, 84
Hormone, 191, 323
HUMASORB-CS™, 309-318
Humates, 47, 98, 101, 114, 205, 222, 223, 238-243, 252, 258, 281, 289-291, 298, 299
Humic acids, 1-3, 5, 9, 37, 42, 44, 46, 47, 53, 54, 65, 66, 84, 92, 99, 111-113,

Subject Index 333

118, 119, 125, 135-140, 143, 146, 153, 169, 173, 177, 178, 183, 184, 188, 192, 195-199, 201, 205, 215, 218, 219, 221-224, 237-239, 244-249, 277, 278, 280, 289, 291, 295, 296, 299, 300, 307, 310-312, 321-325
Humic, 143, 162, 205, 206, 258, 278, 281, 284, 288
   analyte, 114
   colloids, 235-237, 241-243, 245, 246, 250-255, 257, 258
   distribution, 248
   materials, 100, 111, 113, 135, 137, 178, 205, 206, 213, 216, 238, 277
   molecules, 92-94, 96-99, 102, 106, 113, 121, 123, 125, 128, 223
   solubility, 277
   solutions, 113, 114, 120-122, 178
   substances (HSs), 1-7, 9-19, 21, 37, 38, 46, 49, 50, 54, 57, 59, 61, 63, 65, 79, 80, 83, 84, 87, 88, 91-98, 100-109, 111-113, 115, 117, 121-125, 127, 133, 135, 140, 141, 143-146, 151-154, 161, 174, 177-188, 205, 206, 213-216, 219, 223-225, 227, 235, 237, 239, 242, 246, 254, 256, 257, 259, 264, 274, 275, 277, 284, 287-290, 298-300, 307, 315, 319, 325
Humification, 4, 5, 9, 10, 11, 14, 15, 19, 21, 34, 37, 39, 41, 49, 56, 79, 193, 202, 295, 296, 299
Humin, 1, 2, 92, 97, 100, 101, 106, 135, 143, 289, 299
Humps, 301
Humus, 1, 2, 79, 91, 288
Hydrodynamic (Stokes) diameter, 216
Hydrogen bonding, 88, 93-100, 102, 121, 137
Hydrolysis, 4, 5, 265
Hydrolyzable tannins, 4
Hydrophilicity, 10, 11, 102, 113, 115, 154, 177, 185, 278
Hydrophobic
   acid fraction (HPOA), 65-67, 69, 70, 74
   organic compounds, 169, 177
Hydrophobicity, 10, 11, 15, 66, 96, 98, 100, 102, 106, 112, 113, 120, 122, 137, 140, 153, 154, 162, 165-170, 172, 173, 185, 187, 193, 284
Hydroquinone, 11, 37, 42
Hydroxyethylcellulose, 301-303
Hydroxyl groups, 47, 93, 181
Hydroxylation, 4
3-Hydroxypicolinic acid, 13, 14
Hyperchromism, 118
Hyphan, 244
Hypochromism, 118
Hysteresis, 160, 161, 199, 200, 202

Ideal tracers, 252, 256
IHSS standard, 44-46
Immobilization, 153, 192, 205, 206
Impurities, 37, 39, 42, 46, 145, 179-181, 300
Inductively coupled plasma-mass spectrometry (ICP-MS), 215, 216, 218-220, 222, 223
Infrared spectroscopy (IR), 4, 15, 58, 237
Inner
   coordination shell, 227, 228
   sphere, 172
Inorganic
   carbon, 14, 254
   contaminants, 96, 180, 237, 310, 311, 316
Insecticides, 177, 188, 191
Insulin, 146, 290
Integrated areas, 69, 70, 71, 75, 77
Integrity, 218, 223
Intensity standard, 64
Interferences, 218, 219
Intermolecular forces, 10, 121
Internal transmission, 9
International Humic Substances Society (IHSS), 38, 41, 44, 45, 92, 93, 99, 102, 103, 107, 108, 109, 133, 146-150, 219, 281, 290, 291, 294-297, 299-301, 305, 307
Ion
   distributions, 143, 144, 147, 148, 150
   exchange capacities, 278
   intensities, 22, 23, 32, 146, 221
Ionic bonding, 277-280
Ionic strength, 105, 111, 114, 135, 137, 139, 154, 162, 173, 215, 223, 235, 239, 244, 247, 263

Ionization, 13, 24, 25, 29-31, 99, 143, 144, 145, 148, 290
Iron
  concentrations, 65
  Fe(III), 37, 66, 246
  FeO, 237
  ferric hydroxide, 172
  oxides, 153
Irreversibility, 194, 199, 200, 202
Irrigation inputs, 322
Isoelectric focusing, 105
Isolation, 66, 91-93, 96, 97, 100, 101, 103, 105, 106, 193, 261
Isotachophoresis, 105

Johnston Island, 315, 317, 318

Kaolinite, 154, 237
K_{DOM}, 177, 179, 183-187
Keto group, 87
Keto-enol tautomerism, 71
Ketone carbons, 44
Ketones, 68, 186
Kinetic, 153, 157-161, 172, 192, 193, 195, 236, 240, 242, 244, 246, 251, 255-258
  of metal binding and release, 242
  processes, 251
  rate, 257, 258
Kinetically controlled availability model (KICAM), 242, 243
K_{OW}, 165-171, 173
Kubelka-Munk functions, 55

Labile components, 53
Lake Skjervatjern, 261, 263-274
Langmuir
  behavior, 159
  equation, 154, 159, 194, 200
  isotherm, 158, 159, 162
Laser
  ablation, 143, 215
  ablation mass spectrometry, 143
  desorption, 143, 144, 146
  desorption/ionization (LDI), 290, 304
  energy, 291-293, 295
  light scattering (LLS), 135, 136
  power, 148
  shots, 146, 290, 300

  induced fluorescence spectroscopy, 239
Leaching, 4, 287, 289, 315
Leaf, 4
  leachates, 4, 5, 277
  tissue, 4
Leonardite, 112, 219, 221-224, 300, 301, 307, 321, 322
  HA standard, 290
Lewis
  acids, 37, 49
  base, 49
Ligand exchange, 154
Light scattering, 111, 135, 136, 139, 140
Lignin, 4, 5, 10, 15, 26, 32, 33, 35, 53, 71, 87, 91, 265, 266, 272
Lignite based carbon, 277
Line broadening, 67, 113, 127
Linear
  isotherms, 165
  relationship, 71
Lipids, 4, 5, 22, 23, 26, 33, 35, 71, 106
Liquid scintillation, 179
Local
  equilibrium, 255
  structure, 232
Long chain hydrocarbons, 5, 32, 92, 105
Long-lived radionuclides, 251, 255
Long-term stability, 235
Luminescence fluorescence spectra, 207, 262, 268-270, 272, 273

Macromolecular species, 3, 9, 63, 96, 105, 112, 118, 122, 183, 216, 219
Macromolecules, 2, 37, 46, 47, 49, 56, 91, 96, 105, 111, 112, 118, 121, 145, 177, 195, 215-219, 221
Magic angle spinning, 22, 38, 83
Magnesium, 278
Maltose, 301
Manganese(IV), 37
Markoff processes, 11
Marshland, 252
Mass
  range distribution, 144-146
  spectrometry, 14, 22, 135, 143, 215, 218, 289, 290, 295
  to charge distribution, 143, 145, 150
  to charge ratio, 143, 144
  transfer, 160

Subject Index 335

Matrix assisted laser desorption time of flight mass spectrometry (MALDI-TOF-MS), 13, 14, 143-146, 148-150, 289, 296, 299, 307
Maximum capacities, 313
Melanoidins, 19
Membrane
　filter, 101
　separation, 309
Mesityl oxide, 289, 299
Mesomeric forms, 9
Metal
　binding, 3, 137, 215, 216, 227, 240
　impurities, 37, 42, 46
　ion concentrations, 66, 238
　ion humic colloid complexation, 237
　oxygen vibrations, 46
Methoxyl carbons, 44, 56
Methyl
　cyanide, 95, 99, 102
　ether groups, 4
　isobutylketone, 97
Micellar-like structures, 135
Microbial
　action, 205
　biomass, 4
　carbon source, 57
　growth, 12
　metabolites, 37
　populations, 21
Microbiological turnover, 254
Micro-column test, 288
Micronutrients, 315, 325
Microorganisms, 2, 4, 38, 92, 254
Microtox®, 262
Migration
　case studies, 236, 256, 258
　lengths, 249
　time, 251, 296, 298
Mineral
　grains, 4
　(hydr)oxides, 106
　surfaces, 153, 154, 246
Mixed waste, 309, 310
Mixtures, 2-5, 11, 21, 26, 32, 35, 55, 91, 94, 95, 99, 100, 103, 144-146, 150, 161, 193, 194, 221, 280, 299, 301, 304, 310
Mobile
　humic colloids, 246
　phase, 13, 15, 113-115, 117, 122, 193
Mobility, 87, 153, 162, 192, 215, 224, 235-237, 251, 255, 258
Models, 1, 3, 5, 6, 11-15, 38, 39, 67, 68, 71, 76, 79, 111, 121, 135, 144, 153-156, 159-162, 166, 169, 177, 186-188, 193, 200, 206, 219, 227, 228, 236, 237, 240, 241, 255, 256, 262, 279, 290
Moisture content, 22, 99
Molar volume, 95, 100, 166, 169
Molecular
　analysis, 143
　clusters, 14
　conformation, 47, 216, 218
　formula, 19
　ions, 144, 146, 150
　masses, 15, 146, 148-150, 223, 241, 242
　size, 11, 14, 15, 96, 105, 111, 112, 115, 117, 120, 127, 132, 146, 149, 215, 218, 267, 269, 274, 278
　species, 4, 5, 215, 218, 219
　structure, 19, 266
　weight (M_w), 2, 12, 14-16, 34, 35, 39, 57, 68, 92, 96, 105, 111, 112, 114-117, 120-122, 126, 130, 144, 154, 158, 161, 165, 177, 180-183, 215, 216, 218, 219, 221-224, 242, 243, 262, 295
Monomeric units, 3
Monosilicic acid, 38
Montmorillonite, 278
Multidimensional NMR analysis, 3
Multipulse sequences, 84
Municipal solid waste, 21
Muscovite, 237
Mycorhizal fungi, 324

N contents, 39, 98, 101
N,N-dimethylformamide, 95, 98
$Na_2B_4O_7$, 101, 289, 296, 299, 300
$Na_4P_2O_7$, 54, 97, 101, 112, 145, 219
National Priorities List (NPL), 309
Natural
　aquifer systems, 236, 251, 254, 258
　buffering capacity, 274
　humic colloids, 258
　organic compounds, 1, 289
　organic matter, 1, 64, 135, 261, 277

Natural, *continued*
  products, 9, 269
  waters, 1, 3, 4, 71, 205
Nematicides, 191
NH$_4$NO$_3$, 54, 261, 263
Ni(II)-humate, 243
Nitroesters, 309
Nitrogen laser, 146, 290, 300
N-methyl-2-pyrrolidinone, 95, 99
NMR spectroscopy, 11, 53, 125, 205
  ^{13}C chemical-shift range, 85
  ^{13}C CP-MAS NMR spectra, 44-46
  ^{13}C NMR, 3, 5, 22, 33, 38, 45, 53, 63-65, 71, 75, 77, 78, 93, 113, 127, 130-132, 179
  ^{13}C/^{12}C ratio, 254
  CP cross-relaxation time constants (T_{CH}), 64
  CP/MAS ^{13}C NMR spectroscopy, 63-65, 67, 68, 71-75, 77, 78, 83, 113, 130
  CP-TOSS, 55, 59
  2D heteronuclear ^{13}C-1H NMR (HETCOR), 83-88
  ^{1}H-^{13}C correlation analysis, 84
  liquid state ^{13}C-NMR spectroscopy, 63, 65, 67, 68, 178
  ^{1}H-NMR spectra, 13
  solid-state ^{13}C NMR, 4, 5, 32, 34, 44, 63, 65, 67, 68, 71, 83, 84, 87, 125
  spectrum, 32, 38, 44, 77, 87
NOM, 1, 2, 4, 5, 227, 261, 262, 264-274, 277
Nominal molecular weight, 12, 268
Non-aqueous phase liquids (NAPLs), 162, 309
Non-humic substances, 2, 92, 102, 103, 105, 106, 264
Non-hydrolyzable tannins, 4, 71
Non-ionic surfactant, 278
Nordic aquatic FA, 281
Normal
  mode, 216
  polarity mode, 290, 300
Np(IV), 238, 241, 242, 246, 247
Np(V), 238, 246, 247
Np(V)→Np(IV), 246
Nuclear Overhauser enhancement, 65, 68
Nucleic acids, 91, 118

O/R ratios, 57, 59
O/C atomic ratio, 264
O-C-O, 56
Octahedral
  coordination, 47
  particles, 40, 47
O-demethylation, 4
O-dichlorobenzene, 169
Oligonucleotides, 13
Oligosaccharides, 301
Onions, 324
Open
  genesis, 9
  transport code, 255
Optical absorptivity, 130
Optimum Contact Time, 113
Organic
  coatings, 4
  compounds, 1, 2, 4, 47, 63, 71, 166-168, 170, 173, 184, 187, 270, 277, 278, 284
  matter (OM), 1, 22, 23, 26, 32, 35, 53, 57, 63, 66, 91, 97, 100, 102, 154, 165, 166, 192, 219, 261, 262, 264, 268
  nitrogen, 213, 263, 273
  solvents, 2, 11, 97, 99, 100, 102, 106, 173
  wastes, 21
Organoclays, 277-282, 284, 285, 287, 288
Origin, 127, 132, 144-146, 148-150, 177, 183, 186, 187, 193, 196, 236, 251
Outer sphere, 172
Oven drying, 312, 313
Overlaying soil zone, 254
Oxidation, 4, 11, 13, 26, 34, 37, 42, 45, 46, 57, 68, 97, 254, 265, 273, 289
  states, 37, 311
Oxihumolite, 289-291, 296-298
Oxygen, 9, 11, 15-18, 47, 48, 57, 93, 172, 180-182, 185, 264
Oxygenated organic molecules, 273
Oxygen-containing functional groups, 57, 85
Ozonation, 277

PAHs, 11, 166-169, 177, 183, 185, 206, 278
Paper electrophoresis, 105

Paradigm, 1, 3-5
Paramagnetic centers, 64
Paraquat, 206, 207, 212, 213
Partial least square regression, PLS-R, 178, 186
Particle
    morphology, 135, 138-140
    size, 11, 137, 139, 146, 313, 318
Particulate organic matter (POM), 91
Partition, 165-167, 169, 170, 183, 277, 278, 280, 281
    coefficients, 154, 167, 169, 171-173, 177, 179, 183, 184, 187
Partitioning, 154, 165-167, 169-171, 173, 186, 215, 277, 278
Pb^{2+}, 224
^{208}Pb, 220, 221
PCBs, 166-168, 191
peak
    intensities, 119, 120, 127, 128
    positions, 268
    shifts, 269
Peat
    FA standard, 294, 297
    HA standard, 290, 294-297, 301
    moss, 277
Pectins, 10
Pentoses, 26
Peptides, 4, 32, 68, 77, 91, 92, 104, 146
Peptization, 97
Perchloroethylene (PCE), 309, 316, 317
Permeable reactive barrier, 318
Perrhenate, 311, 312, 318
Perylene, 169
Petiole analyses, 322
pH effects, 98, 211, 227, 230
pH variations, 228, 230
Phenanthrene, 172, 173
Phenolic
    carbons, 44, 56
    esters, 26, 35
    groups, 87, 100
Phenols, 26, 93, 97, 171
Phosphate
    availability, 321
    buffer, 15, 208-212, 262
    fertilizers, 325
Phosphoric acid, 14, 281
Phosphorus
    pentoxide, 98

    uptake, 322-324
Photon correlation spectroscopy, 237
Phthalates, 191, 205
Phthalic acid, 262, 265, 272, 273
Phyllite, 237, 248, 249
Physical protection mechanisms, 53
$\pi - \pi$ interactions, 213
Plant, 1-5, 37, 68, 76, 91, 177, 205, 257
    degradation, 2
    exudates, 4
    fragments, 4
    growth, 321
    litter, 4, 71
    nutrition, 12
    polymers, 2, 3, 5
    remains, 91
    residues, 53
    tissue, 3, 4
Pleistocene, 252, 253
Plutonium distribution, 257
Polar
    FA, 281
    forces, 95
    functionalities, 93, 98
Pollutants, 11, 167, 205, 215
Polyacrylamide gel electrophoresis, 105, 215
Polyacrylate, 171
Polycondensation, 56, 193, 202
Polydispersities, 114, 120-122, 221, 224
Polyelectrolyte, 96, 97, 115, 117, 215, 218, 223, 299
    contribution, 242
    microgels, 241, 243
Polyethylene, 167
Polyethyleneglycol, 115
Polyflavonoid, 5
Polymerization, 9, 11, 26, 37, 47, 49, 50, 56, 76, 221
Polymer, 40, 46, 47, 76, 83, 112, 114, 115, 118, 120, 121, 166, 167, 169, 173, 177
    solution theory, 100
    model, 111
Polynitroaromatics, 205
Polynuclear aromatic rings, 56
Polyphenols, 2, 4, 9, 15, 37
Polypropylene, 12, 38, 206
Polysaccharides, 23, 26, 46, 53, 87, 112,

Polysaccharides, *continued*
  114, 115, 117, 122, 146, 301
Polystyrenesulfonates, 13, 15, 112, 114,
  116, 121
Polystyrenesulfonic-co-maleic acid,
  PSM, 227-231
Pores, 102, 115, 165, 265, 277, 278, 287
Positive ion, 146, 149
  spectra, 144
Potassium, 146, 281, 287, 289-291, 296,
  297, 325
  humates, 291, 296, 297
Potato yields, 321
Potentiometric titration, 263
Precursors, 4, 5, 9
Predictive modeling, 246, 255, 258
Preferential sorption, 154
Primary structures, 15
Progestogen, 192
Propionaldehyde, 68, 76, 77, 78
Proteins, 2, 32, 113, 118, 218
  standards, 146, 149
Proton
  acceptor, 96
  donor, 96
Pullulans, 126
Pulse
  delays, 65, 67, 69, 71, 127
  sequence, 65, 84, 113
Pulsed ion extraction capabilities, 13
Purification, 54, 84, 108, 193, 300, 301,
  307
Py-FI mass spectra, 23, 26, 32
Pyrene, 112, 169, 171, 173, 184, 185
Pyrethroids, 177, 178
Pyridine, 98, 99
Pyrolysis
  mass spectrometry (PY-MS), 3, 143
  gas chromatography-mass
  spectrometry, 105, 265
  field ionization mass spectrometry,
  22

Quantitative liquid-state measurements,
  65
Quantitative reliability, 63
Quartz, 22, 126, 172, 237, 262, 289
Quaternary amines, 278, 287
Quenching, 112, 171, 179, 206-213
Quercetin, 76

Quinine sulfate, 126, 262
Quinones, 68, 77, 98, 179, 205, 206

R_1 ratio, 57
Radial pair-distribution, 228
Radicals, 9, 15, 19, 47, 64, 65
Radioactive waste disposal, 235, 236
Radionuclide-humate interaction, 237
Ramp CP/MAS pulse sequence, 65
Random coil, 106, 111, 216, 219
Randomized intervention analysis (RIA),
  263
Rate constant, 157, 245
Re(VII), 312, 318
Reaction
  mechanisms, 5, 34
  sequence, 11
Reactive
  groups, 63
  sites, 9, 195
Reactivity, 10, 57, 63, 111, 123, 196, 215
Real system kinetics, 258
Recycling, 21
Redox states, 246
Reduced runoff, 325
Reducing sugars, 77
Reflector, 146, 149
Refractive index, 95, 113
Refractory, 92, 156
Regenerated resin, 102
Regeneration, 287, 288
Relative fluorescence intensity (RFI),
  126, 129, 130, 132, 268, 269, 274
Relaxation delay time, 13
Release of water molecules, 241
Remediation, 153, 154, 161, 162, 191,
  309, 315
Remote carbon atoms, 71
Removal mechanism, 278
Reproducibility, 115
Reproductive function, 191
Resolved subspectra, 83
Resource Conservation and Recovery
  Act (RCRA), 309
Respiration, 323
Reverse osmosis, 102, 261, 264, 265, 277
Reversibility, 194, 199, 202
RI detectors, 114, 115, 117, 120-122
Ring cleavage, 4, 37, 48, 49
Rock pile, 256

Subject Index

Root initiation, 323
Rye, 54, 56, 57, 59

Saccharides, 2, 104
Saccharose, 301
Safety assessment, 236
Salicylic acid, 205-207, 211-213, 273
Sample
    acquisition, 148, 149
    exclusion, 114
Saturation, 145, 159, 238, 240, 241, 247, 313
SAXS, 15, 135, 136, 138, 139
Scatchard, 240
Scattering behavior, 138, 139
Schubert, 35, 240
Second derivative, 269
Secondary synthesis reactions, 2
Sediment, 1, 79, 106, 111, 154, 169, 170, 173, 205, 236, 237, 246, 249, 255, 258, 289
    dissolution, 258
    surface, 154, 247, 254
Sedimentary carbonate, 254
    humic acid, 264
    organic carbon (SOC), 253, 254
Selective-ion electrode, 263
Self associations, 92, 94
Semiquinone, 47
Senescence, 4, 5, 71, 77
Separation, 2, 54, 105, 112, 113, 126, 132, 171, 215-218, 223, 224, 287, 296, 300
Sephadex G-25, 125-127, 132
Sewage, 21, 66, 191, 192, 202
    sludge, 21, 66, 191
Shape, 47, 135, 149, 150, 210, 239, 314
Side band, 44, 113
Signal suppression, 55
Silica gel, 37, 42, 113, 237
Silicic acid, 37-40, 42, 43, 47, 49
Silicon, 37, 39, 47
Simulated waste streams, 310, 312
Simulations, 155, 159-162
Sinapinic acid, 146, 148, 290
Single-crystal X-ray diffraction, 3
Size, 12, 15, 18, 39, 98, 100, 102, 103, 105, 106, 111-114, 118-120, 122, 125, 128, 135, 137, 143, 149, 156, 162, 177, 179-181, 187, 218, 223, 241, 265, 267, 269, 274, 325
    distribution, 122, 149, 155
    exclusion chromatography (SEC), 13, 15, 16, 111, 112, 114, 124, 125, 144, 146, 149, 178, 179, 181-183
Sludges, 191-193, 195, 196, 200, 202
Small angle
    X-ray scattering, 15, 111, 135, 140
    neutron scattering, 15
Sodium, 5, 13, 15, 38, 68, 97, 98, 101, 114, 145, 154, 173, 219, 227, 278, 281, 287-290, 300, 315, 317, 318
    azide, 173
    humates, 290-292, 296, 297
    tetraborate, 295
Soil, 1-5, 37, 53, 57, 64, 79, 91-93, 96, 97, 100, 101, 105, 106, 125, 128, 130, 153, 154, 159, 169, 170, 173, 192, 200, 205, 218, 289, 324
    aggregates, 92
    amendment, 191, 192
    biomass, 2, 4
    conditioners, 280
    fertility, 1
    HAs, 38, 41, 42, 44-48, 125, 127, 130, 132, 193, 195, 199, 200, 202, 221
    matrix, 5, 100, 149
    moisture, 323
    organic matter (SOM), 2, 4, 34, 53-55, 57, 59, 91, 97, 100, 106, 202
    productivity, 53, 59
    solution, 323
    systems, 5
Solid phase extraction (SPE), 170-173, 194
Solids effect, 154, 155
Solubility, 10, 91, 95, 96, 100, 113, 135, 138, 143, 145, 166, 167, 169, 171, 173, 196, 200, 202, 277, 278
    parameters, 95, 166-171, 173
Solubilization, 93, 95, 96, 98, 101, 137, 140
Solvation, 94-98, 103, 106
Sorption, 92, 102, 153-162, 165, 167, 169, 171-173, 177, 186, 205, 246-249, 251, 277, 278, 281, 287
    mechanism, 247
    parameters, 157
    process, 153, 157, 158, 247

Sorption, *continued*
  sites, 157, 159
Source decay processes, 149
Specific UV absorbance, 66
Spent Decontamination Solution (SDS), 317
Spherical shape, 312
Spherocolloids, 183
Spheroidal particles, 40, 41, 47
Spin counting experiments, 65
Spin lattice relaxation, 55, 64, 65, 83
Spinning rate, 64, 67, 127
Spin-spin relaxation time, 87
Sr^{2+}, 224
Stability, 34, 35, 49, 53, 56, 121, 237, 238, 251, 318
  constant, 49, 238
Stagnant flow conditions, 252
Stern-Volmer
  model, 206-208, 210-213
  plot, 208, 210, 212
Sterols, 23, 26, 32-35
Stoichiometry, 264
Stokes-Einstein relationship, 216
Storage, 191, 310, 313-315
Strontium, 309
Structural
  diagram, 3
  groups, 53, 59, 63, 65, 69, 71
  group quantitation, 63, 64, 71
  models, 143
  unit, 3, 265
Structured groups of peaks, 296
Structures, 2-5, 15, 19, 26-28, 32, 44, 47, 53, 59, 63, 71, 77, 83, 84, 87, 91, 93, 96, 98, 100, 102, 105, 112, 115, 122, 125, 130, 143, 165, 177, 179, 180, 185, 193, 205, 206, 227, 242, 265, 272, 299
Suberins, 91, 105
Subsurface remediation, 153
Sulfolane, 98
Sulfur, 12, 219, 230
$^{34}S/^{32}S$ ratio, 254
Supercritical fluids, 106
  CO_2 drying, 312, 318
Surface
  areas, 169, 311-313, 318, 323
  bound HA, 237
  tension, 136, 138

  waters, 63, 68, 277
Suwannee River, 3, 66, 80, 102, 145, 169, 227, 281
Swamps, 252
Sweep width, 68
Swelling, 173, 278
Synchrotron radiation, 15
Syringic acid ($C_9H_{10}O_5$), 265

Tc(IV), 238, 246
Tc(VII), 246
Ternary systems, 246, 248
Tetrachloroethylene, 311
Tetragonal distortion, 230, 232
Th(IV), 238, 244, 245, 248, 249
Theoretical modeling, 228
Thermal
  degradation, 143
  diffusion, 216
Three-dimensional excitation emission matrix (3-D EEM), 125, 126, 128, 130
Titration, 93, 114, 207, 263, 267
TOSS (total sideband suppression), 55, 84
Total
  carbon recoveries, 127
  organic carbon (TOC), 12, 14, 159, 262-264, 280, 281, 284, 287, 288
  permeation volume, 114
  reactive Al, 273
Toxic metals, 309-312, 315, 317
Toxicity, 153, 177, 191, 192, 261, 262, 267, 274
Trace elements, 215, 219, 246, 257, 258
$1s \rightarrow 4p$ transition, 232
Transfer distance, 84
Transport calculations, 257
Tricaprylin, 170
Trichloroethylene (TCE), 309, 311-314, 316, 317
Triethylamine, 146
Trifluoroacetic acid, 145, 146
2,4,6-Trihydroxybenzoic acid, 68, 76-79
(TRIS)-nitric acid buffer, 219
Tritium, 251, 309
  tracer pulse, 251
Trona deposit, 281

Ubiquitin, 146
Ultracentrifugation, 15, 237

Subject Index

Ultrafiltration, 12, 19, 102, 111, 215, 218, 267-270, 274
Ultrasonification, 12, 38
Ultraviolet-visible spectra, 38, 125, 179, 219, 262
Un-curled state, 187
Unpaired electrons, 64, 65
UO_2^{2+}, 224
Uranium, 238, 246, 249, 252, 256, 309
    humate, 239, 240
    mine, 252
    U(VI), 238-240, 244, 248
UV-detected, 118, 119, 120

van der Waals forces, 96-98, 102
Vanadium, 12, 230
Vanillic acid ($C_8H_8O_4$), 265
Vanillin, 26, 265
Vanillyl phenols, 265
Vapor pressure osmometry, 111
Variable contact time, 113
Vegetable crops, 324
Vibrational spectroscopy, 143
Vibrio fischeri, 262
Viscosity, 95, 111, 216, 287
Vitamins, 269
Void volume, 114, 126

Volatile matter, 22, 34

Waste
    repositories, 252, 257
    sites, 205, 309
    streams, 310, 315, 318
Water, 1, 66, 91, 92, 101, 103, 105, 264, 277, 281, 289
    chemistry, 103, 263
    holding capacity, 1
Weak anion exchange resins, 102
Wet beads, 310, 311, 313-315, 318
Wheat plants, 323

XAD resin, 104, 106
XAD-4, 102-104, 227
XAD-8, 55, 66, 92, 101-104, 106
XANES, 227, 229-232
Xenobiotic compounds, 5, 191, 192
X-ray
    absorption, 227
    microscopy, 15
    powder diffraction, 38, 46

z-Average particle radius, 136
^{64}Zn, 220, 221
Zone electrophoresis, 105